Elements of ANATOMY AND PHYSIOLOGY

STANLEY W. JACOB, M.D., F.A.C.S.

Associate Professor of Surgery, School of Medicine,
University of Oregon Health Sciences Center
First Kemper Foundation Research Scholar,
American College of Surgeons
Markle Scholar in Medical Sciences

CLARICE ASHWORTH FRANCONE

Medical Illustrator, Formerly Head of the Department of Medical Illustrations
University of Oregon Medical School

W. B. SAUNDERS COMPANY • Philadelphia • London • Toronto

W. B. Saunders Company: West Washington Square
Philadelphia, PA 19105

1 St. Anne's Road
Eastbourne, East Sussex BN21 3UN, England

1 Goldthorne Avenue
Toronto, Ontario M8Z 5T9, Canada

Library of Congress Cataloging in Publication Data

Jacob, Stanley W

Elements of anatomy and physiology.

1. Human physiology. 2. Anatomy, Human. I. Francone, Clarice Ashworth, joint author. II. Title. [DNLM: 1. Anatomy. 2. Physiology. QS4 J16e]

QP34.5.J28 612 75–28795

ISBN 0–7216–5088–0

Elements of Anatomy and Physiology ISBN 0-7216-5088-0

Last digit is the print number: 9 8 7 6 5 4 3

Dedication

To my father, without whose encouragement this would not have been possible.

Stanley W. Jacob

To my grandchildren

Clarice Ashworth Francone

PREFACE

The human body is more complex than the most involved computer mechanism ever produced. Moreover, scientists from all parts of the world are constantly adding to our knowledge of physiology and anatomy. It is not possible for any one person to absorb all of these facts or even to understand the immense ramifications of this explosion in knowledge. It is, however, possible for even a beginning student to master enough information so as to appreciate the major components and activities of the body. In this text we have tried to describe the structure and function of the human body in a manner that the student with no prior preparation in the sciences will find interesting and comprehensible.

To help the student understand and retain new terms and ideas we have incorporated a number of special features. For example, because most students using this text will be participating in some way in the care of patients, we have taken pains to mention the clinical implications of disordered physiology. Pronunciations are provided for unfamiliar terms, and Latin and Greek roots are included as memory cues. Major headings in the text are presented as questions—What Does a Neuron Look Like? How Does It Work?—that focus attention on what is to be learned.

Recently there has been a dramatic increase in basic knowledge of the cell and of the role of DNA and RNA in heredity and control of cellular life functions. Because these functions are so intimately involved in the workings of the body in both health and illness, all persons who contribute to patient care need to know about them. We have therefore dealt with cellular physiology more extensively than is common in a beginning textbook, but have presented the information as simply as possible.

In the firm belief that visual displays contribute to comprehension and help to develop an overall perspective of the field of anatomy and physiology, the authors have prepared new halftone and line drawings as well as new tables and charts.

We wish to express our indebtedness to our colleagues as well as to the many teachers and students who have read the manuscript for this text and have contributed valuable and constructive criticisms. We would like also to express special appreciation to Terry Bristol and Beverly Methvin.

The authors wish to acknowledge the editorial assistance of the W. B. Saunders Company. A particular note of gratitude goes to Robert Wright, whose advice and encouragement contributed significantly to the completion of this work.

STANLEY W. JACOB
CLARICE A. FRANCONE

AUDIOVISUAL AIDS

To supplement this text, certain illustrations have been made into audio-visual teaching aids which can be purchased directly from the W. B. Saunders Co., West Washington Sq., Phila., Pa., 19105.

These aids consist of ten 35 mm. filmstrips in full color. They are based on illustrations from the text, *Structure and Function in Man,* by Jacob and Francone, and cover the structure and function of important body systems. Each filmstrip includes about 60 frames; each is accompanied by a narration on a tape cassette, and a printed script of the narration. Filmstrips can be projected in class with any standard filmstrip projector or slide projector with a filmstrip attachment. They can also be used for individual study in a desk-top viewer. The narration is also available on long-playing records.

CONTENTS

Chapter 1
INTRODUCTION TO ANATOMY AND PHYSIOLOGY 1

Chapter 2
THE CELL 18

Chapter 3
TISSUES AND THE SKIN 30

Chapter 4
THE SKELETAL SYSTEM 37

Chapter 5
THE MUSCULAR SYSTEM 59

Chapter 6
THE BLOOD 88

Chapter 7
THE CIRCULATORY SYSTEM, LYMPHATIC SYSTEM, AND ACCESSORY ORGANS 98

Chapter 8
THE RESPIRATORY SYSTEM 131

Chapter 9
THE URINARY SYSTEM 147

Chapter 10
THE DIGESTIVE SYSTEM 158

Chapter 11
THE NERVOUS SYSTEM 177

Chapter 12
THE ENDOCRINE SYSTEM 210

Chapter 13
THE REPRODUCTIVE SYSTEM 225

Index 241

CHAPTER 1

INTRODUCTION TO ANATOMY AND PHYSIOLOGY

STUDYING THE HUMAN BODY

The human body may be viewed as a collection of thousands of billions of minute living units called *cells*. These cells are marvelously combined and organized to operate as a harmonious whole—the living body. The simplest structures formed by the cells are known as *tissues*, and these tissues combine to form larger more sophisticated structures known as *organs*. The organs work together as *systems*, such as the respiratory system, the digestive system, and the circulatory system.

When looked at in this way, the human body is an awe-inspiring phenomenon. Its highly organized design and the fine balance and interdependence of the various parts are the underlying bases of the tremendous adaptability and diverse capabilities of human beings. An understanding of the structures and functions of the human body will greatly expand the student's appreciation of the magnificence of life.

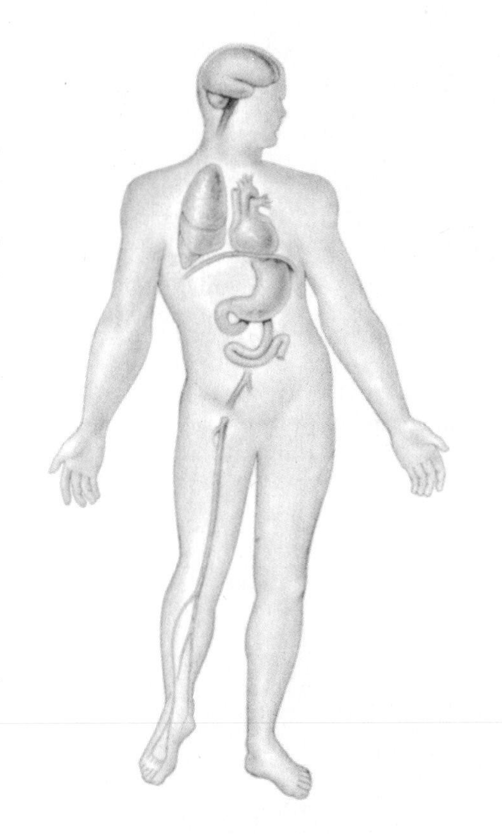

What Is Human Anatomy?

Anatomy is the science that studies the shape and structure of living organisms and their parts. *Human anatomy* is the study of the shape and structure of the human body and its parts.

Gross anatomy, or gross human anatomy, deals with the large, macroscopic structures that can be observed by normal dissection and with normal, unaided vision.

Microscopic anatomy deals with structures that can be seen only with the use of the light microscope. The modern electron microscope, which was developed during the 1940's, is 400 times more powerful than

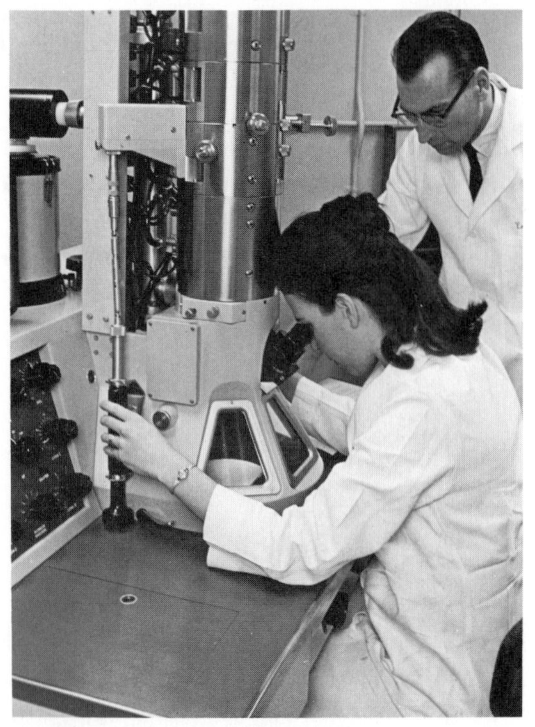

Figure 1. The electron microscope is now being employed to study structures at a magnification of 200,000×.

the light microscope (Figs. 1 and 2), and has become an important instrument of research in current anatomical investigation.

What Is Human Physiology?

Physiology (from the Greek, *physis*, meaning natural science, and *-logy*, meaning study of) is the science that deals with the functions of living organisms and their parts. It is the study of what the parts do and why. *Human physiology* is the study of the functions of the human body and its parts. Some physiologists specialize in the study of the functions of systems, such as the circulatory system, whereas others may center their attention on a particular organ, such as the heart or kidneys.

Cellular physiology, the most prominent branch of modern physiology, is the study of the *how* and *why* of the activities of individual cells and their internal parts.

How Are Anatomy and Physiology Related?

The structure and the function of any part of a living organism are always related. For instance, the heart is a specially designed muscular pump whose main function is to keep the blood moving continuously throughout the body. The design of the heart is very much different from the

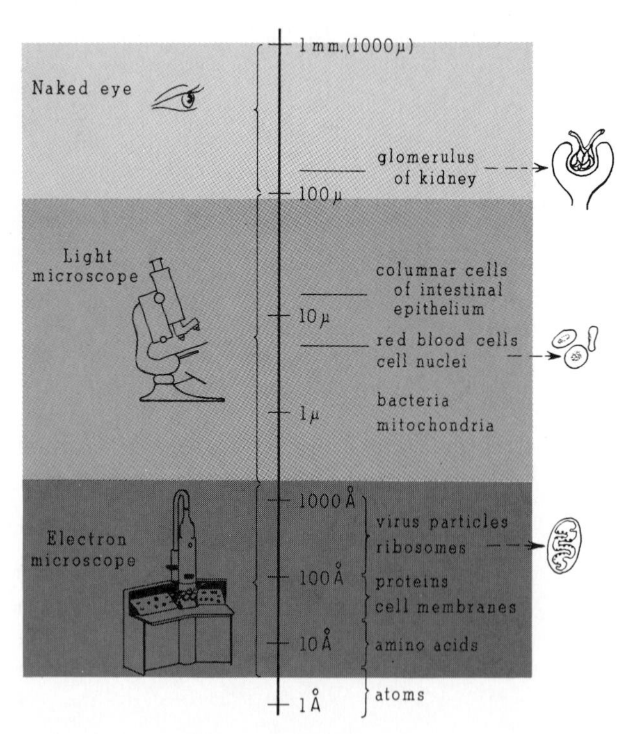

Figure 2. The limits of resolution of the electron microscope as contrasted to those of the eye and the light microscope. A millimeter, mm., is equal to 1/1000 meter, or 0.03937 inch; a micrometer, μm, 1/1000 millimeter; an angstrom, Å, 1/10,000 micrometer.

design of a lung, whose function is the exchange of carbon dioxide for oxygen between the outside environment and the blood.

The structure of a part can provide a clue to its function. Likewise, knowing the function of a part helps us to properly understand its size, shape, and structural organization. Just as structure and function are related in the practical study of the

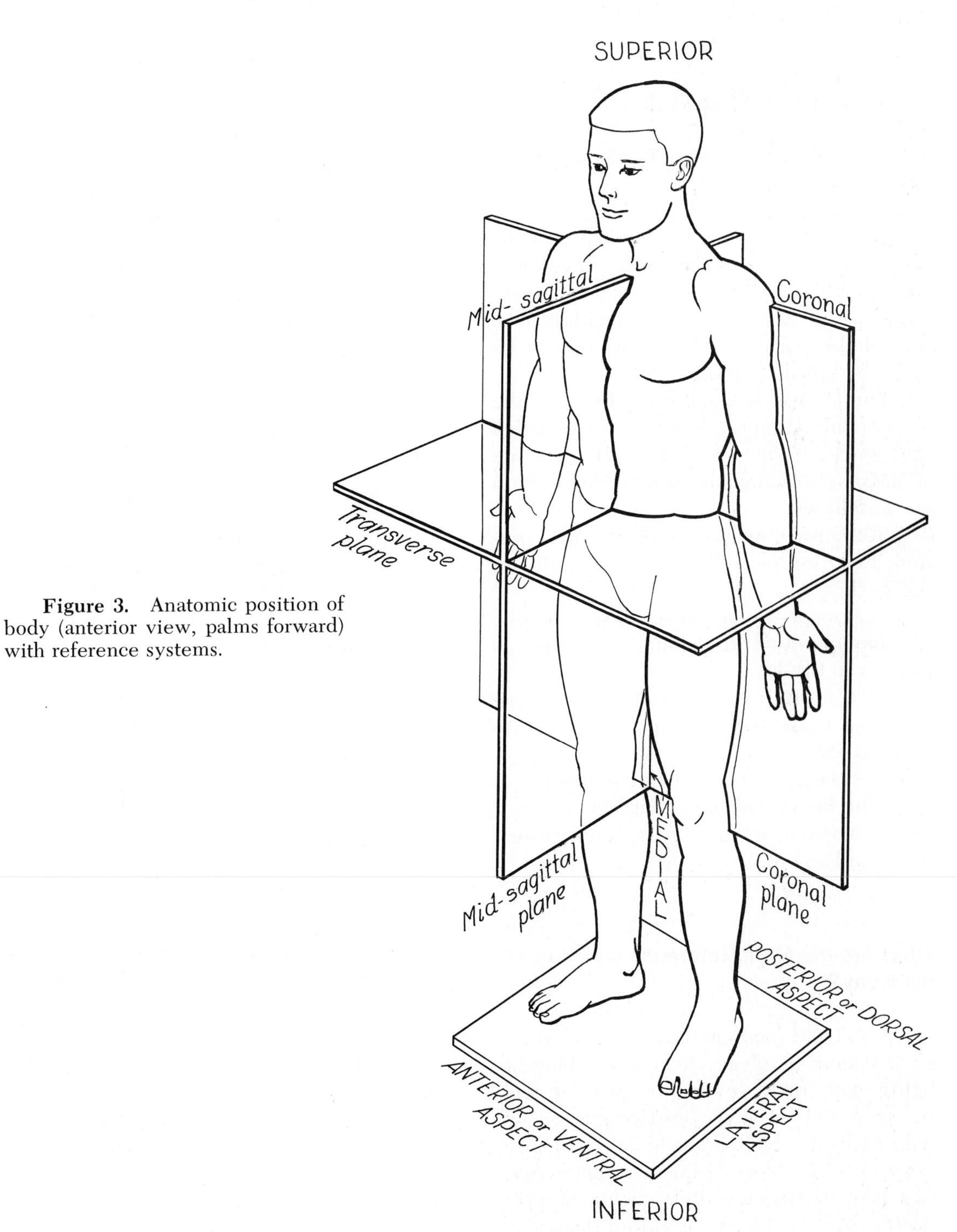

Figure 3. Anatomic position of body (anterior view, palms forward) with reference systems.

body and its parts, so are anatomy and physiology always closely related. Consequently, the use of the terms "anatomy" and "physiology" is best thought of as indicating an emphasis, rather than a sharp division in what is being studied.

SYSTEMS OF REFERENCE

How Are the Structures of the Body Accurately Identified?

Because of the great complexity of the human body, anatomists have developed several *systems of reference* over the years to aid in the rapid and accurate identification of the part or area of the body to be described or discussed.

Directions in the Body. Perhaps the most common approach involves the use of the general terms of anatomical direction (Fig. 3). The four main terms of direction are as follows:

SUPERIOR—means above or upper portion. For example, the head is superior to the neck.

INFERIOR—means lowermost or below. The foot is inferior to the ankle; the neck is inferior to the head.

ANTERIOR (OR VENTRAL)—means toward the front; the breasts are on the anterior chest wall.

POSTERIOR (OR DORSAL)—means toward the back; the spinal cord is posterior to the internal organs (heart, lungs, intestines, and so on).

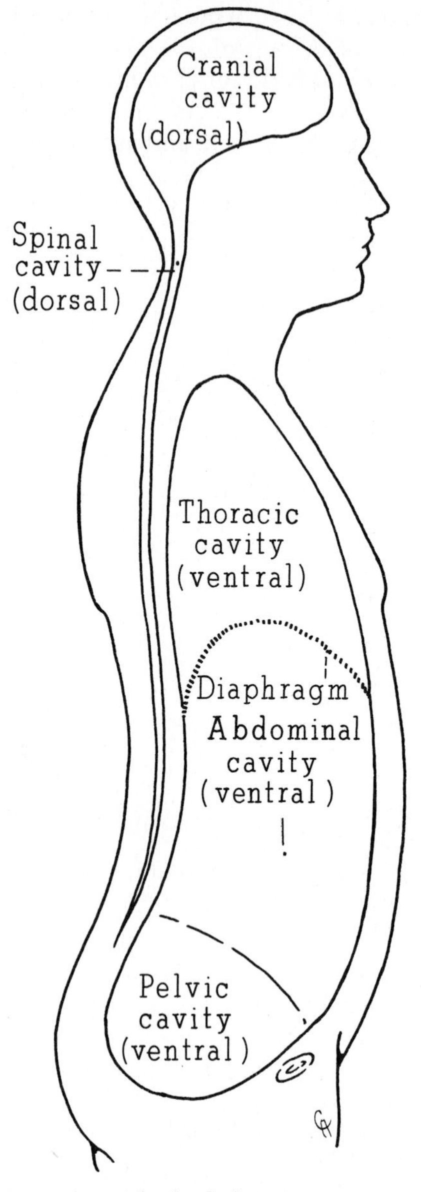

Figure 4. The body has two major cavities, dorsal and ventral, each subdivided into two lesser cavities. For convenience the abdominal and pelvic cavities pictured here are referred to simply as the abdominopelvic cavity.

What Are the Main Reference Cavities of the Body?

A second common anatomical reference system involves viewing the body as being made up of several "rooms" or *cavities* (Fig. 4). The body has two major cavities, each of which is made up of smaller cavities. The larger of the two major cavities is usually referred to as the *ventral cavity*, and the smaller is called the *dorsal cavity*. How might these be identified by alternative terms?

VENTRAL CAVITY. The ventral cavity contains the internal organs, which maintain the basic life processes. These include the heart, lungs, kidneys, stomach, and intestines, as well as several others. The ventral cavity is frequently distinguished into three smaller cavities.

The uppermost (superior) of the smaller cavities contains the heart and lungs, and is referred to as the *thoracic* (tho-ras′-ik) *cavity*, or, more commonly, as the chest. The middle cavity is the *abdominal cavity,* which contains the liver, kidneys, and most of the digestive tract, including the stomach. The lowermost (inferior) cavity is referred to as the *pelvic cavity,* since it is bounded on the outside by the pelvic bones; it contains the lower digestive tract and the bladder.

DORSAL CAVITY. The dorsal cavity contains the brain and spinal cord. It is commonly distinguished into the *cranial cavity,* containing the brain, and the *spinal cavity,* containing the spinal cord.

What Is the "Structural Unit" Reference System?

In studying the physiology of an organism, the most useful system of anatomical reference is that based on the *structural units* approach. In this approach, the body is described in four levels of detail—the cellular level, the tissue level, the organ level, and the systems level. As previously mentioned, these four levels are related in an organized fashion, as follows: The cells combine to form tissues; the tissues combine to form organs; and the organs work together to form systems.

The Cellular Level. All living matter is composed of cells; tiny microscopic units that are the smallest living parts of all organisms. All cells are made up of three basic parts—the *cell membrane,* which envelops the other parts; the *cytoplasm* (si′to-plazm″), which is the body-substance of the internal portion; and the *nucleus* (nu′kle-us), the control center, located near the center of the cell. However, beyond this basic common structure, different cells vary in size, shape, and composition, according to what they do. For instance, muscle cells differ from nerve cells, and both of these differ from bone cells, although all (muscle, nerve, and bone cells) have a cell membrane, cytoplasm, and nucleus. Cytoplasm is derived from the Greek *cyto-*, meaning cell, and *-plasm,* meaning formed material. The word nucleus is derived from the Latin word *nucis,* meaning kernel.

The Tissue Level. Collections of similar cells combine to form particular types of tissues. The term "tissue" is derived from the Latin word *textura,* meaning texture, or something woven together. Cells hold themselves together in tissues by weaving a network or matrix of soft gluelike substance, known as the intercellular matrix. The prefix "inter-" means between. Anatomists usually refer to four main categories of tissues that make up most of the organ structures of the body. These are epithelial (ep″i-the′le-al) tissues, connective tissues, muscular tissues, and nervous tissues. These four types also correspond to the four main types of cells—epithelial, connective, muscle, and nerve. *Epithelial tissues* are found covering all surfaces of the body, both internally and externally; for instance, the coverings on the organs and digestive tract, and the skin covering the whole body. *Connective tissues* are harder—containing much more intercellular matrix substance—and function in holding things together; for example, bones, tendons, and cartilage.

Muscular tissues make up muscles, but are also found wherever movement occurs; for instance, besides arm and leg muscles, muscular tissue is found in the stomach and digestive tract, as well as in association with the skin. *Nervous tissues* make up the brain and spinal cord and all the millions of nerves that carry signals throughout the body.

The Organ Level. An organ is a somewhat independent structure within the body, made up of several types of tissues, *all serving a common function.* For instance, the stomach is an organ made up of layers of epithelial, connective, and muscular tissues, all working together as a whole to break up food into tiny particles for digestion. The stomach is connected to

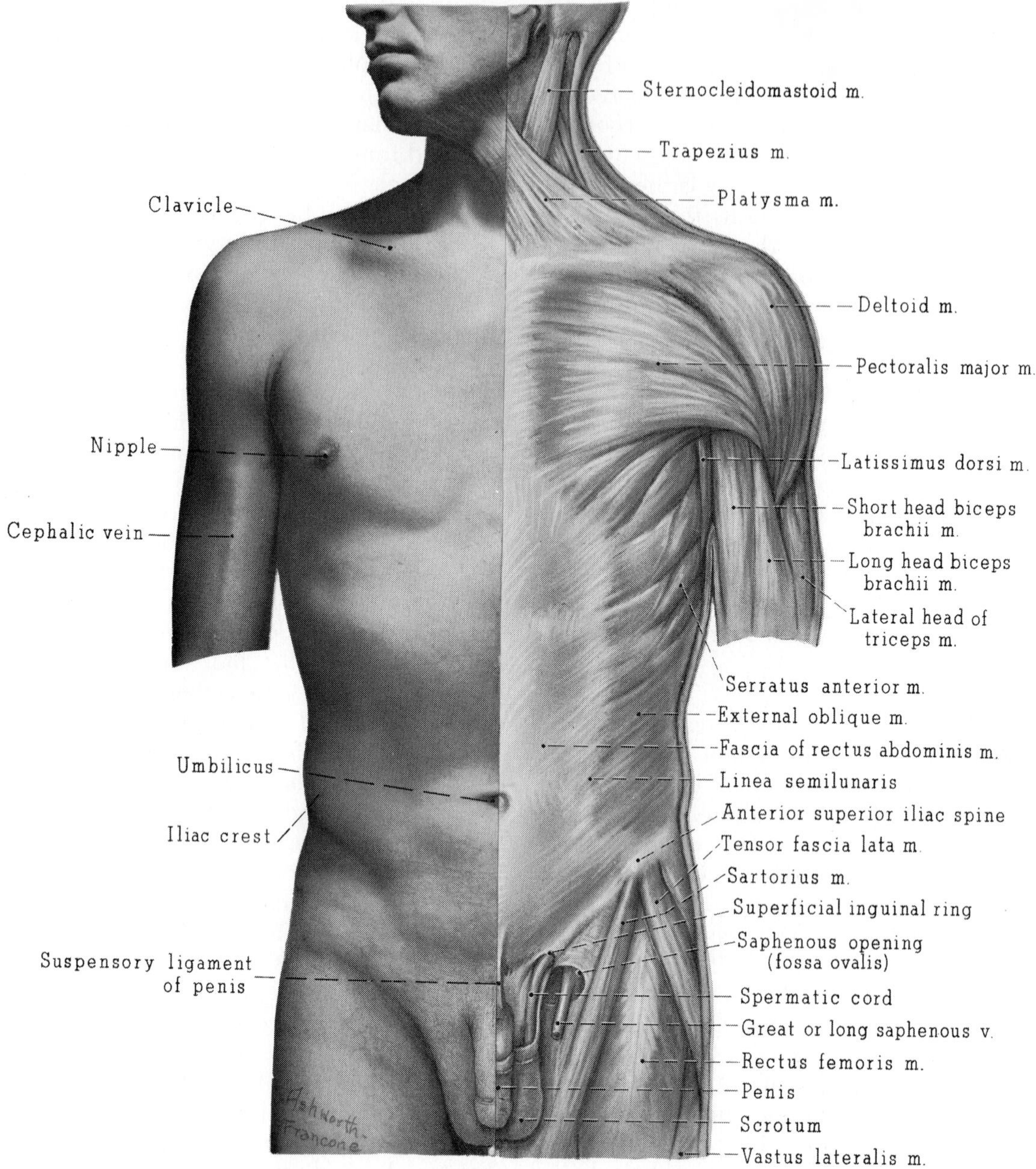

Figure 5. Anterior surface of male, left half with skin removed to expose first layer of muscles.

the brain centers by nerves (nervous tissue). Other organs are the liver, heart, lungs, and skin.

The Systems Level. The highest level of structural unit that is used for purposes of reference is the system. A system is a group of organs acting together to accomplish some overall bodily function. For instance, the circulatory system is made up of the heart, arteries, veins, and capillaries, all serving to circulate the blood throughout the body. The skeletal system is made up of bones, tendons, and cartilage, all serving the functions of support and movement. Figures 5 through 14 are representative anatomical diagrams showing a variety of bodily structures with their identifying names.

Text continued on page 16.

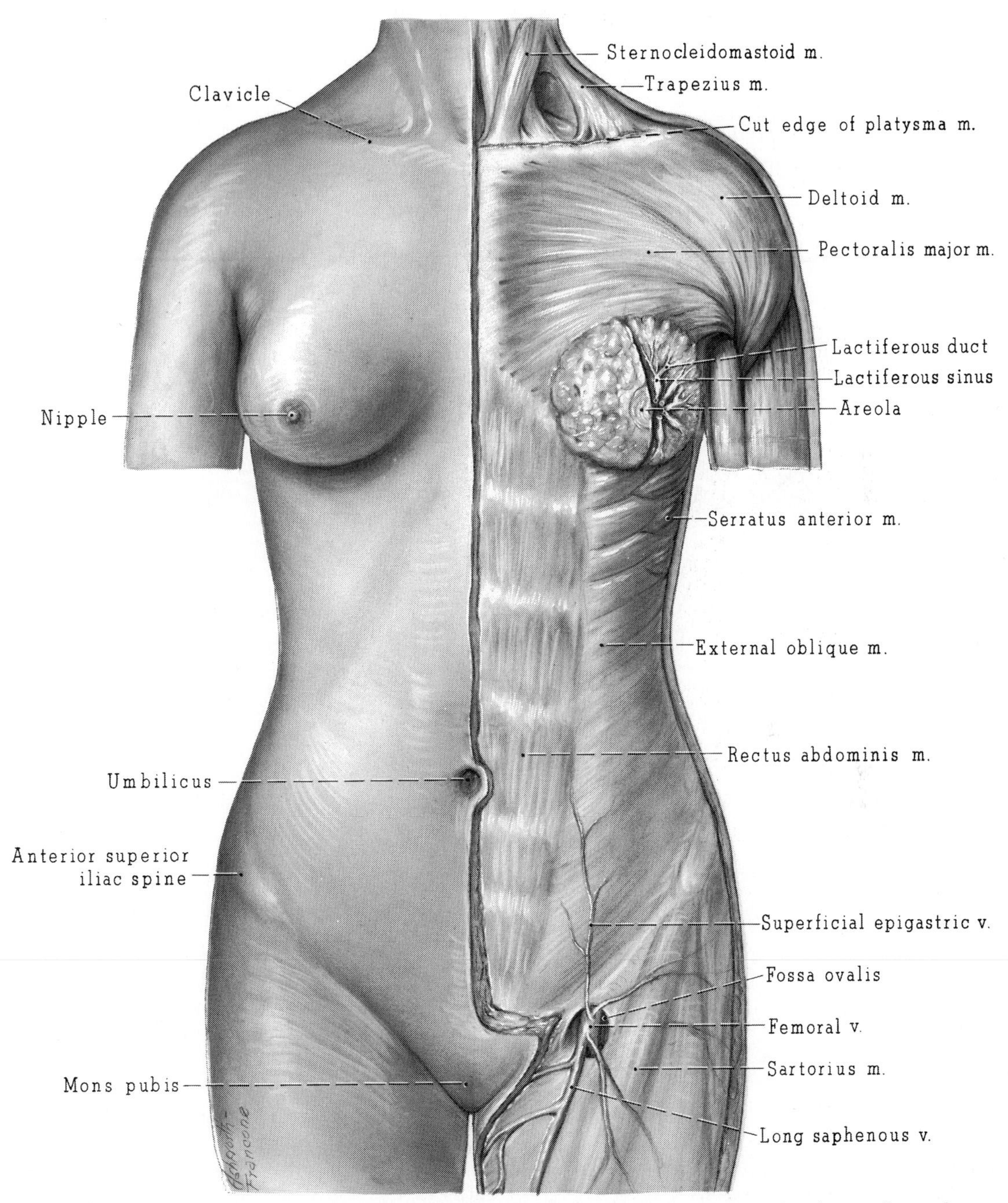

Figure 6. Anterior surface of female, left half with skin removed to expose first layer of muscles.

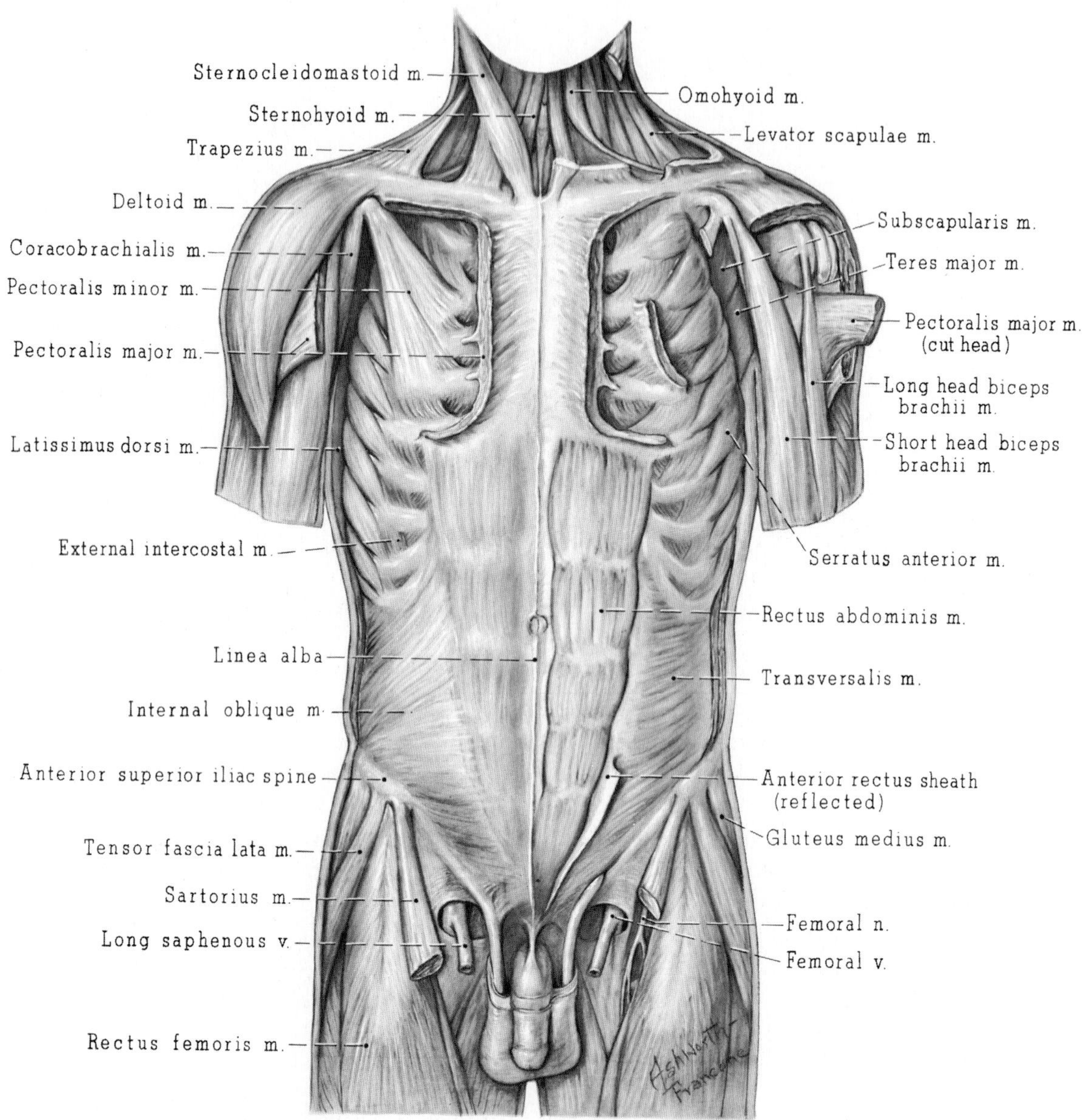

Figure 7. Pectoralis major muscle removed on right side, pectoralis minor on left side; second and third layers of abdominal muscles exposed.

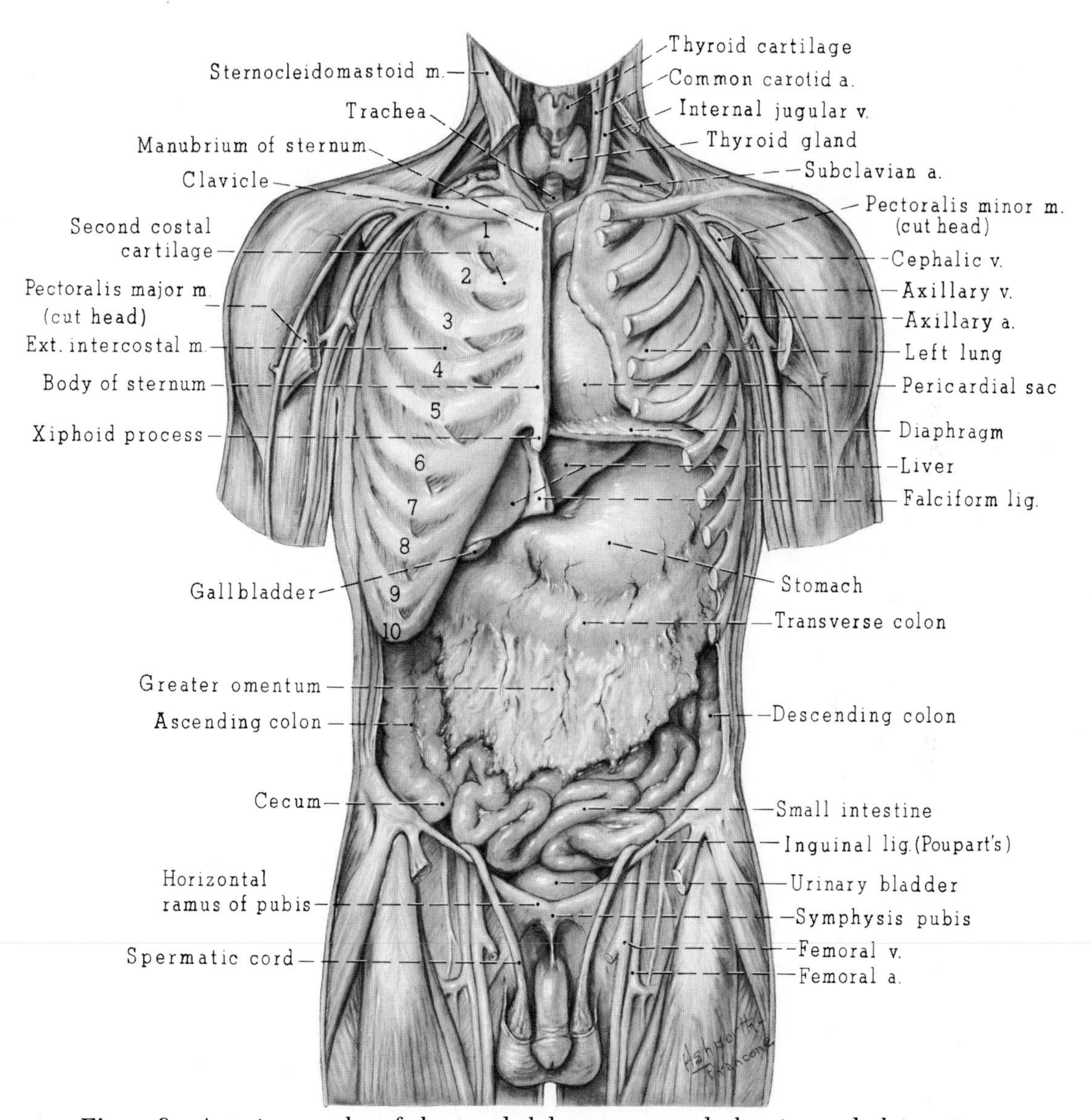

Figure 8. Anterior muscles of chest and abdomen removed, showing underlying viscera.

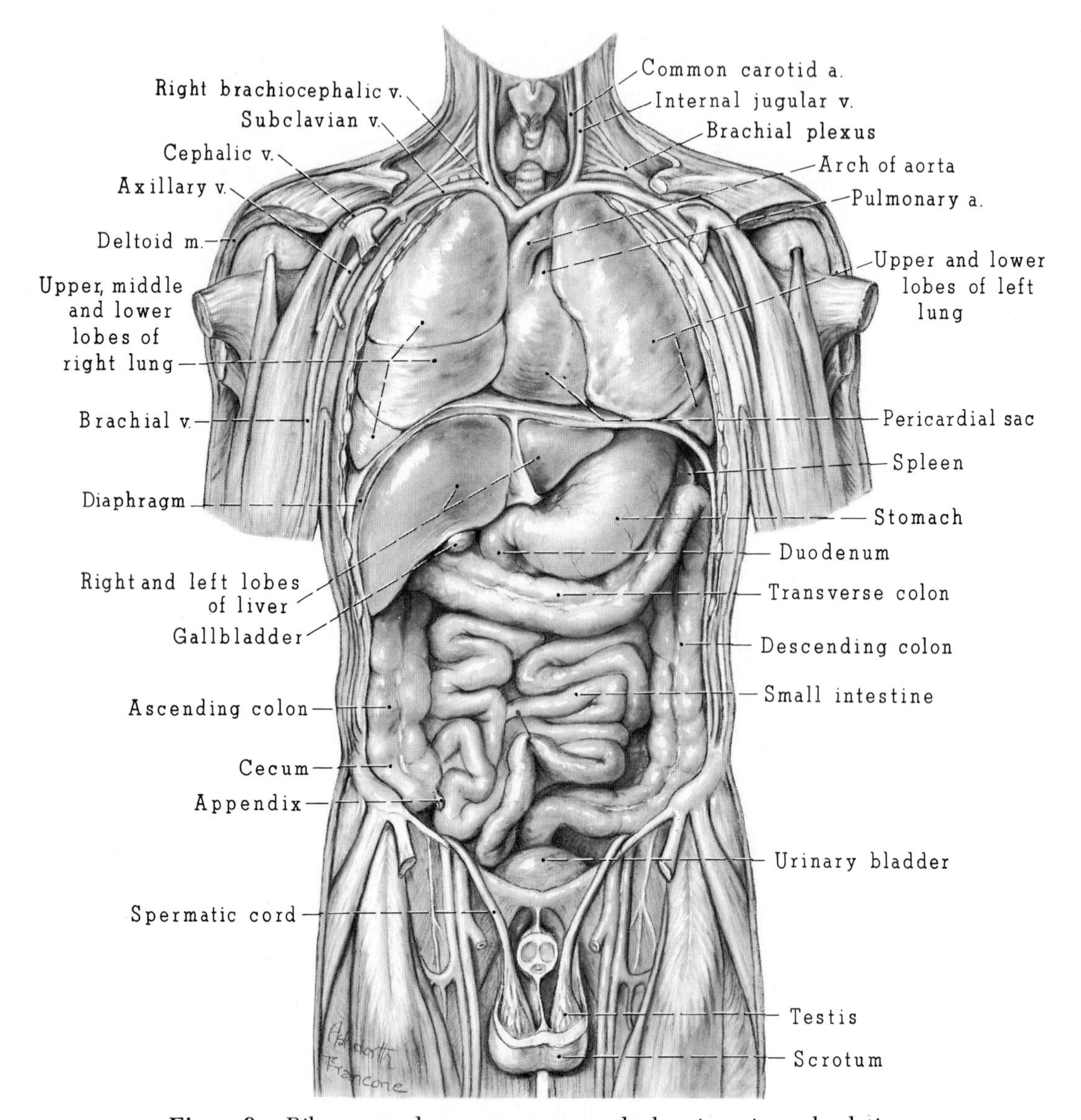

Figure 9. Rib cage and omentum removed, showing visceral relations.

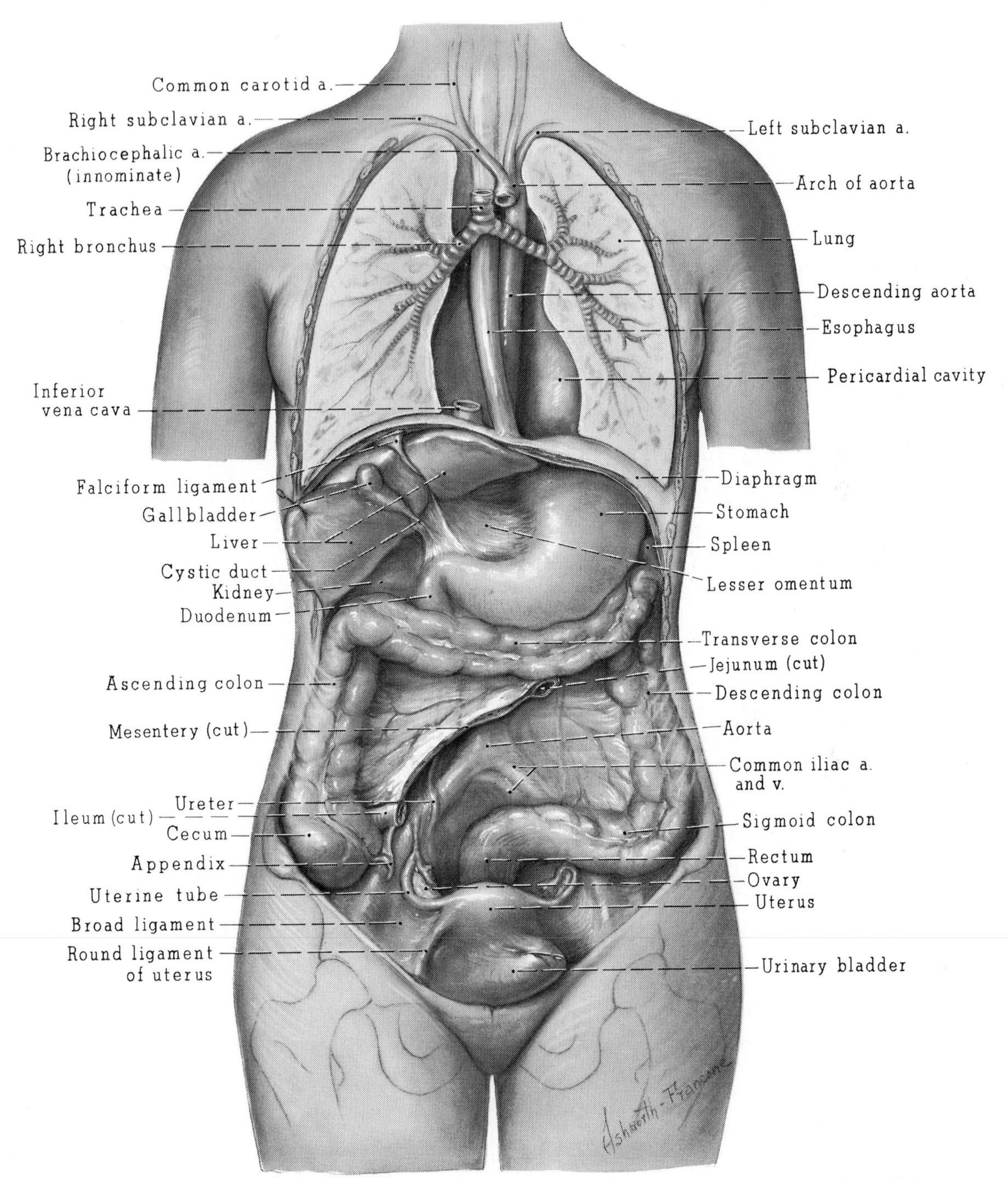

Figure 10. Female, demonstrating visceral relations; lungs sectioned, heart and small bowel removed.

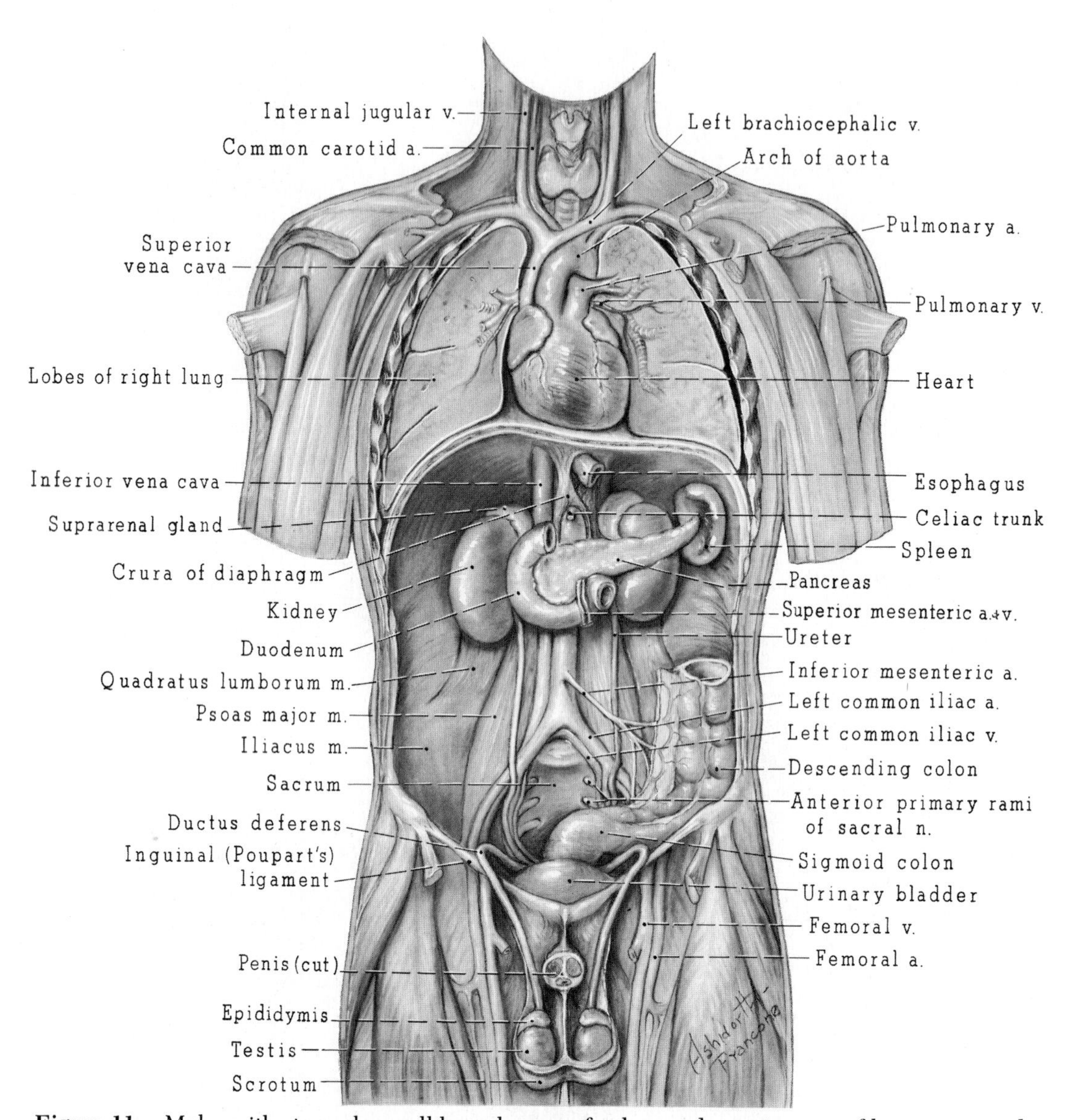

Figure 11. Male, with stomach, small bowel, most of colon, and anterior part of lungs removed.

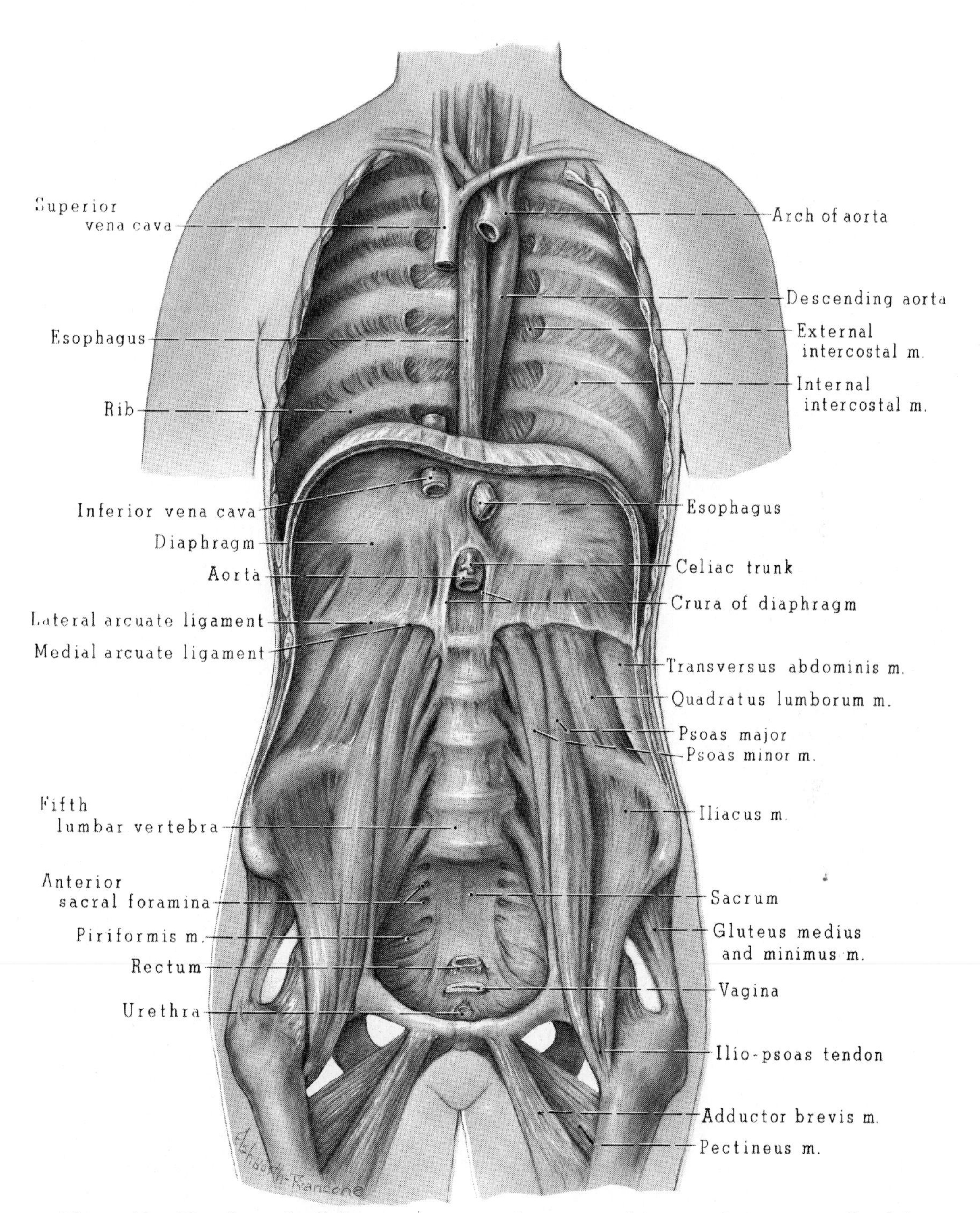

Figure 12. Female, with all the viscera removed, exposing the internal posterior walls of chest and abdominal and pelvic cavities.

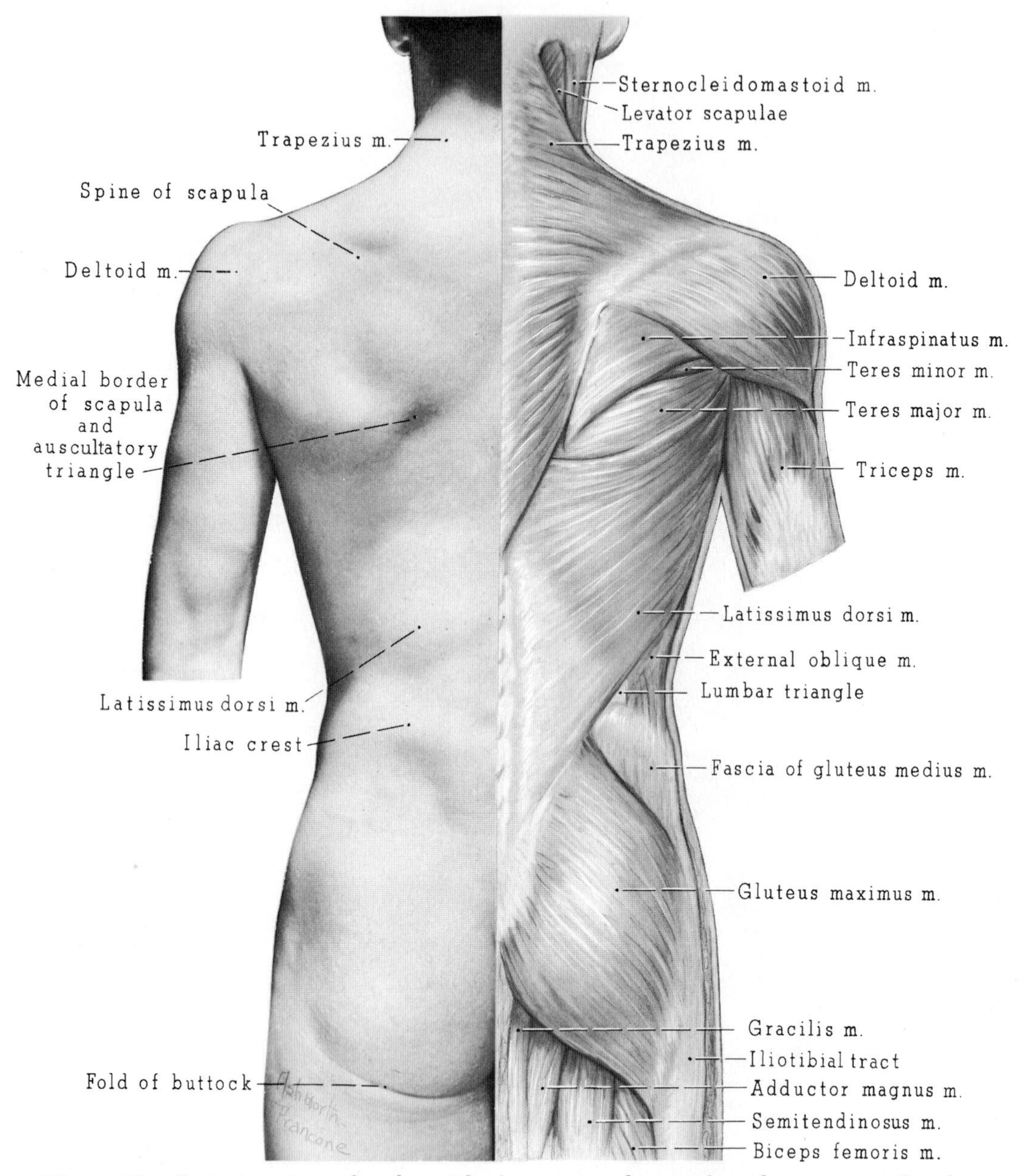

Figure 13. Posterior view of male, with skin removed on right side to expose first layer of muscles.

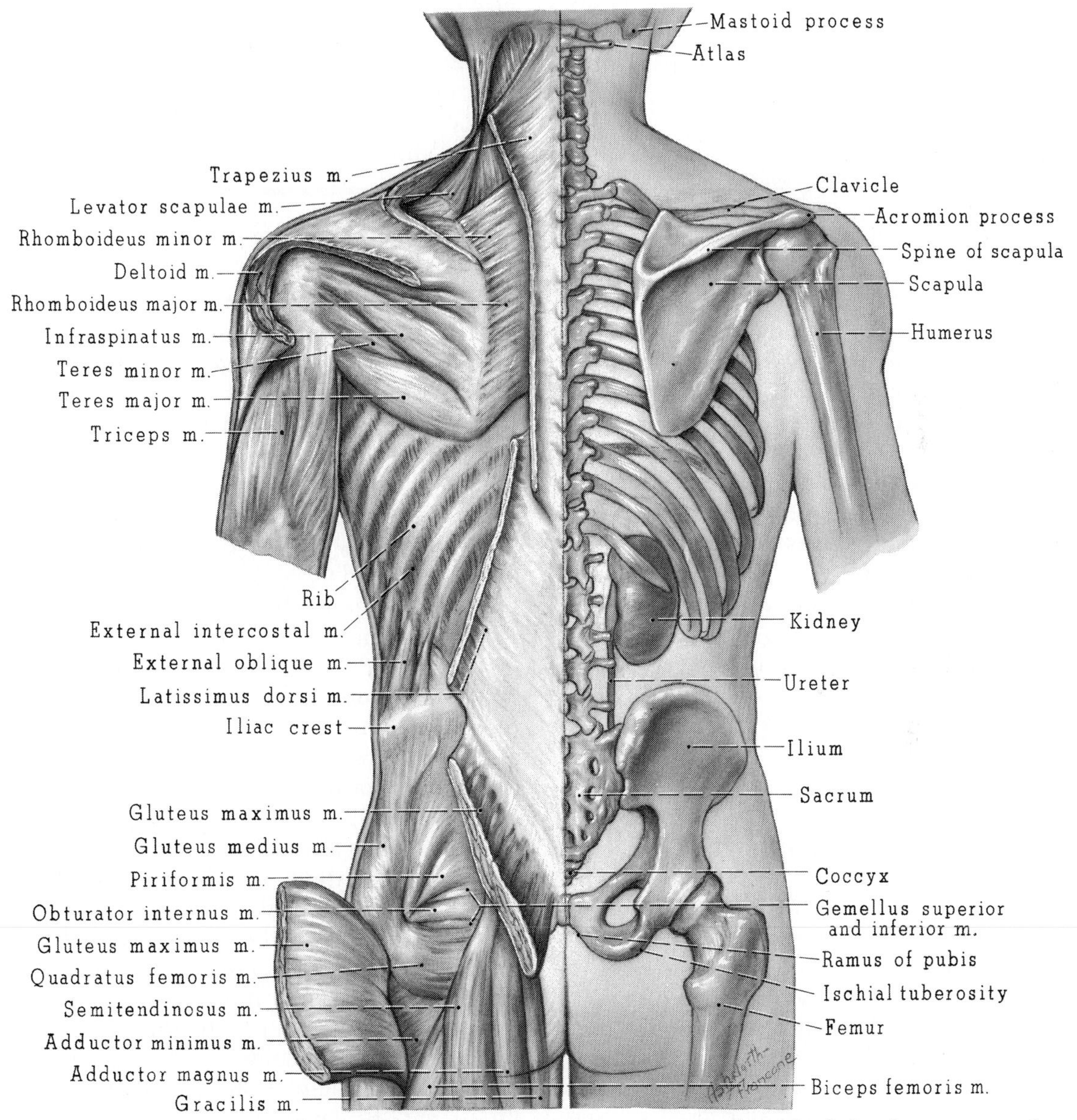

Figure 14. Most of the superficial muscles have been removed on the left side to expose the deep layers. All the muscles have been removed on the right side, exposing the skeletal framework.

THE COORDINATED WHOLE

How Are the Parts of the Body Functionally Related?

The human body is not simply a collection of substances or parts; it is a highly organized and precisely coordinated whole that functions as an harmonious unity. Each cell (or tissue, or organ, or system) benefits from its own activity as well as helping and benefiting from the activity of the cells of the other organs and systems. They all work to *help* themselves and one another, as well as to *control* themselves and each other.

The overall coordination of the body and its parts is so important that two special organ systems have evolved to handle this special function. These are the nervous system, which operates by sending and receiving electrical signals throughout the body, and the endocrine system, a system of eight glands, each of which keeps track of chemical substances in the body and sends out, when needed, stimulating chemical messengers, known as *hormones* (from the Greek *hormaein,* meaning to set in motion). These two systems are responsible for maintaining the symphony of activities of the individual cells, tissues, organs, and systems.

What Is the Internal Environment?

A cell can live within a certain range of conditions or environments. There must be a certain amount of food, water, and oxygen present, and waste products must be taken away regularly. This situation is quite similar to that for each human being. However, the individual person can move around and seek environments in which he can survive, unlike the cell, which is limited in its movement.

To understand the problem of cellular environment, you might ask yourself what you would need to survive if you were sealed in a large jar. How would you obtain fresh air? Where would you obtain your food? How would you get rid of your waste products? How would you protect yourself from heat and cold?

Living organisms solve these problems partly by their movement within the external environment, and partly by maintaining a fairly constant *internal environment* for each cell. If this internal environment changes too much, either over the whole body or in particular areas, then the cells in that area are going to begin to act abnormally and, eventually, die.

What Is Homeostasis?

The process of keeping the internal (cellular) environment fairly constant, or within a "normal" range, is known as *homeostasis* (ho″me-o-sta′sis). The term homeostasis was derived from the Greek words, *homoios,* meaning to keep the same or to maintain and *stasis,* a standing still.

In order to accomplish homeostasis, the various systems of the body must be able to respond to the important changes in the environment, before these changes begin to harm the cells. Four major organ systems are responsible for maintaining a homeostasis of the primary factors in the internal environment of each cell. These are the *respiratory system,* which operates to maintain an oxygen supply and remove carbon dioxide; the *digestive system,* which supplies all solid nutrients and salts; the *urinary system,* which regulates the water content and the concentration of many substances in the fluid surrounding each cell; and the *circulatory system,* which connects each individual cell with the other organ systems. In this way, the circulatory system holds a central place of importance in maintaining the internal environment of each cell.

STUDYING FOR TESTS

Here is a suggestion to help you discover how much you are learning. It will be useful to follow it each time you complete a chapter.

Take a piece of notebook paper and fold it in half. Use this to cover the text under each question in the chapter. Then *write down* on another sheet as much as you can in answer to each question. Go through the entire chapter before checking your answers.

LEARNING EXERCISES

1. Construct a diagram showing the major cavities of the body; label.
2. Following Figure 3, construct a drawing of the human body in anatomical position; with the book closed, label the diagram.

CHAPTER 2

THE CELL

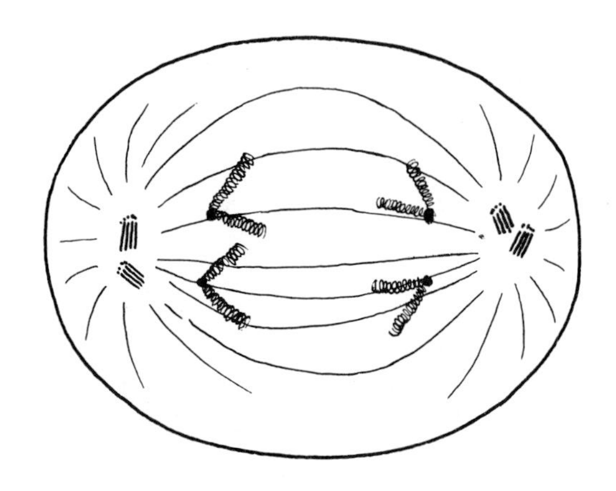

WHAT IS A CELL?

Cells are the tiny building blocks of the body. Unlike simple substances or collections of substances, cells are *living units.* This is shown by their ability to *grow* and *reproduce,* their need for *nourishment* and disposal of *waste products,* and their ability to *respond* and *adapt* to changes in the environment.

Human cells are so small that 2000 of them would have to be lined up to measure one inch; each cell being about 1/2000th of an inch across. This is so small that individual human cells cannot even be seen without the aid of a microscope. The microscope was developed only about 135 years ago, and only since then have we known that large living organisms are made up of cells.

Cells come in many different shapes. Some are shaped like long hot dogs, some are like square bricks, some are like flat tiles, and many are of various, irregular shapes.

What Are the Three Main Divisions of the Cell?

For study purposes, the three main divisions of each cell are the cell mem-

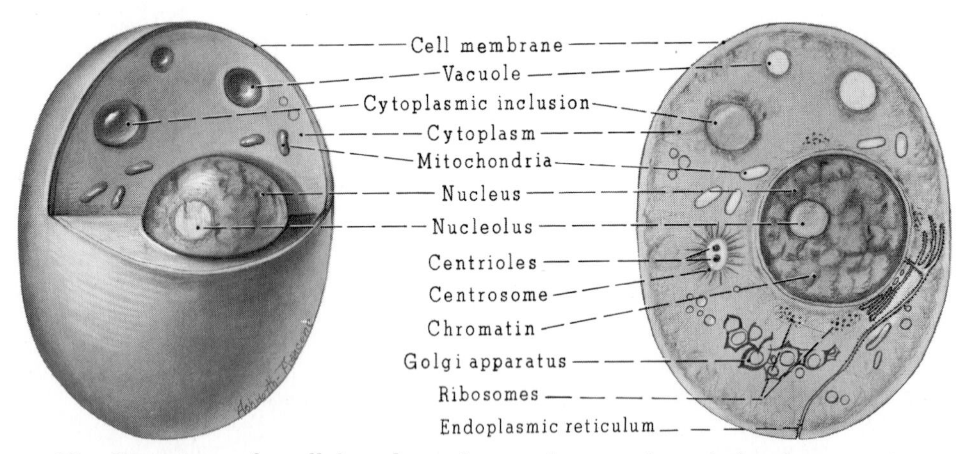

Figure 15. Two views of a cell, based on what can be seen through the electron microscope.

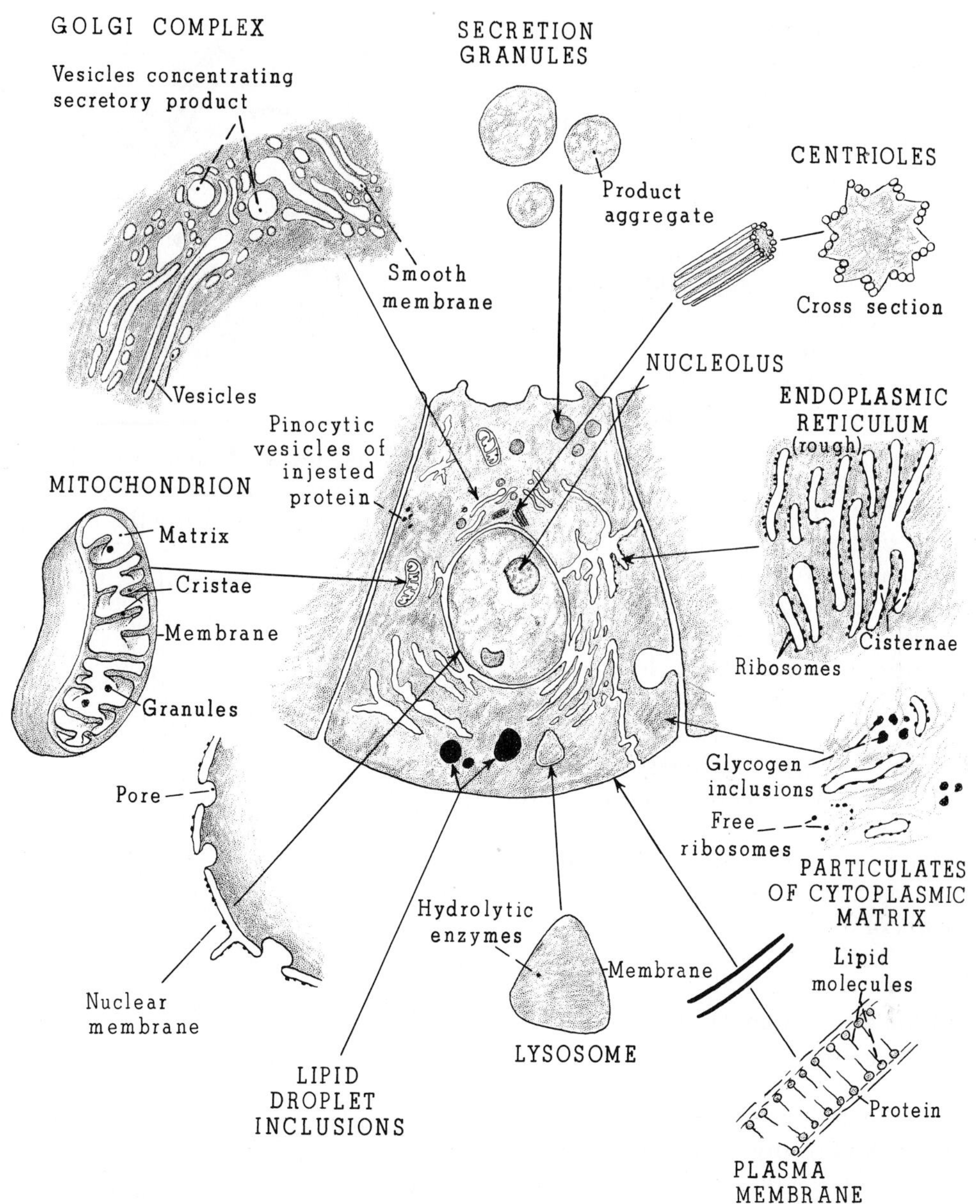

Figure 16. Parts of a cell as seen through the electron microscope. (After Bloom and Fawcett: *A Textbook of Histology*, eighth edition. Philadelphia, W. B. Saunders, 1962.)

brane, the cytoplasm, and the nucleus (Figs. 15 and 16).

The cell membrane is the extremely thin covering of the outer surface of each cell. It operates in both cellular ingestion (eating) and cellular excretion (of waste products), and generally serves to control the exchanges of air, water, and chemical substances between the cell and its environment.

The cytoplasm is the main body substance of the cell. It is made up of water and contains many microminiature "factory" units, discussed below.

The *nucleus* is contained in its own separate membrane, which floats in the cytoplasm, usually toward the center of the cell. The nucleus is the control center of the cell, directing and coordinating the activities of all parts of the cell.

How Does the Cell Membrane Function?

The cell membrane controls what substances pass into or out of the cell. This control is selective and is very important for maintaining the proper makeup of the cellular material. Tiny holes in the cell membrane allow some substances, such as oxygen and carbon dioxide, and water, to pass in either direction fairly freely. But larger substances are restricted from moving into or out of the cell. When the cell wants to ingest some larger nutrient (food) substances, such as proteins or sugars, it forms little incuppings and "swallows" these substances into the cell. *Pinocytosis* (pi″no-si-to′sis) (pino- from the Greek *pinein*, meaning to drink, + *kytos*, cell + *osis*, process) is the process in which the cell membrane forms a channel and "swallows" nutrient molecules (into little packets, called vacuoles) in its environment (Fig. 17). Phagocytosis (fag″o-si-to′sis) is the process of ingesting large particles, in which the cell surrounds and engulfs the particles. Figure 18 shows a white blood cell (neutrophil) engulfing some bacteria. (Phago- is derived from the Greek *phagein*, to eat).

What Are the Cytoplasmic Organelles? What Do They Do?

The cytoplasm of all cells has been found to be highly organized, containing several types of functioning "factory" units, which are like miniature cellular organs. These are referred to as *organelles* (or″ganels′); this term means, literally, "little organs". These include the mitochondria, lysosomes, ribosomes, and the endoplasmic reticulum.

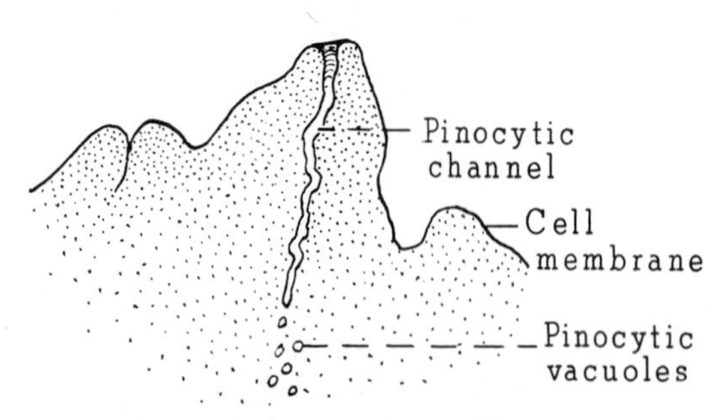

Figure 17. Pinocytosis.

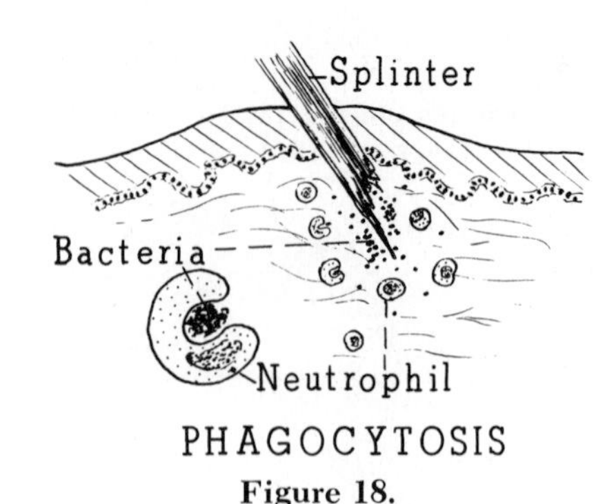

Figure 18.

Mitochondria (mit″o-kon′dre-ah). These are commonly referred to as the "powerhouses" of the cell. Their main function is to change small food molecules into other forms of energy that can be used by the cell for a variety of activities. Mitochondria look like tiny, sausage-shaped bodies, and are made up of a double membrane that is formed into several "inner folds."

When simple food molecules, such as the sugar glucose, enter the mitochondria they are broken into small pieces. This process releases energy. The energy is then stored in a very important, special molecule called *ATP*. The abbreviation ATP stands for *adenosine* (ah-den′o-sin) *triphosphate*. The ATP molecules, with their stored energy (they can be thought of as storage batteries or universal energy transporters) move out of the mitochondria and supply energy wherever it is needed in the cell (Fig. 19).

The number of mitochondria in a cell depends on its energy needs. Muscle cells use a great deal of energy in their daily activity, and each cell contains several thousand mitochondria. Skin cells often have less than a hundred.

Lysosomes (li′so-soms). Sometimes referred to as the digestive organs (or organelles) of the cell, (probably because of their Greek origin; *lyso-* meaning to dissolve, and *-some*, meaning body), lyso-

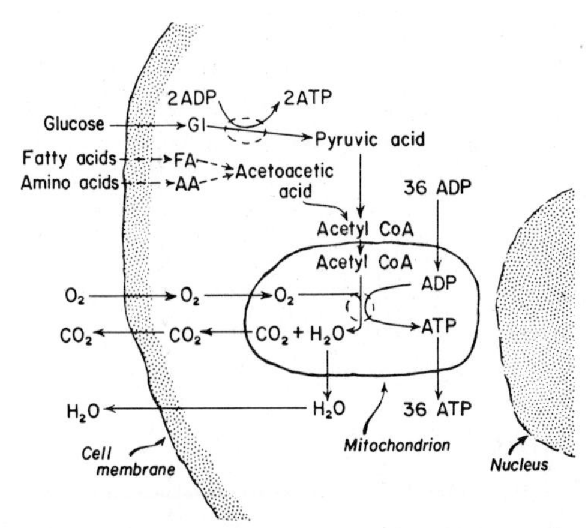

Figure 19. Formation of ATP in the cell. (From Guyton, A. C.: *Basic Human Physiology*. Philadelphia, W. B. Saunders Co., 1971, p 17.

somes are tiny units, even smaller than mitochondria, containing several powerful enzymes. (An *enzyme* is a chemical substance that causes a chemical reaction to take place.) When large, complex food molecules are taken into the cell, the lysosomes act to break down the particles into simple substances. These simple food molecules can then enter the mitochondria for energy processing.

Ribosomes (ri'bo-soms). These are still smaller granules of ribonucleic acid, or RNA, and protein, and are usually attached to the endoplasmic reticulum. The ribosomes are the specific sites at which cells construct new protein molecules. These new molecules are used for many purposes within the cell, and some are passed out of the cell to act in other parts of the body.

Endoplasmic Reticulum (re-tik'u-lum). This is one organelle which is spread through the cytoplasm. It is a cellular membrane arranged to form complex networks of channels and surfaces throughout the cytoplasm. In some parts of the endoplasmic reticulum one finds thousands of ribosomes; here is where new proteins are made. The channels of the endoplasmic reticulum serve also to carry substances to different parts of the cell, and sometimes function in storage.

How Does the Nucleus Function?

The nucleus of the cell controls and coordinates cellular activity by controlling the chemical makeup of the cytoplasm.

How Does the Nucleus Control and Coordinate Cellular Activities?

When some process or activity of the cell needs to be increased, the nucleus (or, more specifically, the DNA molecules in the nucleus) send out directions by messenger (messenger-RNA molecules) to produce specific *proteins* that will then act to speed up the process or activity. In order to slow down some cellular activity, the DNA can either stop sending out directions for production of the key substances or else send out new directions for construction of a protein blocking substance, or inhibitor, which rapidly slows down the activity.

How Does the Nucleus Direct Protein Manufacture?

The easiest way to understand the function of the nucleus is by using an analogy. Imagine the cell to be a large carpentry work shop. The director of the workshop sits in a separate office—this office would be the nucleus. The director himself has all the plans and designs needed to construct objects in the workshop; the director, with all the different plans and designs, corresponds to the special *DNA molecules* in the nucleus. Since the director is always very busy deciding which products should be made in the work area, he never leaves his office. So, in order to get the plans and designs from his office (the nucleus) to the working area (cytoplasm), a messenger is needed. The messengers in the cell are special *RNA molecules,* known appropriately as *messenger RNA.* These messenger RNA molecules take the plans or designs for the specific products the director wants manufactured and carry them from the director's office (nucleus) to the work benches. The work benches, where the products are to be assembled, correspond to the *ribosomes.* The ribosomes are composed of another, very large type of RNA known simply as *ribosomal RNA.* So far, we have the plans and designs (brought by or contained in the messenger RNA) at the work benches (ribosomes) but we lack the materials needed to construct the product specified in the plans.

The materials needed to manufacture the specified product must be gathered and brought to the work benches. This job requires another worker. In the cell, this worker is another type of RNA, known as *transfer RNA,* since it functions to transfer the basic building materials from around the working area (cytoplasm) to the ribosomes (work benches). All the objects produced in the cell's workshop, at the direction of the director (DNA) are *proteins;* proteins are made up of *amino acids.* The amino acids, then, are the basic building materials that need to be gathered by the transfer RNA's, and brought to ribosomes. At the ribosomes, the amino acids are assembled into proteins, according to the plans brought by the messenger RNA.

Just how the materials (amino acids) are assembled to form the products (proteins), is still somewhat of a mystery. Some scientists think the ribosomes do it, and therefore, according to our analogy, should be thought of as automatic work benches, which receive design plans and materials and then automatically construct the right objects. Others feel that the messenger RNA and transfer RNA participate in the construction process. In any case, we can

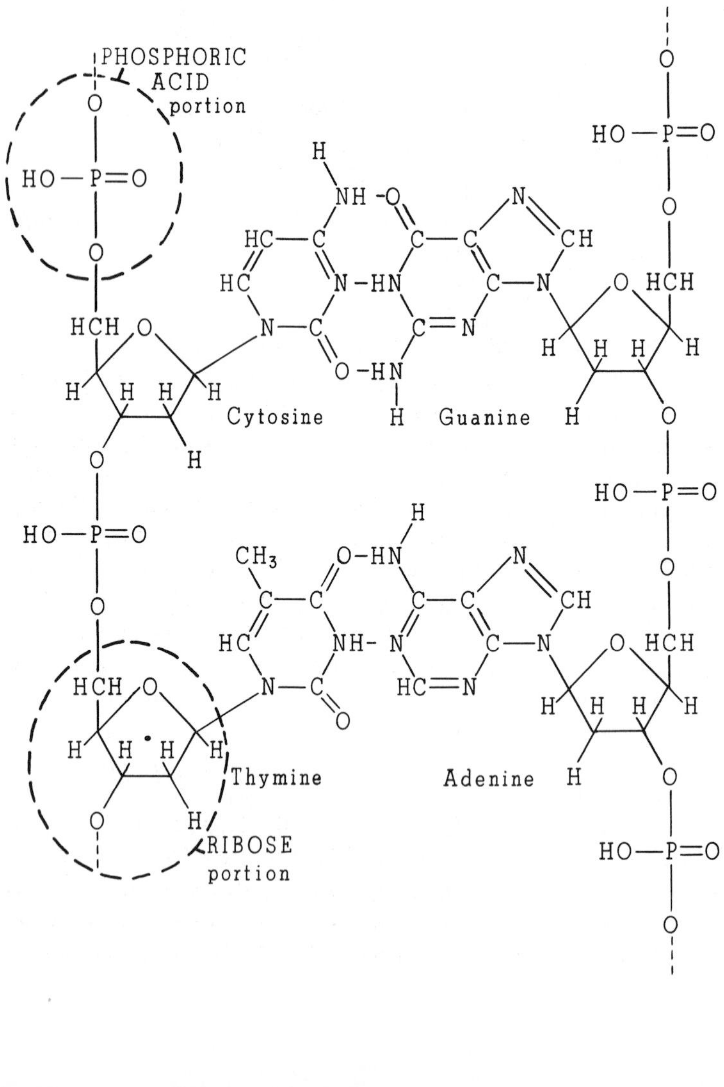

Figure 20. Chemical structure and diagrammatic representation of DNA molecule and its components (Richard Lyons, M. D.).

generalize and say that the DNA, which always remains in the nucleus, is the ultimate director—it selects the plans to be sent, and the number of products of that type to be manufactured. The RNA's (messengers, transfers, and ribosomal) are the workers; they operate to carry the plans, gather the materials (amino acids), and construct the products (proteins).

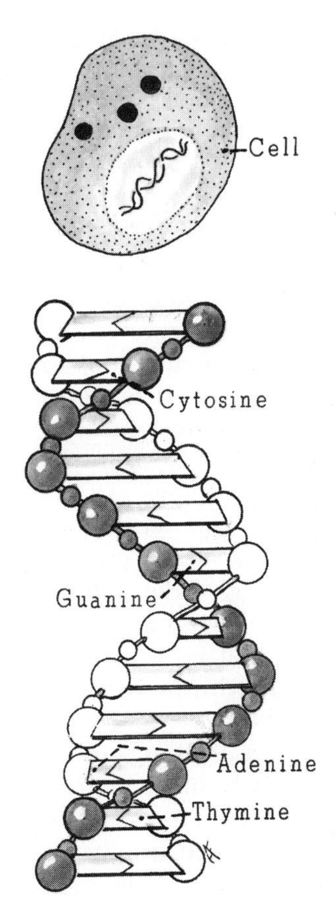

Figure 21. Diagrammatic representation of DNA helix.

What Is DNA?

DNA is the main substance found in the nucleus of every cell; it is rarely if ever found outside the nucleus. It is a very large molecule whose full name is *deoxyribonucleic* (de-ok″se-ri″bo-nu′kle′ik) *acid*, abbreviated DNA. There are thousands of these large DNA molecules in the nucleus of every cell. Some of the DNA in the nucleus simply holds information or directions for production (referred to as *synthesis*) of protein molecules, whereas other DNA's operate in coordinating and selecting which directions should be sent out (by messenger) and when. The protein molecules are made in the cytoplasm (at the ribosomes), but the directions on how to synthesize them are kept stored in the nucleus, in the DNA molecules.

The DNA molecules are extremely long and thin, and are composed of surprisingly few basic chemical compounds (Fig. 20). These include *phosphoric acid*, a sugar called *deoxyribose*, and four nitrogenous bases: *adenine*, *guanine*, *thymine*, and *cytosine*. The structure of the molecule is easily described by first noting that it has two very long strands of alternating deoxyribose and phosphoric acid (d-p-d-p-d-). The number of each of these molecules in each strand is well into the thousands. Attached to each deoxyribose molecule is one of the four nitrogenous bases. The specific sequence of bases along the strands makes up a code—the genetic code.

The nitrogenous bases of each strand face each other, and are loosely bonded in opposite matching pairs. The deoxyribose and phosphoric acid sequences serve as the sides of a ladder-like structure. The two strands (on the ladder) are twisted, to form the double spiral, or helix (Fig. 21). The opposite pairing of the bases is quite specific; guanine (gwan′in) and cytosine bond only with each other; likewise, adenine and thymine are exclusive complements. Therefore, if there is a sequence of bases on one strand: G-A-A-C-T-G, then matching it respectively on the opposite strand is the complementary sequence: C-T-T-G-A-C.

What Is the Language of DNA?

Within the last decade, biochemists have managed to decipher the coding system of the DNA molecule—the code by which DNA molecules store the information on how to make all the various proteins produced by cells. This code, most interestingly, is the same for all living organisms. The code itself turns out to be extremely simple and straightforward. The letters of the code are represented by the nitrogenous bases, which we can abbreviate as above: C, for cytosine; G, for guanine; T, for thymine; and A, for adenine.

These letters go to make up words, or code words. The key to understanding the language of DNA turns out to be the fact that each code word contains only three letters. For instance, CAT, or TAC, or GAA, and so forth. Each of these words names, in the language of DNA, one of the amino acids. The amino acids are the building blocks of the proteins. So, if the DNA wants to give the code or information for making some protein, it simply sends out protein; that is, it sends out the messenger (RNA) with the words, or sentence, that specifies that particular protein. For instance, CAT-CAT-TAC-GAA-CAT, would specify one very small protein made up of five amino acid units (only three different kinds of amino acids, however). Most proteins are larger than this, ranging from 25 or more amino acid units up to several thousand for complex proteins. The protein hemoglobin, which operates in red blood cells to carry oxygen, is known to be made up of about 575 amino acid units. Insulin, the protein involved in the maintenance of the blood sugar level (and given artificially to diabetics) contains about 51 amino acid units.

Table 1. SOME BIOLOGICALLY IMPORTANT AMINO ACIDS AND THEIR RNA CODE WORDS

AMINO ACIDS	RNA CODE WORDS			
Alanine	CCG	UCG		
Arginine	CGC	AGA	UCG	
Aspargine	ACA	AUA		
Aspartic acid	GUA			
Cysteine	UUG			
Glutamic acid	GAA	ACU		
Glutamine	ACA	AGA	AGU	
Glycine	UGG	AGG		
Histidine	ACC			
Isoleucine	UAU	UAA		
Leucine	UUG	UUC	UUA	UUU
Lysine	AAA	AAG	AAU	
Methionine	UGA			
Phenylalanine	UUU			
Proline	CCC	CCU	CCA	CCG
Serine	UCU	UCC	UCG	
Threonine	CAC	CAA		
Tryptophan	CGU			
Tyrosine	AUU			
Valine	UGU			

What Is RNA?

The RNA molecules, or ribonucleic acids, are a group of molecules found mostly in the cytoplasm; however, their functions are to be understood in relation to the organization and coordination activities of DNA.

The basic building blocks of RNA molecules are quite similar to those of DNA; however, RNA contains the sugar *ribose*, instead of deoxyribose, and the base *uracil* (u′rah-sil), in place of thymine, and RNA is usually *single stranded;* in all other ways, it resembles DNA in molecular structure.

There are three different types of RNA found in the cell, each with a specific function related to protein synthesis. These are *messenger RNA* (abbreviated mRNA), *transfer RNA* (tRNA), and *Ribosomal RNA* (rRNA).

Messenger RNA. The name is quite appropriate, since this molecule carries the information from the DNA molecule in the nucleus out into the cytoplasm, where the actual protein synthesis takes place. mRNA is formed when the two strands of the DNA molecule separate and the paired building blocks for RNA match up against the code of the exposed DNA. Then an enzyme stimulates the building blocks to combine into one RNA molecule. The mRNA that has been formed then contains the information in the gene (genetic sentence) of the DNA (see Table 1). The messenger RNA then moves to the cytoplasm.

Transfer RNA's. These are the smallest RNA molecules. They circulate around the cytoplasm, where they capture stray amino acids and bring them to a specific site in the cytoplasm, to be combined into protein molecules. There is one type of transfer RNA for each type of amino acid.

Ribosomal RNA. The site of the actual protein synthesis is at the ribosomes, which are made up of ribosomal RNA and special ribosomal proteins. *Ribosomal RNA* is a very stable molecule of high molecular weight, which probably determines the alignment of the messenger RNA when it reaches the ribosome from the nucleus.

How Is the Protein Synthesized?

The process by which a particular protein comes to be synthesized within the cell is most easily viewed as beginning with the uncoiling of a specific DNA molecule in the nucleus. RNA parts then move to line up along the code strand of the uncoiled DNA molecule, according to the base-pair rule. By action of an enzyme, the messenger RNA molecule is formed. The

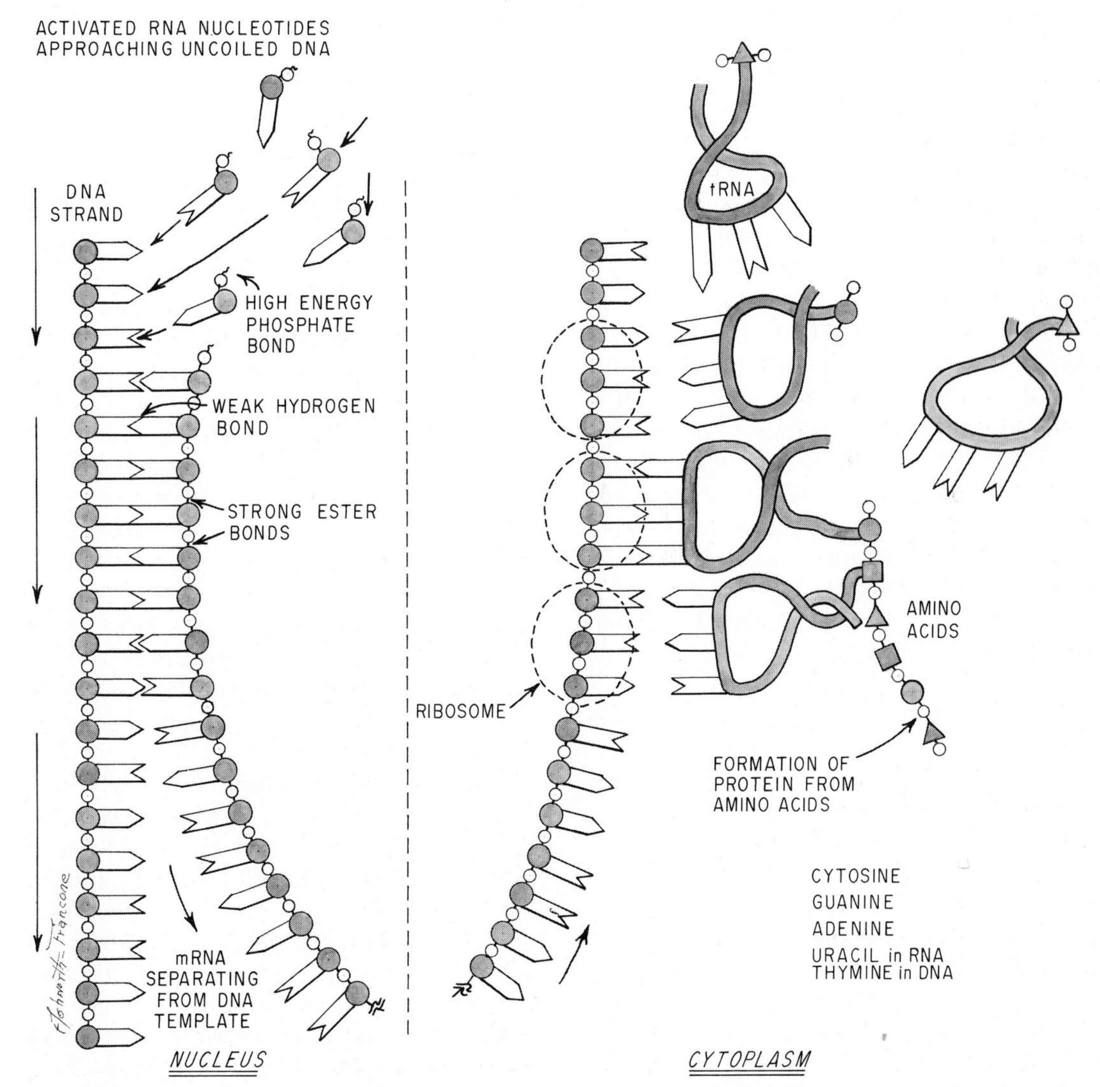

Figure 22. Messenger RNA duplicates the information encoded in the DNA molecule and attaches itself to the ribosomes. Amino acids, carried by loops of transfer RNA, are bonded together, move away, and become a three-dimensional protein.

mRNA then migrates out of the nucleus to the ribosomes in the cytoplasm (Fig. 22).

In the meantime, transfer RNAs have collected the amino acids and carried them to the ribosomes for synthesizing the protein molecule.

As the messenger RNA arrives at the ribosome, it is sequentially "read," or "translated," and the amino acids are properly lined up, ready for formation of the protein.

In some unknown manner, the ribosome then stimulates the now properly ordered amino acids, to form the indicated protein—the protein whose code was contained in the DNA and mRNA.

What Are Relative Sizes in the Cell?

In order to help visualize the relative sizes of the parts of a cell, we can imagine a cell enlarged to a point at which its diameter is 100 yards—the length of a football field.

Each of the *mitochondria*, or energy factories of the cell, would be about the size of an automobile. You could fit 30 of these, end-to-end, across the length of the football field—100 yards. Glucose molecules, which in our enlarged cell would be about 1/8 inch in diameter, enter the mitochondria and are processed. ATP molecules (1/2 inch), carrying the stored energy, migrate out of the mitochondria to various parts of the cell, where they release the energy as needed for cellular activities.

The *lysosomes*, which are the cell's digestive organs, would be about the size of small sports cars.

The *ribosomes* would be about the size of baseballs (3.6 inches in diameter), and they would be scattered throughout the cytoplasm.

The *cell membrane* would be only 1 1/2 inches thick, or about twice as thick as this textbook. Similarly, the endoplasmic reticulum would be constructed of a membrane about 1 1/2 inches thick.

An average *bacterium* that might invade the cell would be about the size of a mitochondrion (the size of an automobile).

On the *molecular level*, the water molecules would be only 1/18th inch in diameter, or smaller than the ball in a ballpoint pen. You could fit 780,000 of these balls into a teacup. The large protein molecules, those that make up many of the structures of the cell, and take part in most cellular chemical processes, would be 1 to 2 inches in diameter.

Perhaps the most interesting features of our enlarged cell are the long DNA molecules. They would appear as lengths of relatively thin rope (1/3 inch in diameter). There is so much DNA in the nucleus that if we put all the DNA end-to-end it would make a thin rope 5500 *miles* long. When this is wound-up, as it is in the nucleus, it would make a tight ball about 10 yards in diameter.

The whole *nucleus*, which contains some special nuclear proteins as well as the DNA, would be about 17 yards in diameter.

How Do Cells Reproduce?

All body cells reproduce by a process of division called *mitosis* (mi-to'sis). In the process of mitosis the original cell self-duplicates its internal constituents and then divides, producing two *daughter* cells, which are identical to the parent type.

It is through the self-duplication of the material in the nucleus—the DNA molecules—that it is possible for the parent cell to divide into two identical daughter cells. The DNA is the crucial hereditary (or genetic) material.

DNA self-duplication begins when the DNA molecules split apart lengthwise into two long strands (just as in mRNA formation). Next, DNA building blocks line up along the two strands to form, in each case, the complementary strand. The result is the production of the two new, identical DNA molecules. When the cell divides, one goes to one daughter cell and the other to a second daughter cell. Each new DNA (and each new daughter cell) contains one strand from the original DNA of the mother cell.

What Are the Four Main Stages of Mitosis?

Normal mitosis is usually described in four stages (Fig. 23). The normal, functioning state of the cell is the *interphase,* or "resting," state. During this phase the DNA molecules are somehow stimulated to self-duplicate. Also, during this phase the cell stores up many cytoplasmic constituents that will later be divided between the two daughter cells. (The factors that stimulate a cell to divide in an adult are not well understood. However, this is the subject of intensive research, since it is believed to involve the key to understanding the uncontrolled growth found in cancerous tumors.)

In the *prophase,* the DNA complex becomes tightly coiled; forming the rod like *chromosomes* (kro′mo-soms). Microtubules (mi-kro-tu′buls) begin to develop around each of the *centrioles* (sen′-tri-ol), like the spokes of a wheel. Some of these microtubules penetrate the nucleus, and apparently attach to the individual, coiled chromosomes at their midpoint, or *centronucleus.* During this process the nuclear membrane disintegrates and the nucleoli disappear.

Next, in *metaphase* the chromosomes move toward the center of the cell and arrange themselves in a plane perpendicular to the line connecting the centrioles. The structure formed is usually referred to as the *equational plate.*

Then, in *anaphase* the duplicated chromosomes begin to separate; one set moves to one pole and the duplicate set moves to the opposite pole. One chromosome from each of the 46 chromosomes found in human cells finds its way into each daughter cell, giving each cell an identical set of chromosomes.

Finally, in *telophase* the chromosomes reach the general location of the centrioles. Then the chromosomes uncoil, the nuclear membrane reappears, and an equatorial membrane appears at the former site of the metaphase plate, dividing the cytoplasm into two parts.

By this process, one cell with 23 pairs of chromosomes (or 46 individual chromosomes) becomes two cells, each of which has 23 pairs, or 46 individual chromosomes.

CLINICAL CONSIDERATION

What Is a Disease? An Injury?

The most general definition of a disease is any impairment (hindrance) of nor-

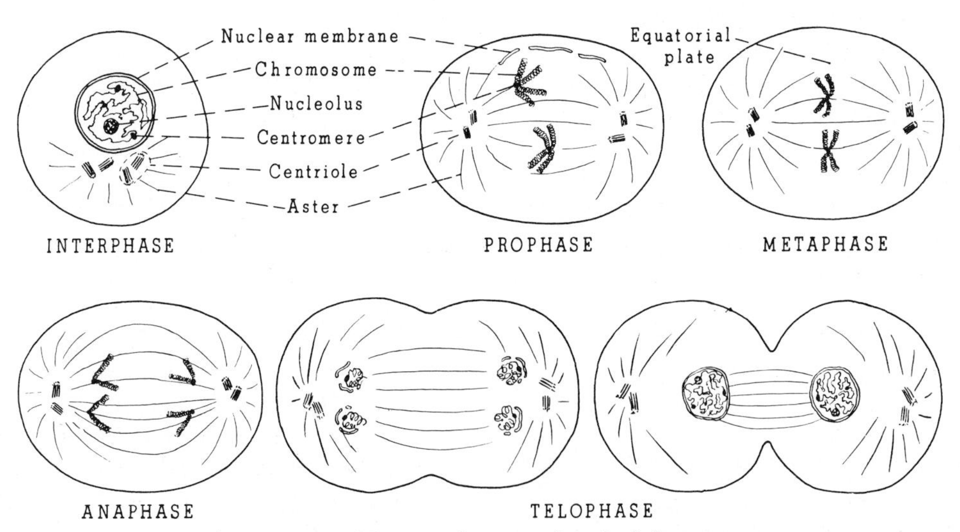

Figure 23. Mitosis, showing detail of division.

mal bodily activity, which affects the performance of the vital functions. In diagnosing a disease, it is most important to gain an understanding of the origin or cause of the abnormal conditions or symptoms of the body. An *acute disease or illness* is one with a sharp onset, and having a short and relatively severe course. A *chronic disease or illness* is one that persists over a long period of time.

Hereditary diseases are abnormalities that are usually handed down from generation to generation. It is generally agreed today that these problems can be traced, in the final analysis, to some abnormality of the chromosomes or the DNA that composes them.

Infectious diseases are those that can be transmitted from one individual to another. These diseases are due to the invasion and growth in the body of either viruses or bacteria. The invasion of larger parasitic organisms can also disrupt normal functioning.

Injuries can be classified into two general types. Immediate traumatic injuries, or *traumas,* result from some physical or chemical interference. This includes physical forces that result in broken bones and bruises, as well as contact with or ingestion of poisons. *Long-term injuries* can occur from many different aspects of our relation to our environment. Included in this category are injuries or abnormalities resulting from long-term usage of cigarettes, alcohol, and a variety of dietary excesses or deficiencies.

Table 3. **CHEMICAL COMPOSITION OF BODY FLUIDS**

	EXTRACELLULAR FLUID	INTRACELLULAR FLUID
Na^+	137 mEq/l	10 mEq/l
K^+	5 "	141 "
Ca^{++}	5 "	0 "
Mg^{++}	3 "	62 "
Cl^-	103 "	4 "
HCO_3^-	28 "	10 "
Phosphates	4 "	75 "
SO_4	1 "	2 "
Glucose	90 mgm%	0 to 20 mgm%
Amino acids	30 "	200 "
Cholesterol, Phospholipids, Neutral fat	0.5 gm%	2 to 95 gm%
PO_2	35 mmHg	20 mmHg ?
Pco_2	46 "	50 " ?
pH	7.4	7.1 ?

BODILY FLUIDS

What Is the Distribution of Water in the Body?

In the human body, about 60 to 70 per cent of the lean body weight is water (Table 2). About two thirds of this is contained inside the individual cells, as part of the cytoplasm. This is called the *intracellular* (inside the cell) water. The remaining

Table 2. **PERCENTAGE COMPOSITION OF REPRESENTATIVE MAMMALIAN TISSUES***

TISSUE	WATER	SOLIDS	PROTEINS	LIPIDS	CARBOHYDRATES
Striated muscle	72–78	22–28	18–20	3.0	0.6
Whole blood	79	21	19	1	0.1
Liver	60–80	20–40	15	3–20	1–15
Brain	78	22	8	12–15	0.1
Skin	66	34	25	7	present
Bone (marrow-free)	20–25	75–80	30	low	present

* Source: White, A., Handler, P., Smith, E., and Stetten, D.: *Principles of Biochemistry.* New York, McGraw-Hill, 1954.

one third is called the *extracellular* (outside of the cells) water (Table 3).

This extracellular water is very important. A large part of it is a major component of the blood. What is not contained in the blood is found slowly circulating between the cells as part of their environment.

What Is the Extracellular Fluid?

The extracellular water, both in the blood stream and between the cells, contains many dissolved substances. This water, together with its dissolved substances, is called the *extracellular fluid.*

The extracellular fluid is made up of water, salts, dissolved sugars (such as glucose), proteins, and a variety of substances in smaller amounts. It is very important that the concentrations of substances in this extracellular fluid remain fairly constant. This is a part of the homeostatic environment of the cells. If the concentration changes, so as to contain too much or too little of these substances, the cells will become sick and act abnormally, and eventually begin to die.

LEARNING EXERCISES

1. Construct a diagram of a cell and label the following parts: the cell membrane, cytoplasm, nucleus, mitochondria, ribosomes, lysosomes, and endoplasmic reticulum.
2. Draw a double-stranded DNA molecule with a G-A-A-C-T-G base sequence on one strand and the complementary base sequence on the other strand. Use a symbol D for deoxyribose, and a symbol P for phosphoric acid.

CHAPTER 3

TISSUES AND THE SKIN

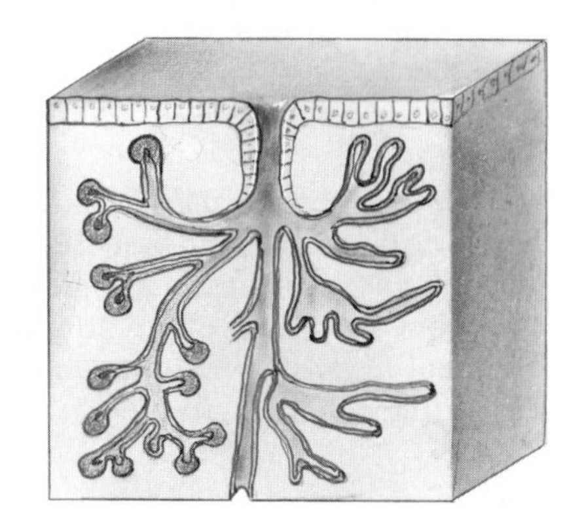

TISSUES

What Is a Tissue?

A tissue is a network of similar cells woven together. The cells hold themselves together by producing an intercellular (between the cells) substance, to which they all attach themselves. The precise amount and composition of these intercellular "cementing" substances varies, depending on the type of tissues, as will be seen.

These are four main types of tissues in the body: *epithelial, connective, muscular, and nervous.*

Epithelial Tissues. These tissues function in protection, absorption, and secretion. When serving a *protective* function, epithelial tissue is found in sheets, covering a surface such as the skin (Fig. 24). In the absorptive function, the epithelial tissues form specialized surfaces, as in the tiny air sacs of the lung tissue, where oxygen is absorbed into the blood. In the functions of secretion there are two specialized epithelial tissues—mucous membrane and glandular epithelium. *Mucous membrane* lines the digestive, respiratory, urinary, and reproductive tracts, secreting mucus as well as other special substances. Glands are incuppings of special epithelial tissue; they function to produce and secrete a wide variety of important substances. The glands include simple sweat glands, mammary glands, and large salivary glands. (Fig. 25), as well as the endocrine gland group, which includes the thyroid, pituitary, ovaries, and testes. (The endocrine glands are discussed in Chapter 11.)

Connective Tissues. These tissues allow movement and provide support. In this type of tissue, there is an abundance of fiber-like intercellular material, which the cells produce and deposit in order to give great strength to the connective tissues (Fig. 26). The two main types of *loose connective tissue* are known as areolar (ah-re′o-lar) tissue and adipose tissue. *Areolar tissue* is the most widely distributed connective tissue; it resists tearing and is somewhat elastic. Its main function is to connect other tissues and to provide support when needed. *Adipose tissue* is specialized areolar tissue, with many fat-containing cells. Its functions are in fat storage and as a cushion between organs and other structures in the body.

Dense connective tissue makes up the main parts of tendons, ligaments, and various tough tissue sheaths, which hold various parts of the body in position. For instance, dense connective tissue holds joints together and holds muscles and organs in position.

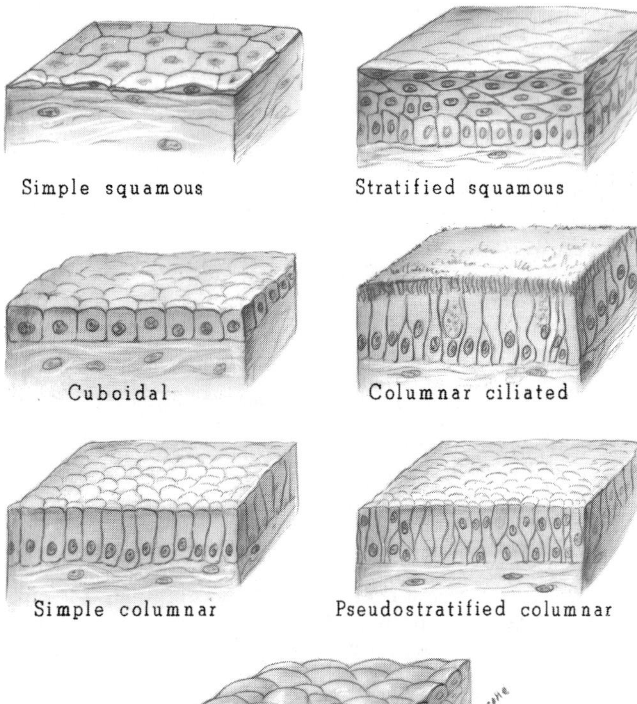

Figure 24. Types of epithelial tissue, classified according to shape and arrangement of cell layers.

There are a number of specialized connective tissues. *Cartilage,* which is a special connective tissue, is the flexible material found on the outer ear and nose; it is also found between bones at the joints, where it serves to cushion shocks to the body. *Bone,* which is also a special connective tissue, has the important property of rigid strength. This property arises from the fact that bone cells surround themselves with large stores of inorganic salts – primarily *calcium phosphate. Blood* and blood-forming tissue are also considered to be special connective tissues. This is because blood cells are born and grow to maturity in the bone marrow, which is the tissue in the center of most bones (Fig. 27). Blood is a fluid tissue that circulates

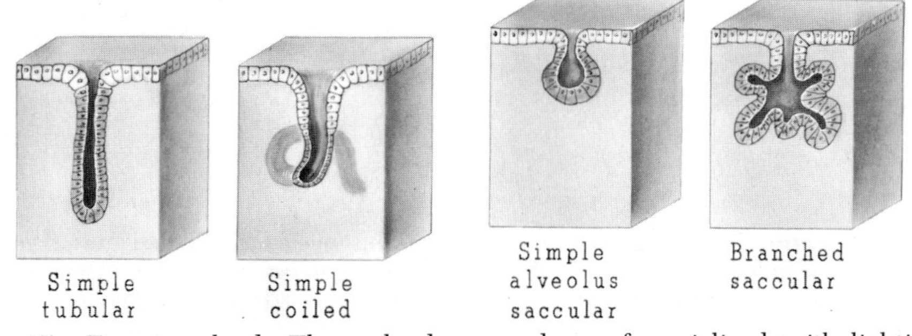

Figure 25. Exocrine glands. These glands are made up of specialized epithelial tissues.

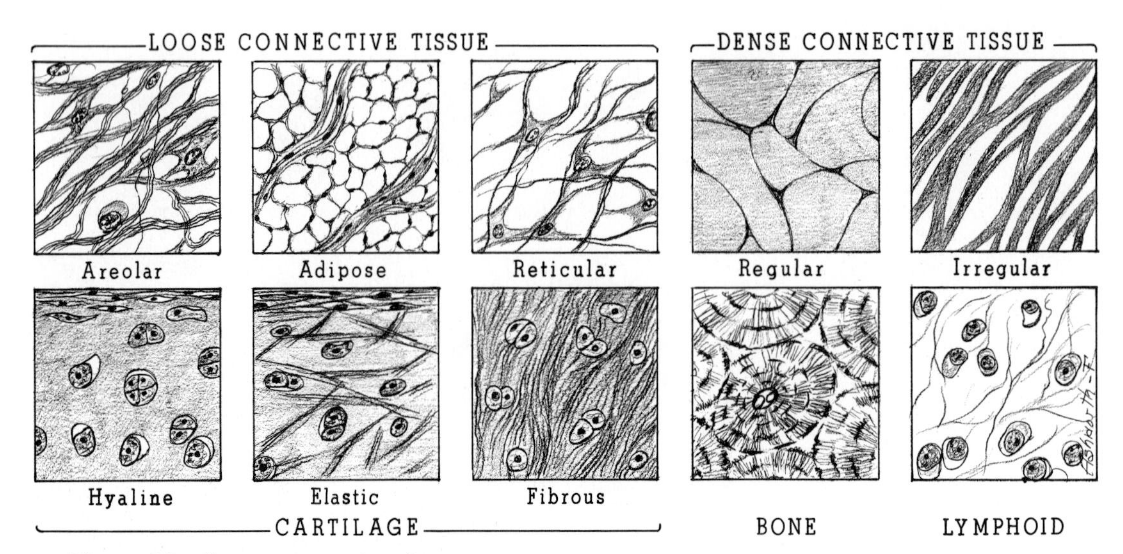

Figure 26. *Loose connective tissue:*

Areolar: loosely arranged fibroelastic connective tissue.

Adipose: regions of connective tissue dominated by aggregations of fat cells.

Reticular: makes delicate connecting and supporting frameworks, enters into the composition of basement membranes, produces lymphocytes and macrophages, and plays important roles as scavenger and agent of defense against bacteria.

Dense connective tissue:

Regular: fibers that are oriented so as to withstand tension exerted in one direction.

Irregular: fibers that are arranged so as to withstand tensions exerted from different directions.

Cartilage:

Hyaline: the most fundamental kind of cartilage, consisting of a seemingly homogeneous matrix permeated with fine white fibers.

Elastic: specialized cartilage with elastic fibers in the matrix.

Fibrous: specialized cartilage emphasizing collagenous fibers in its matrix.

Bone: a tissue consisting of cells, fibers, and a ground substance, the distinguishing feature of which is the presence of a ground substance of inorganic salts.

Lymphoid tissue: a tissue consisting of two primary tissue elements – reticular tissue and cells, chiefly lymphocytes – intermingling in intimate association in the reticular interstices.

through the body, carrying nutrients to cells and removing waste products.

Muscular Tissue. The main function of muscle tissue is to produce movement through contraction and relaxation. The voluntary muscles are made up of a particular type of muscle tissue. The internal, involuntary muscles are made up of an-

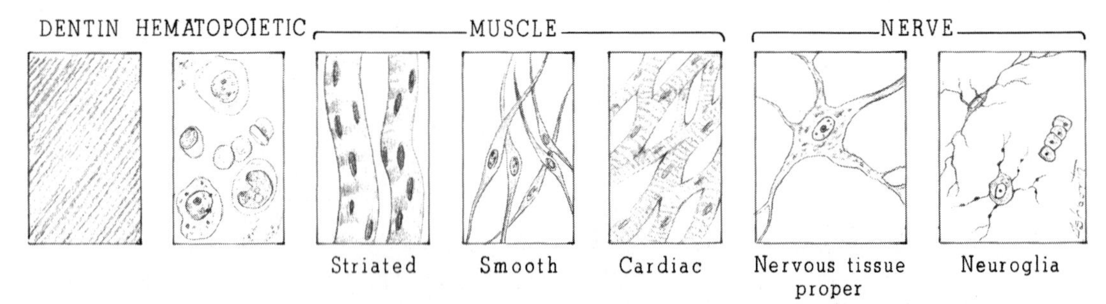

Figure 27. *Dentin,* like bone, consists of a collagenous mesh and calcified ground substance; unlike bone, it contains neither vessels nor total cells. *Hematopoietic tissue:* blood-forming tissues; i.e., lymphoid organs, bone marrow, and spleen. *Muscle tissue* has the properties of contractility and excitability (see Chapter 5). *Nervous tissue* has the properties of excitability and conductivity (see Chapter 6).

other, specialized type of muscle tissue. Finally, the heart, which is the most important muscular organ of the body, contains its own special type of muscle tissue (Fig. 27).

Nervous Tissue. This is the most highly organized tissue in the body. It functions to control and coordinate all the many complicated activities of the body. In nervous tissue, the specialized conducting cells are called *neurons.* Neurons are linked together to form nerve pathways (Fig. 27.).

THE SKIN

What Is the Skin?

The skin is an organ (or group of organs) that covers the external surface of the body. It functions as a protective surface, keeping out bacteria and harmful substances. It also operates in the regulation of bodily temperature and the disposal of water (through sweat). Figure 28 shows a three-dimensional view of the skin.

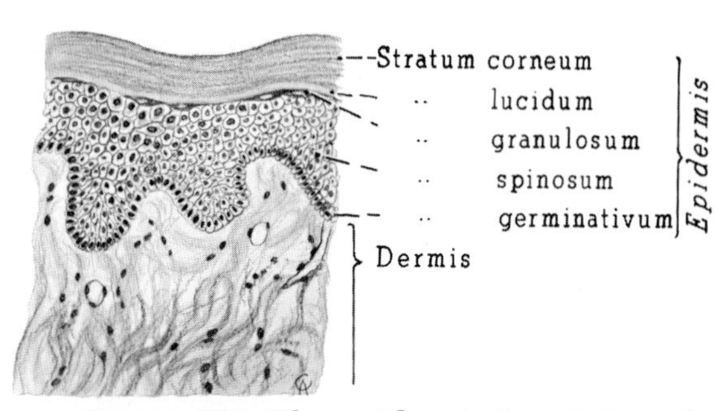

Figure 29. The epidermis (consisting of five distinct layers) and the dermis comprise the protective covering of the body.

What Are the Epidermis and the Dermis?

The skin is composed of two distinct kinds of tissue. The outer surface, known as the *epidermis (Epi* - meaning upon), is made up of several layers of a special epithelial tissue. The deeper layer of tissue is known as the *dermis* (Greek for skin), and consists of a tough connective tissue (Fig. 29).

The innermost cells of the epidermis are continually reproducing and replacing the cells of the outer surface, which flake

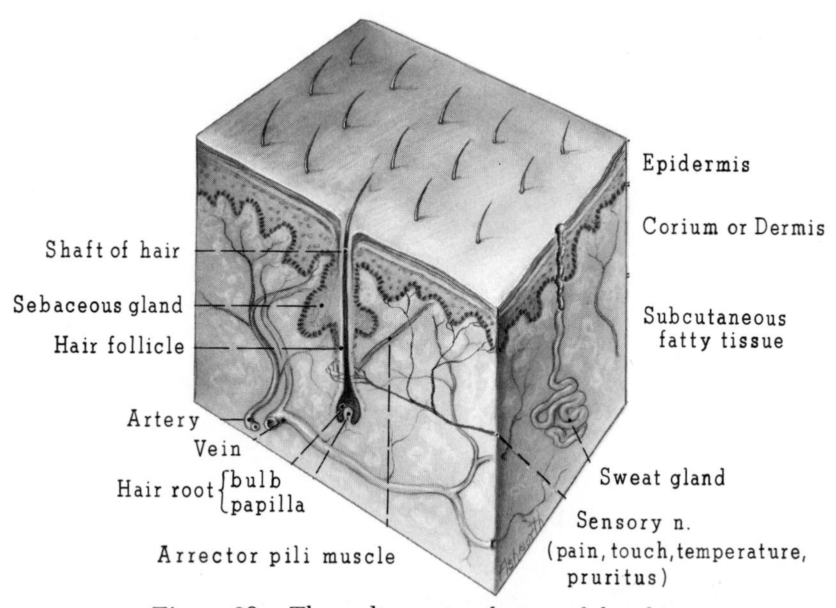

Figure 28. Three-dimensional view of the skin.

off by the millions each day. One of the factors that determine the skin color is the amount of a dark pigment substance called *melanin* (mel'ah-nin), which is produced by the epidermal cells. Melanin production increases with exposure to strong ultraviolet light (suntan).

The dermis (or corium) lies directly beneath the epidermis and is composed of strong fibrous connective tissues. Within the dermis are hundreds of nerve endings, sweat glands, and hair roots. About one third of the body's blood normally flows through the dermis. Heat or motion can increase this flow and give the skin a reddened appearance.

What Are the Appendages of the Skin?

The main appendages of the skin are hair, nails, sebaceous glands, and sweat glands (Fig. 30).

Hair covers the entire body, except the palms, soles, and portions of the genitalia. The visible portion of the hair is the *shaft.* The *root* is situated in an epidermal tube, known as the hair *follicle,* which consists of an outer connective tissue sheath and an inner epithelial membrane. When the *arrector pili* (Latin, "raiser of the hair") muscles contract, the skin assumes a so-called "goose flesh" appearance, and the hairs tend to "stand on their ends," to a certain degree (see Figs. 28 and 30).

Around each hair follicle is a *sebaceous* (se-ba'shus) *gland,* which produces *sebum* (se'bum), the oily substance primarily responsible for lubrication of the surface of the skin.

Sweat glands are simple, tubular glands that are found in most parts of the skin; they are most prominent in the palms and soles. It has been estimated that there are 3000 sweat glands per square inch on the palm of the hand. Sweating leads to loss of heat in the body.

The *nails* are composed of hard *keratin*

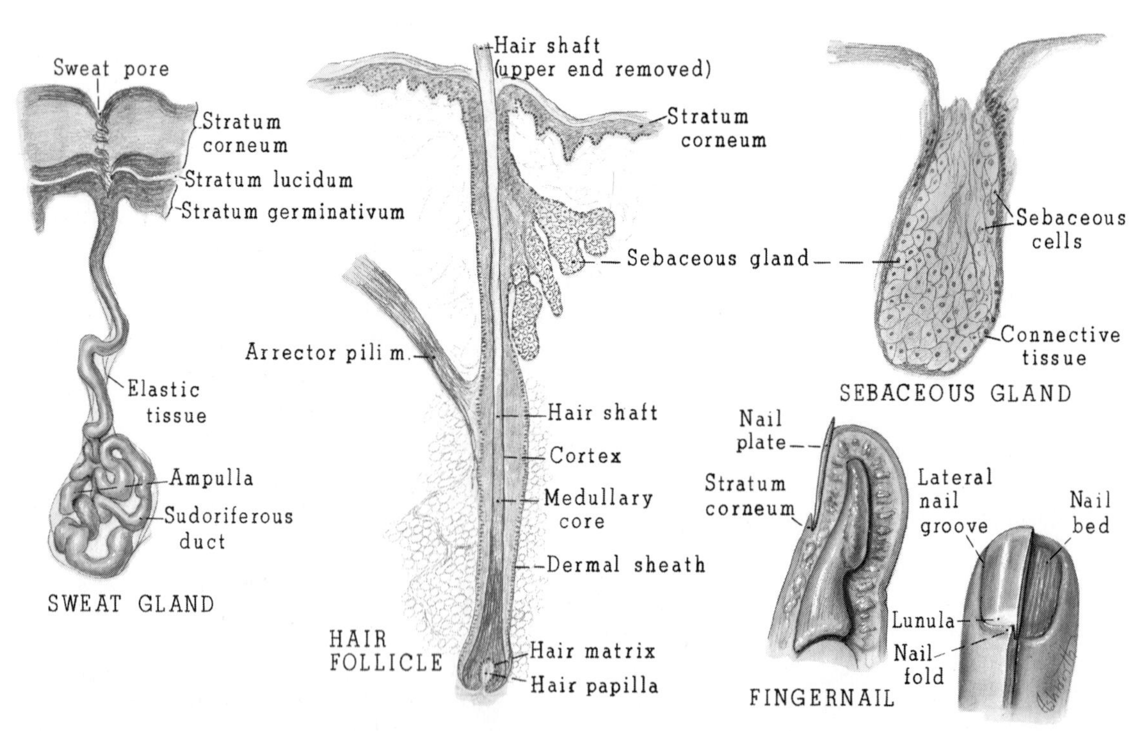

Figure 30. The appendages associated with the skin.

(ker'ah-tin), the same protein substance that makes up a major part of the hair shaft. Air mixed with keratin forms the white crescent, or *lunula* (lu'nu-lah), of each nail. Each nail grows about 1 mm. per week. Regeneration of a lost nail occurs in 4 to 8 months.

What Are the Four Main Functions of the Skin?

The skin functions are *sensation, protection, thermoregulation,* and *secretion.*

Sensation. Located in the skin are specific receptors, sensitive to the four basic sensations of pain, touch, temperature, and pressure. Upon stimulation of a receptor, a nerve signal is sent to the brain, where it is interpreted.

Protection. The skin forms an elastic, resistant covering that protects man from his complex environment. This covering prevents the passage of harmful physical and chemical agents, and prevents excess loss of water and salts.

Temperature Regulation. As the body needs to dissipate heat, blood vessels of the skin dilate (expand), allowing more blood to come to the surface, with a resulting loss of heat; also, the evaporation of sweat from the skin surface leads to a loss of body heat.

Secretion. The skin plays a part in the secretory functions of the body. Sebum, secreted by sebaceous glands, fights fungus and bacterial growth on the skin surface and also helps maintain the texture of the skin. Sweat is a secretion.

Clinical Aspects of the Skin

What Does Skin Appearance Tell Us?

The appearance of the skin can be an important sign in the diagnosis of various disorders. For instance, the skin may be red in hypertension (high blood pressure) and in other conditions in which the blood vessels of the skin are dilated. A pale skin suggests anemia (too few red blood cells or too little hemoglobin). The color of the skin may be blue or purple (a condition known as cyanosis) in severe heart disease and in lung-related diseases such as pneumonia (in which the blood is not being adequately supplied with oxygen). A yellow skin (jaundice) indicates the presence of larger than normal amounts of bile pigments in the blood.

What Are Some Common Skin Problems?

Acne (pimples) is an inflammatory disease of the sebaceous glands that occurs mainly over the face, neck, upper chest, back, and shoulders. It may occur at any time from puberty through the period of sexual development and remain for years.

Hives is a skin condition characterized by the sudden appearance of raised patches that are white in the center and itch severely. Hives often occur after ingestion of certain foods, such as strawberries or seafood, to which the individual is sensitive.

Psoriasis (so-ri'ah-sis) is a chronic inflammatory disease that is neither infectious nor contagious; its cause is unknown. Psoriasis is characterized by patches of dry, whitish scales.

Bed sores are caused by areas of pressure on the body of a bedridden person. Frequent turning and alcohol rubs help to prevent this distressing condition.

Sunburn is a condition in which the skin is swollen and red after excessive exposure to the sun, especially its ultraviolet rays, and can happen even on a cloudy day.

Boils are localized inflammations of the dermis and underlying tissue; they are caused by bacteria that enter through the hair follicles.

How Are Burns Rated?

It is important to be able to determine the extent of a burn injury before any therapy is undertaken.

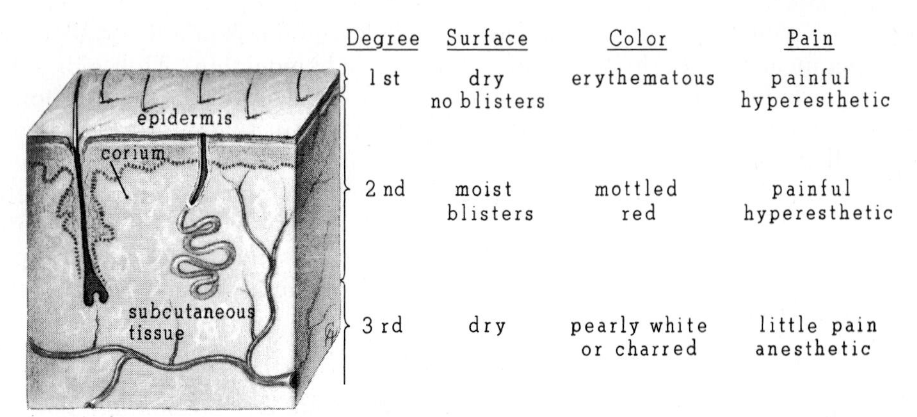

Figure 31. Extent of burn injury—first, second, and third degree. In a first degree burn only the epidermis is injured (as in sunburn); a second degree burn extends into the dermis; a third degree burn involves the full thickness of skin, epidermis, dermis, extending into subcutaneous tissue. (Courtesy of Parke, Davis.)

In a *first degree* burn, only the epidermis is injured (as in sunburn). A red coloration appears on the surface. A *second degree* burn extends into the dermis and forms blisters on the surface. A *third degree* burn involves both the epidermis and dermis, as well as the underlying tissues; complete destruction of all layers of the skin results (Fig. 31).

LEARNING EXERCISES

1. Draw diagrams of various types of tissue, including, at least, the epithelial, connective, muscle, and nervous.
2. For each drawing of a type of tissue make a table listing its distinguishing characteristics. How would you identify each tissue type under the microscope?

CHAPTER 4

THE SKELETAL SYSTEM

WHAT DOES THE SKELETAL SYSTEM DO?

The skeletal system provides the rigid framework that supports and gives shape to the body. It also serves to protect delicate internal organs such as the brain, heart and lungs. Together with the muscular system, the skeletal system enables the body to accomplish a wide range of movements.

Two lesser known functions of the skeletal system are the manufacture of blood cells and the storage of mineral salts, especially calcium.

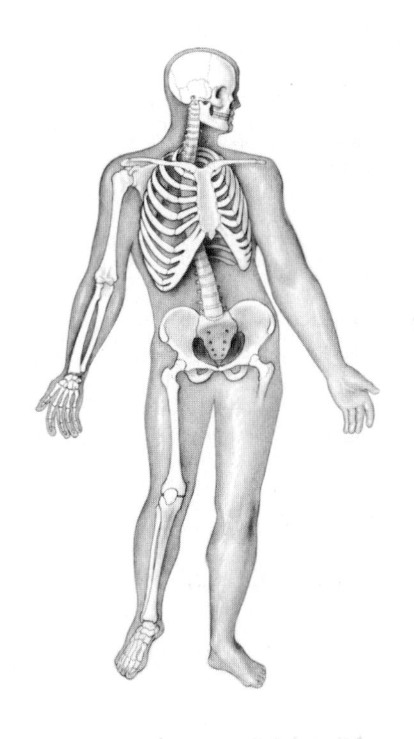

HOW DOES THE SKELETON DEVELOP?

The complete skeleton is first formed in the fetus at the end of the third month of pregnancy. At that time it is composed entirely of cartilage tissue—like the flexible material of your outer ear and nose. The cartilage is gradually replaced by bone tissue. This process, known as *ossification* (os″i-fi-ka′shun), occurs when dormant bone cells are stimulated to mature. During the initial phase the bone cells produce a matrix or membrane system. Later the bone cells and the matrix absorb calcium phosphate crystals in a regular pattern to form the typical hardened network of bone tissue. The early patterns of the solid network are not final but rather are being constantly remade and modified as the bone cells multiply and adapt.

Longitudinal (lengthwise) growth of bones continues in a definite sequence until approximately 15 years of age in the female and 16 in the male. Longitudinal growth should not be confused with bone maturation (the process of maturing) and remodeling, which are processes that continue until the age of 21 in both male and female. This pattern of maturation is so regular that an individual's age can be determined with amazing accuracy from radiologic (using x-ray pictures) examination of his or her bones (Fig. 32).

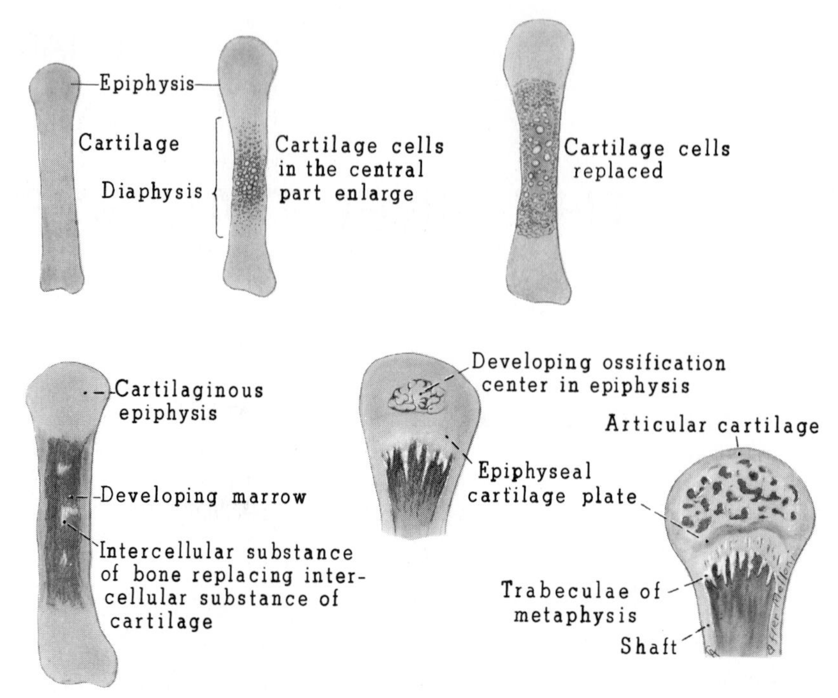

Figure 32. The bony deposit laid down around the diaphysis spreads toward the epiphysis, where ossification is also occurring. Gradual replacement of cartilage by bone occurs, and an increase in lengthwise direction of the bone accompanies this process. Growth in diameter of the bone occurs primarily with the deposit of bony tissue beneath the periosteum.

THE NATURE OF BONE

What Does a Typical Bone Look Like?

The parts and structure of bones in general can be seen by examining a typical long bone, such as the humerus of the upper arm. There are two distinct types of bone tissue; one is referred to as *compact*, and is dense and strong; the other is referred to as *cancellous* (kan′se-lus), and is more spongy and porous (Figs. 33 and 34). A long bone consists of a central shaft referred to as the *diaphysis* (di-af′i-sis), which is composed primarily of compact bone. Careful examination by x-rays (radiology) shows that the central shaft has more-compact bone in the middle. This is appropriate, since mechanical strain is greatest in the middle. The overall strength of the long bone is further insured by a slight curvature of the shaft. The interior of the central shaft is called the *medullary* (med′u-lar″e) *canal*, and is filled, for the most part, with yellow marrow (see below).

The two ends, or extremities, of the long bone, each called an *epiphysis* (e-pif′i-sis), have a thin covering of compact tissue overlying cancellous tissue. The cancellous tissue normally contains *red marrow,* which manufactures blood cells. The extremities of the long bone are generally broad and expanded when compared to the shaft. This allows for easy jointing (articulation) with other bones and provides a larger surface for muscle attachment.

What is the Function of Bone Marrow?

In the normal adult, the ribs, vertebrae, sternum, and bones of the pelvis contain red marrow in cancellous tissue. Red marrow within the ends of the humerus and femur is plentiful at birth but decreases in amount throughout the years.

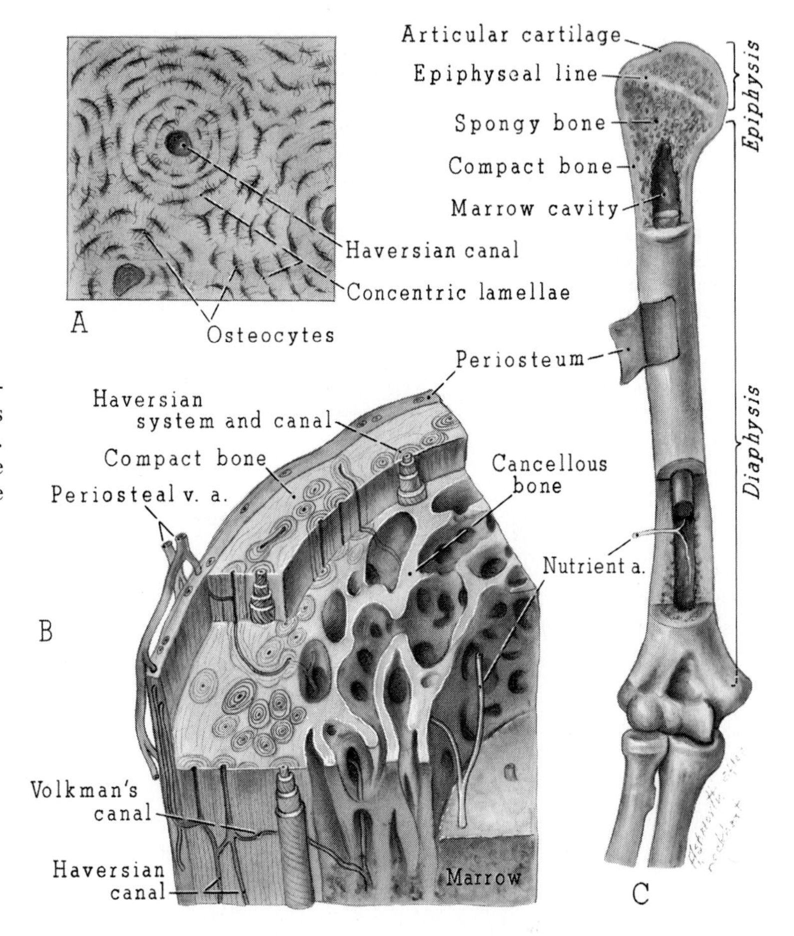

Figure 33. Detailed anatomy of a long bone. Cancellous bone is porous and sponge-like. The Haversian canals carry the small vessels that supply the living bone-cells with blood.

The primary function of red marrow is to form red and white blood cells and platelets. The technical name for this process is hematopoiesis (hem″ah-to-poi-e′sis). The red blood cells, white blood cells, and platelets all originate from the same type of bone marrow cell, the hemocytoblast (he″-mo-si′to-blast) or stem cell. Red blood cells (erythrocytes) and white blood cells (leukocytes), in various stages of maturing, are the main constituents of red bone marrow.

Yellow bone marrow is not very active in blood cell production. It consists chiefly of fat cells and is found primarily in the central shafts of long bones. As a person ages, red marrow is slowly replaced by the less active yellow marrow. Only a small fraction of the original red marrow remains in the bones of elderly persons. This helps to explain the difficulty they often have in replacing lost blood.

CALCIUM

What is the Function of Calcium in the Body?

Ninety-nine per cent of the total calcium of the body exists in solid deposits of calcium phosphate (a salt crystal) in the bone tissue. The small but important remainder is dissolved (in ionized form) in the blood plasma and other body fluids, where it actively participates in vital chemical reactions (Fig. 35).

The solid deposits of calcium salts in the bone tissue serve to give the bones strength and stability of shape. In addition,

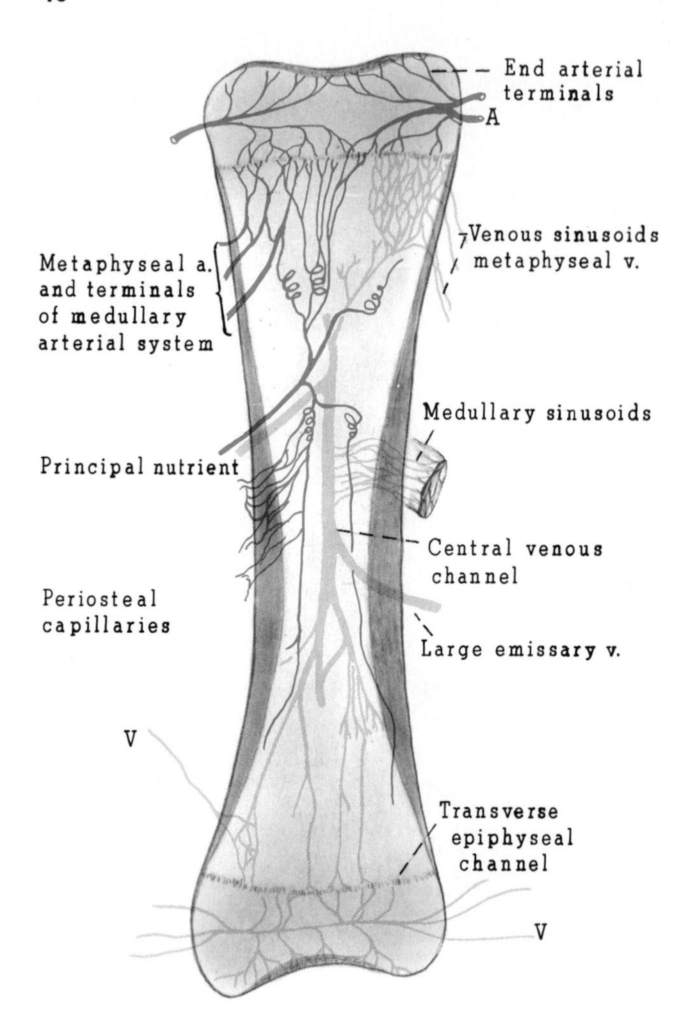

Figure 34. Blood supply to long bone.

these solid deposits help to supply bodily needs during periods, such as pregnancy, when normal mineral intake is deficient or bodily demand is great.

The small amount of dissolved calcium is necessary for the proper operation of both nerve and muscle cells. Because too much or too little dissolved calcium in the body fluids causes serious abnormalities in the operation of both nerve and muscle cells, it is crucial that the concentration be carefully regulated.

How is Calcium Concentration in the Body Fluids Regulated?

The movement of small amounts of calcium between the bone deposits and the body fluids occurs through two opposite processes controlled by the bone cells. They are appropriately referred to as *deposition,* in which calcium from the body fluids is added to bone, and *reabsorption,* in which solid calcium in the bone dissolves back into the fluids.

The total amount of calcium in the body can change a great deal, depending on intake and loss, with no serious health problems developing. During pregnancy, for instance, the total body calcium of the mother can change considerably, but this only leads to problems when her calcium intake is grossly deficient.

Even though such changes may occur in the total amount, the small amount dissolved in the body fluids must remain very nearly constant. This concentration of cal-

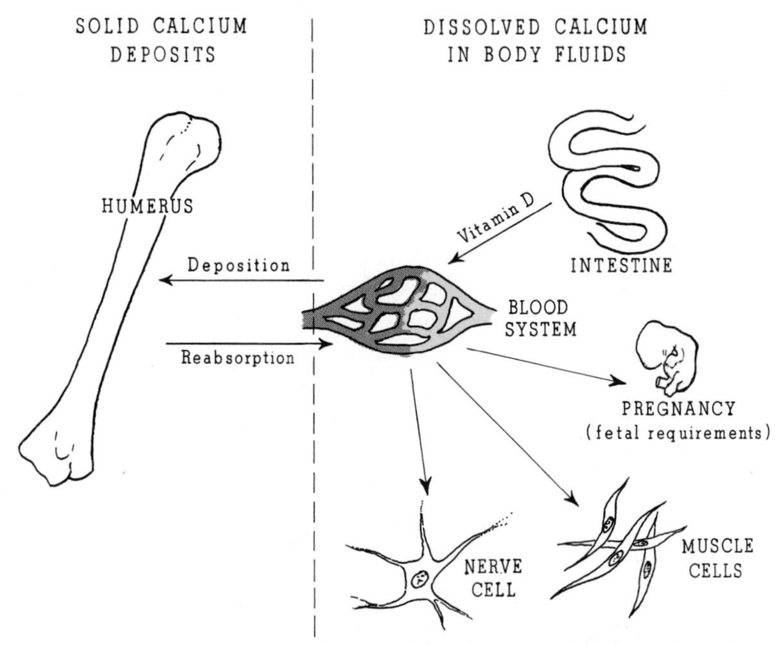

Figure 35. Most of the calcium in the body is in solid bone deposits. The dissolved calcium in the body is regulated by the balancing of deposition and reabsorption. All calcium enters the body through the intestine; vitamin D is essential for uptake into the blood. Calcium is needed in large amounts during pregnancy for the formation of the fetal skeletal framework; if insufficient calcium is ingested by the mother, calcium is removed from the mother's bones to supply the fetus. Nerve cells and muscle cells are extremely sensitive to the calcium concentration in the body fluids, necessitating very sensitive homeostasis of dissolved calcium concentration.

cium in the body fluids is kept at a constant (homeostatic) level primarily through the alternating actions of deposition and reabsorption, controlled by the bone cells.

The parathyroid gland, which will be discussed more fully in Chapter 12, The Endocrine System, is known to be important in stimulating the deposition and reabsorption activities of the bone cells. Generally speaking, the parathyroid gland monitors (keeps track of) the concentration of calcium in the fluids, and when the concentration begins to decrease the parathyroid releases a *hormone* (Chemical stimulant or messenger) that stimulates the bone cells to release more calcium into solution. Less hormone is secreted when reabsorption is called for.

Vitamin D has been shown to be necessary in the process by which ingested calcium is absorbed from the intestines. Parathyroid hormone seems also to regulate the rate of intestinal absorption, but cannot operate if vitamin D is not present.

WHAT ARE THE TWO MAIN SUBDIVISIONS OF THE SKELETON?

The human skeleton has two main subdivisions, known as the axial (ak′se-al) skeleton and the appendicular (ap″en-dik′u-lar) skeleton. The axial skeleton makes up the central axis, or axle, to which the parts of the appendicular (from appendage; addition to main body) skeleton are attached. Together they contain a total of 206 bones (Figs. 36 and 37, and Table 4).

The *axial skeleton* includes the bones of the cranium and face, the spinal column, and the chest (ribs and sternum).

The *appendicular skeleton* includes the shoulder girdle and the upper extremities (arms and hands), as well as the hip

Text continued on page 46.

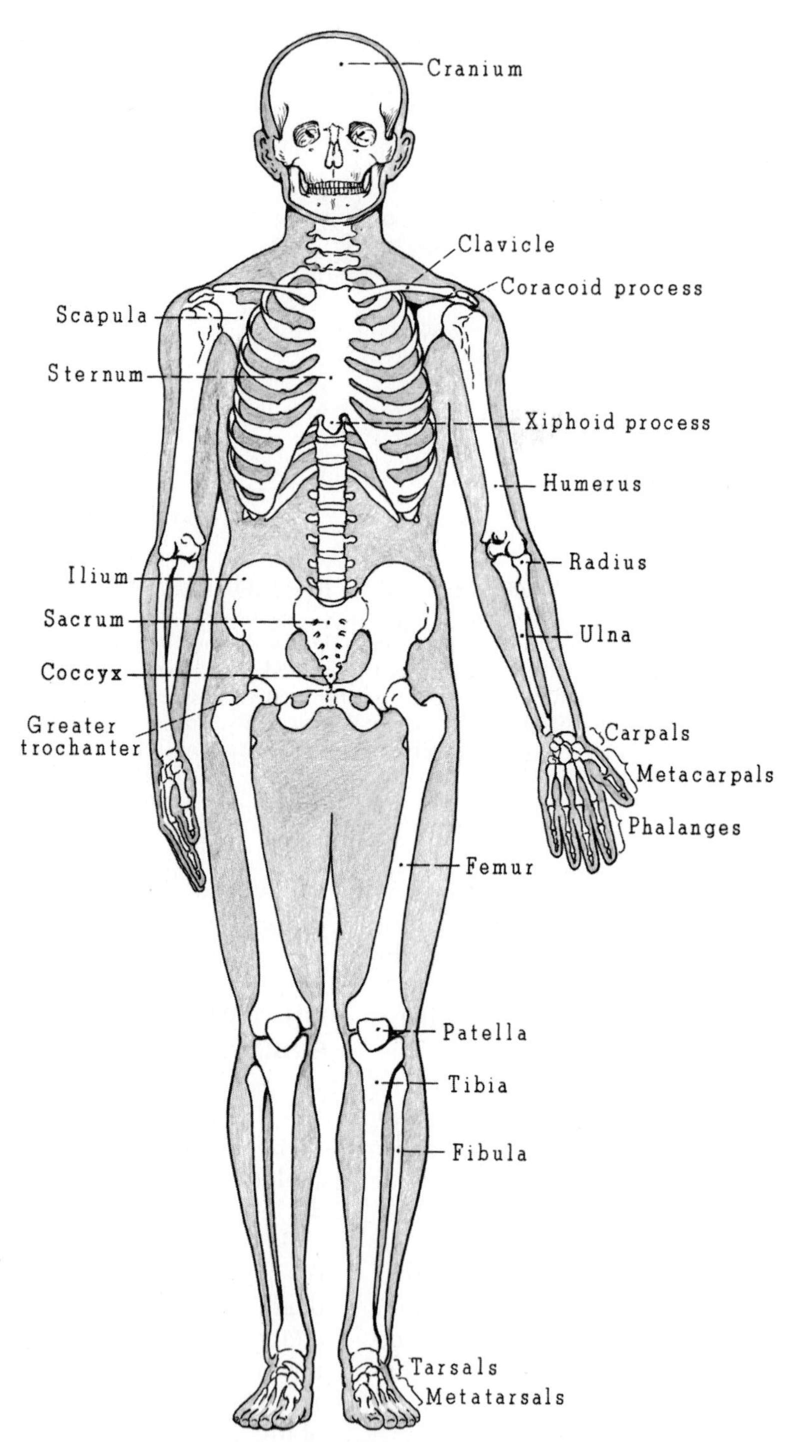

Figure 36. Anterior view of the skeleton.

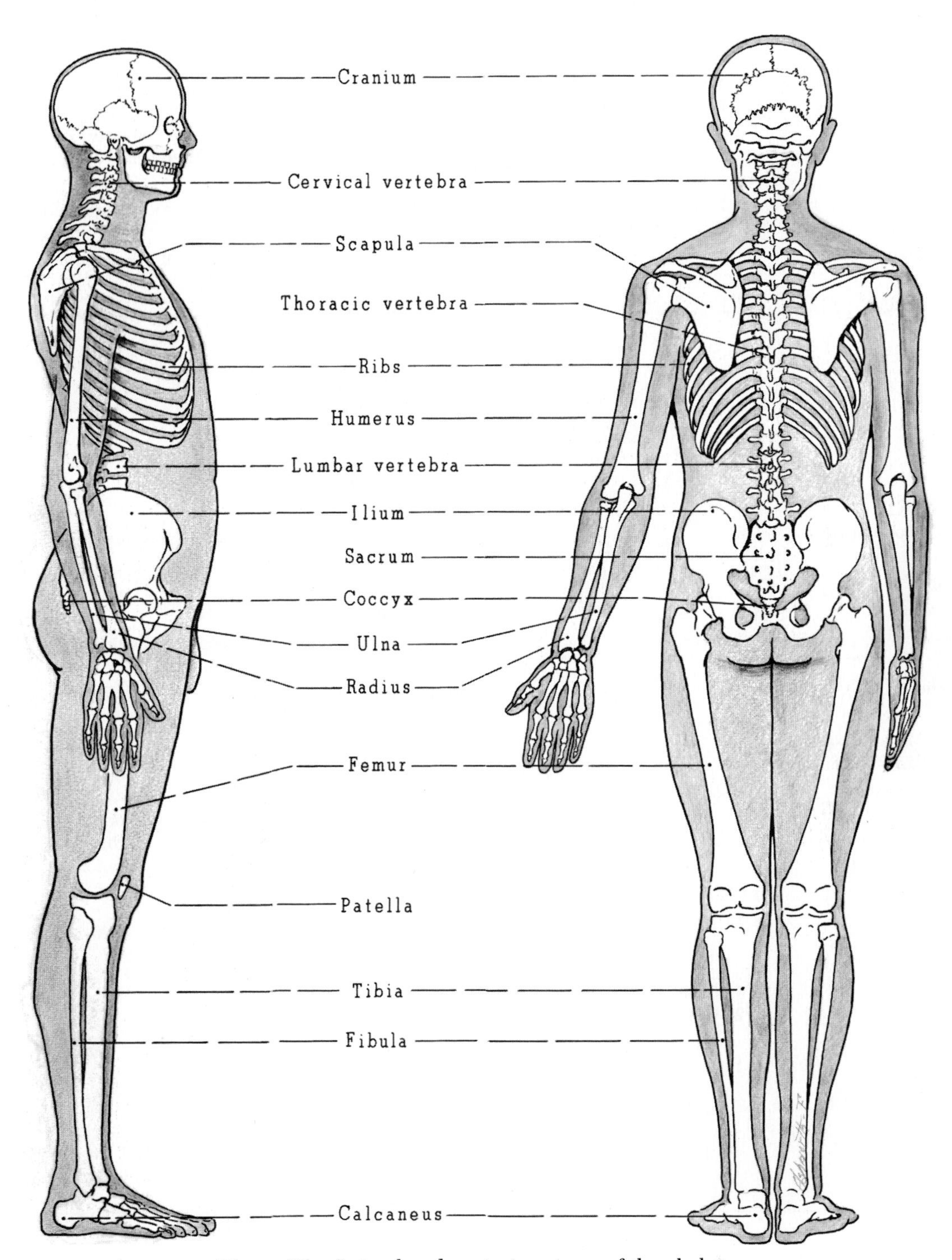

Figure 37. Lateral and posterior views of the skeleton.

Table 4. BONES

BONE	NUMBER	LOCATION
1. *Skull*	28 bones	
Cranium	8 bones	
Occipital	1	Posterior cranial floor and walls
Parietal	2	Forms the greater part of the superior lateral aspect and roof of the skull between frontal and occipital bones
Frontal	1	Forms forehead, most of orbital roof, and anterior cranial floor
Temporal	2	Inferior lateral aspect and base of the skull, housing middle and inner ear structures
Sphenoid	1	Mid-anterior base of the skull; forms part of floor and sides of orbit
Ethmoid	1	Between nasal bones and sphenoid, forming part of anterior cranial floor, medial wall of orbits, part of nasal septum, and roof
Face	14 bones	
Nasal	2	Upper bridge of nose
Maxillary	2	Upper jaw
Zygomatic (malar)	2	Prominence of cheeks and part of the lateral wall and floor of the orbits
Mandible	1	Lower jaw
Lacrimal	2	Anterior medial wall of the orbit
Palatine	2	Posterior nasal cavity between maxillae and the pterygoid processes of sphenoid
Vomer	1	Posterior nasal cavity, forming a portion of the nasal septum
Inferior nasal conchae (inferior turbinates)	2	Lateral wall of nasal cavity
Auditory Ossicles	6 bones	
Malleus (hammer)	2	Small bones in inner ear in temporal bone, connecting the tympanic membrane to the inner ear and functioning in sound transmission
Incus (anvil)	2	
Stapes (stirrup)	2	
Hyoid	1 bone	Horseshoe-shaped, suspended from styloid process of temporal bone
2. *Trunk*	51 bones	
Vertebrae	26 bones	
Cervical	7	Neck
Thoracic	12	Thorax
Lumbar	5	Between thorax and pelvis
Sacrum	1 (5 fused)	Pelvis—fixed or false vertebrae
Coccyx	1 (4 fused)	Terminal vertebrae in pelvis—fixed or false vertebrae

Table 4. BONES *Continued.*

BONE	NUMBER	LOCATION
Ribs	24	True ribs—upper seven pairs fastened to sternum by costal cartilages; false ribs—lower five pairs; eighth, ninth, and tenth pairs attached to the seventh rib by costal cartilages; last two pairs do not attach and are called floating ribs
Sternum	1	Flat, narrow bone situated in median line anteriorly in chest
3. *Upper Extremity*	64 bones	
Clavicle	2	Together, clavicles and scapulae form the shoulder girdle; the clavicle articulates with the sternum
Scapula	2	
Humerus	2	Long bone of upper arm
Ulna	2	The ulna is the longest bone of forearm, on medial side of radius
Radius	2	Lateral to ulna, shorter than ulna, but styloid process is larger
Carpals	16	Two rows of bones comprising the wrist
Scaphoid		
Lunate		
Triangular		
Pisiform		
Capitate		
Hamate		
Trapezium		
Trapezoid		
Metacarpals	10	Long bones of the palm of the hand
Phalanges	28	Three in each finger and two in each thumb
4. *Lower Extremity*	62 bones	
Pelvic	2	Fusion of ilium, ischium and pubis
Femur (thighbone)	2	Longest bone in body
Patella	2	Kneecap; located in quadriceps femoris tendon; a sesamoid bone
Tibia	2	Shinbone; antero-medial side of the leg
Fibula	2	Lateral to tibia
Tarsals	14	Form heel, ankle (with distal tibia and fibula), and proximal part of the foot
Calcaneum		
Talus		
Navicular		
Cuboid		
First cuneiform (medial)		
Second cuneiform (intermediate)		
Third cuneiform (lateral)		
Metatarsals	10	Long bones of the foot
Phalanges	28	Three in each lesser toe and two in each great toe

(pelvic) girdle and the lower extremities (legs and feet).

The Axial Skeleton

What is the Skull?

The skull is made up of both the facial and cranial bones. Among the bones of the face are the maxilla and mandible. The bones of the cranium are those that enclose and protect the brain and its associated structures, including the eyes and ears. The muscles of mastication (chewing), as well as the muscles for head movement, are attached to the cranial bones. At certain locations within the cranial structure are spaces or cavities (cavelike) usually referred to as air sinuses. These are connected to the nasal (nose) cavity (Fig. 38).

Although the cranium looks as if it might be just one large bone enclosing the brain, it is actually several bones that have grown together. During infancy and early childhood, the articulations (joints) between these bones are composed of cartilage, which gradually disappears as the bones grow together, or ossify. The bones actually seem to grow toward each other and eventually meet at what are called juncture (from junction) or suture (from sewn, because of appearance) lines.

These joints between the cranial bones do not allow for any movement.

What Are the Bones of the Torso?

The torso, or trunk portion of the axial skeleton, is made up of the *thorax* (tho′raks), or rib cage, and the vertebral, or *spinal column.*

What Is the Thorax?

That portion of the torso consisting of the sternum, the costal cartilages, the ribs,

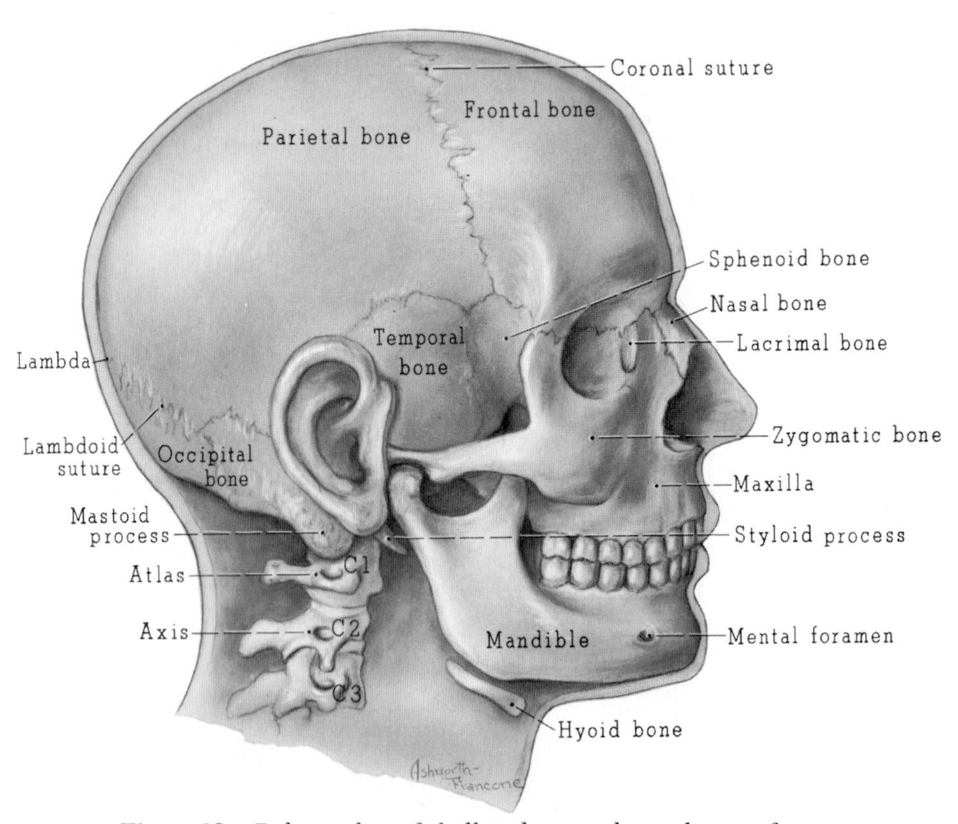

Figure 38. Relationship of skull and cervical vertebrae to face.

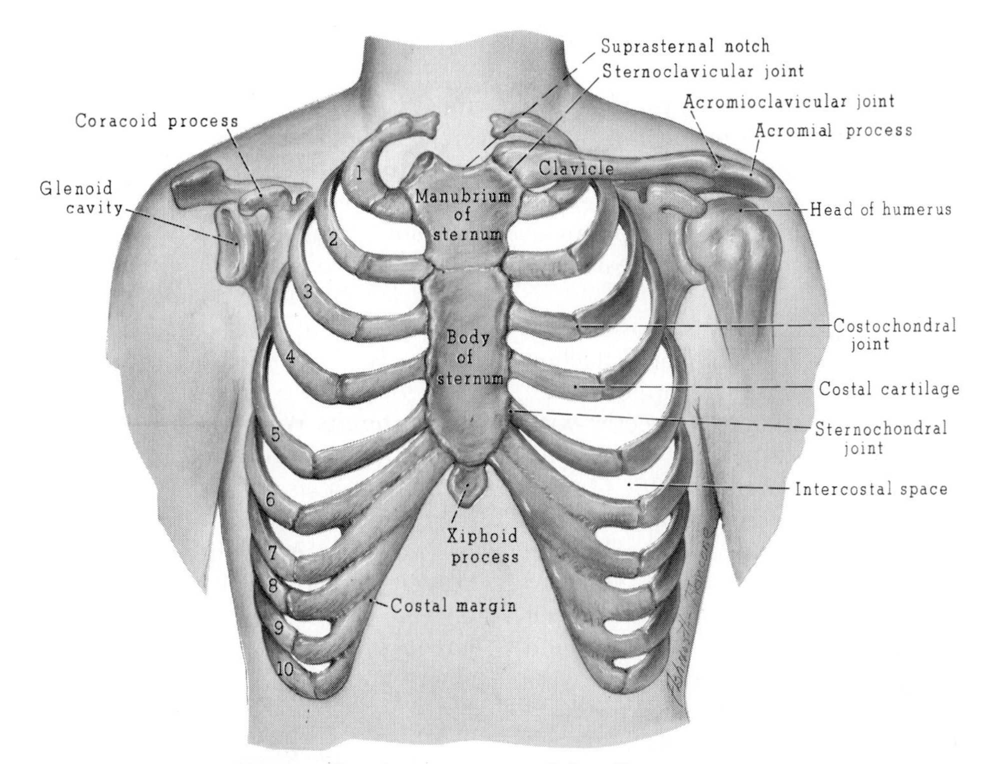

Figure 39. Anterior view of the rib cage.

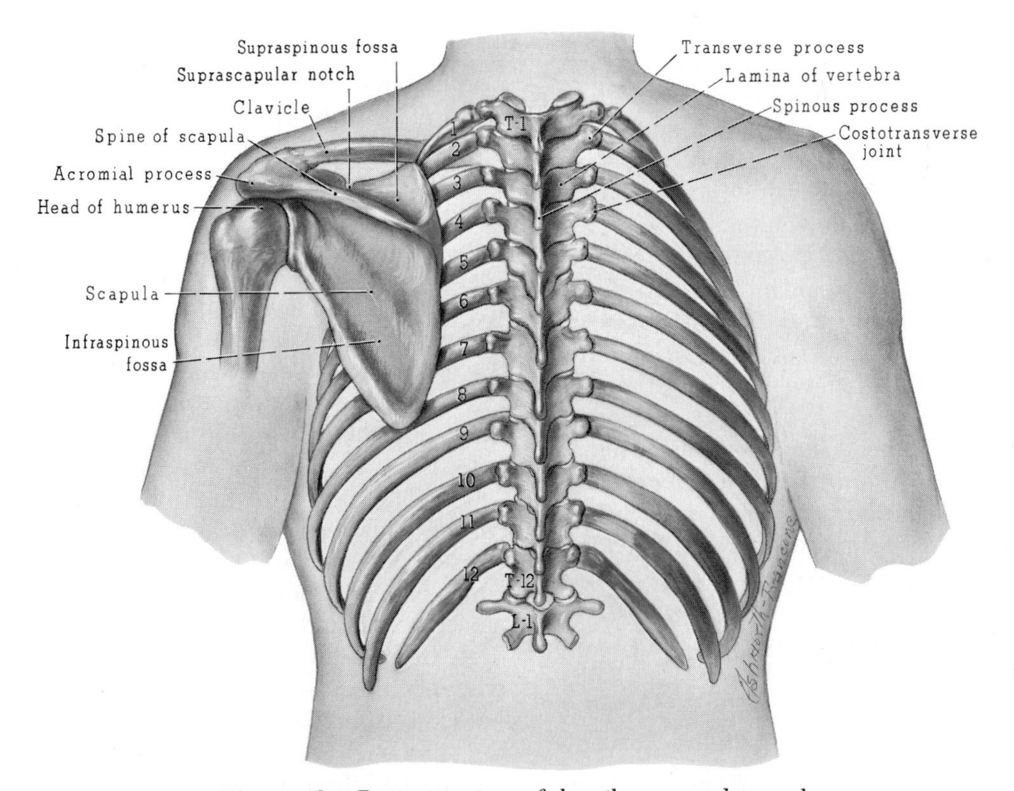

Figure 40. Posterior view of the rib cage and scapula.

and the bodies of the thoracic (from thorax) vertebrae is properly called the thorax. This bony cage encloses and protects the lungs and other structures of the chest cavity. The thorax also provides support for the bones of the shoulder girdle and upper extremities. At birth the thorax is spherical, while in adult life it is more coneshaped, with a broad base.

The twelve pairs of ribs (costae) are named according to how they are attached in front to the sternum. The upper seven pairs articulate (or connect) directly with the sternum and are called *true ribs.*

The lower five pairs (8, 9, 10, 11, 12), which join with the sternum either indirectly or not at all, are called *false ribs.* The eighth, ninth and tenth pairs are attached to the sternum indirectly through the costal cartilages of the above ribs. (Figs. 39 and 40). The eleventh and twelfth "false" ribs are also called *floating ribs,* since their anterior ends are completely unattached.

What Is the Vertebral Column?

The vertebral column provides protection for the delicate and vital spinal cord contained within its jointed channel. It has a remarkable combination of properties, being rigid enough to provide adequate support for the body, yet being highly flexible in many directions, allowing for considerable upper body movement.

The vertebrae are numbered by regions from the top down. There are 7 *cervical,* 12 *thoracic,* and 5 *lumbar* vertebrae. These remain separate throughout life and are called moveable vertebrae. In addition, there are five *sacral* vertebrae, which by adulthood have become fused to form a single *sacrum,* and four *coccygeal* (kok-sij'e-al) vertebrae, which unite firmly into a single *coccyx* (kok'siks). The vertebrae of these last two regions are called fixed and, consequently, the complete vertebrae are referred to as being 26 in number rather than 33 (Fig. 41).

Why Is the Vertebral Column Curved?

In the fetus, the vertebral column shows a single, C-shaped curve. After birth, raising of the head creates the beginning of an S-curve in the neck. Later, the erect posture involved in standing and walking creates the beginning of a similar inward curve in the lumbar region.

The normal curves of the spine can become exaggerated as a result of injury, poor body posture, or disease. When the posterior curvature is exaggerated in the thoracic area, the condition is called *kyphosis* (ki-fo'-sis) or, more commonly, *hunchback.* The word *kyphos* is Greek, and means hump. When the lower anterior curvature in the lumbar region is exaggerated it is known as *lordosis* (from the Greek, meaning bending backward) or, commonly, as swayback. A sideways curve, which involves rotation of some of the vertebrae, is termed *scioliosis* (sko"le-o'sis) (from the Greek word *skolios,* meaning crooked).

Appendicular Skeleton

What Are the Bones of the Upper Extremities?

The bones of the upper extremities can be separated into three groupings.

1. The *shoulder girdle,* or pectoral girdle, is made up of two major bones, the *collar bone,* also called the clavicle; and the *shoulder blade,* or scapula (skap'u-lah) (Fig. 42).

2. The bones of the *arm* proper are the humerus, ulna, and radius. The *humerus* connects at the top with the scapula and at the elbow with the two forearm bones. The forearm bones are the *ulna,* on the little finger side, and the *radius,* on the thumb side (Figs. 42 and 43).

3. The bones of the wrist are called *carpals* (kar-pals), and are situated in two rows of four each. The palm of the hand consists of five metacarpal bones, each with a base, shaft, and head. The metacarpals radiate from the wrist line like spokes from a wheel, rather than being parallel, and join with the proximal phalanges (fa-lan'jez) of the fingers. Each finger (excluding the thumb) has three phalanges—a proximal, a middle, and a terminal (or distal) phalanx.

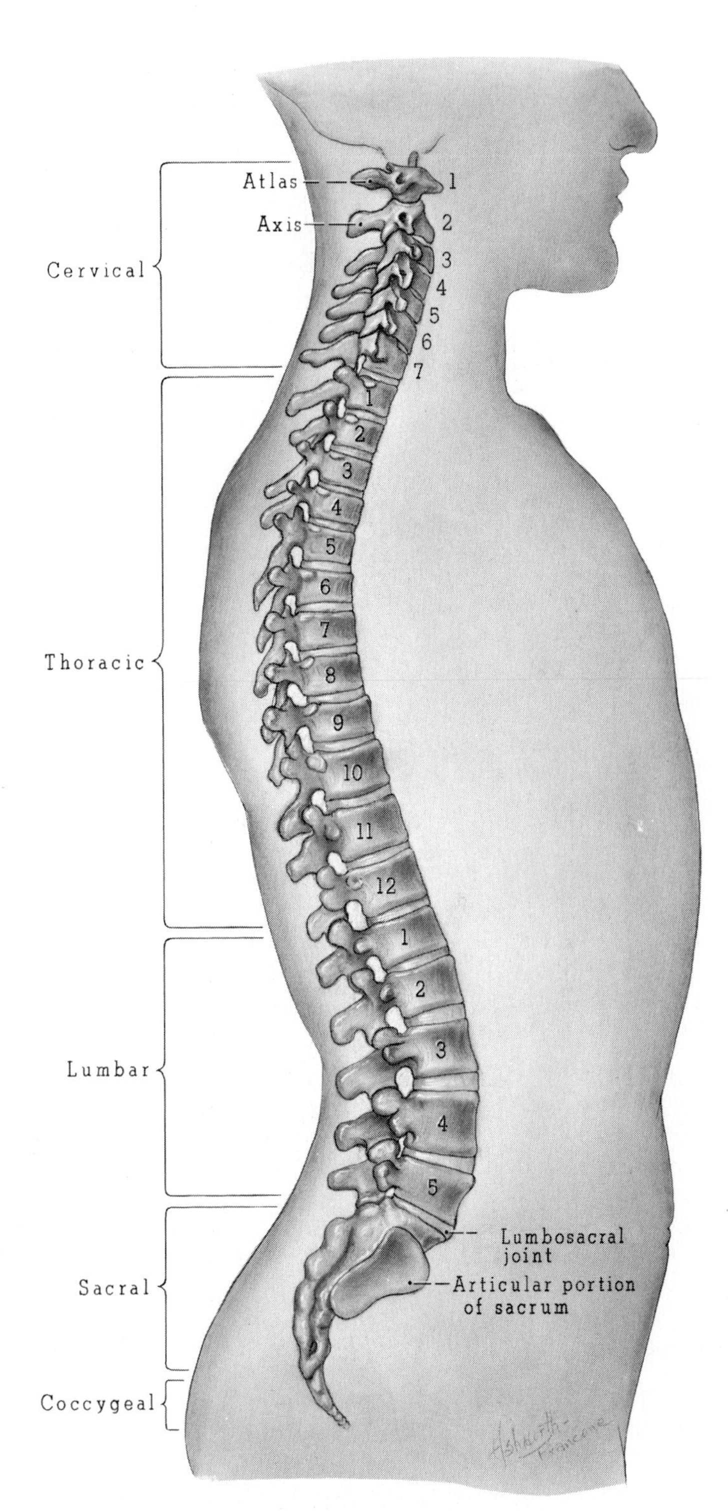

Figure 41. Vertebral column in relation to body outline.

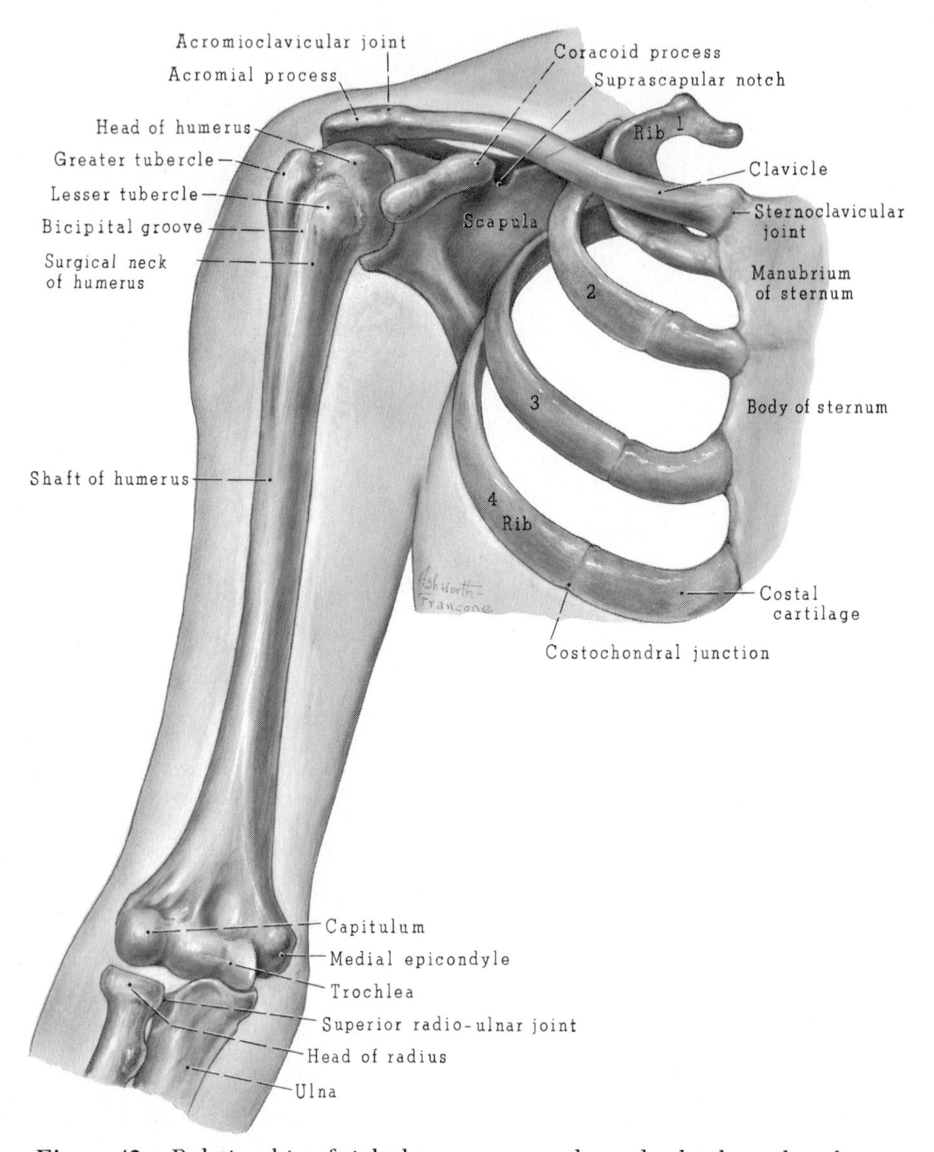

Figure 42. Relationship of right humerus, scapula, and calvicle to the rib cage.

The thumb has only two phalanges (see Fig. 43).

What Are the Bones of the Lower Extremities?

The bones of the lower extremities can also be separated into three groupings.

1. The *pelvic girdle* supports the trunk and provides attachment for the legs. The paired *os coxae* (pelvic bone or "hipbone") originally consist of three separate bones, the ilium, ischium, and pubis. These names are retained as descriptive regions for areas of the fused adult pelvic bone. The male pelvis has bigger bones but forms a narrower structure overall. The female pelvis is shaped like a basin (the word pelvis is from the Greek word *pyelos,* meaning basin) (Fig. 44).

2. The *femur* is the bone of the thigh. It is the largest and heaviest bone in the body. Notice that the femur is *not* in a vertical line with the axis of the erect body. Rather, it is positioned at an angle, slanting downward and inward. From the point of

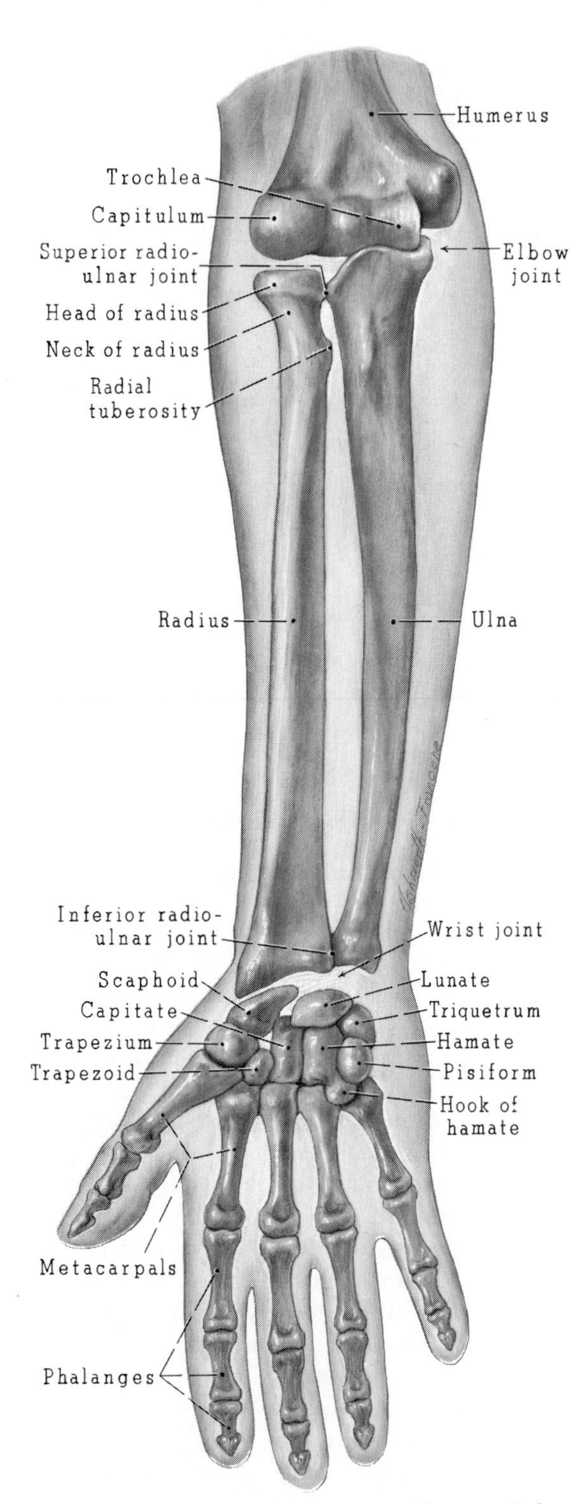

Figure 43. Anterior view of bones of the right forearm and hand.

view of the skeleton, the two femurs appear as a "V." Because of the female's greater pelvic breadth, the angle of inclination of the femurs is greater than in the male (Fig. 45). The *patella* (pah-tel'lah), or kneecap, is a small, flat, and somewhat triangular bone lying in front of the knee joint and enveloped within the tendon that attaches to the large, rectus femoris muscle of the anterior upper leg. The patella is moveable, and serves to increase the leverage of the muscles that straighten the knee (Figs. 49, 51, and 52).

The *tibia* (shinbone) is the larger of the two bones forming the lower leg. The *fibula* (calfbone) is the smaller of the bones in the lower leg and in proportion to its length is the most slender bone in the body, lying parallel with and on the lateral (toward the outside) side of the tibia (Fig. 45).

3. The ankle and foot are composed of the tarsal (tahr'sal) and metatarsal bones, as well as the phalanges. The general design is very similar to that of the wrist and hand. Can you find any differences? You can remember that the tarsals are in the foot and the carpals are in the hand by associating the "t" of tarsal with the "t" of toes (Fig. 46).

ARTICULATIONS

What Are the Three Types of Joints?

The joints, or articulations (ar-tik-u-la'shuns), of the skeletal system may be defined as the region of union of two or more bones. Every bone of the human body, with one exception, articulates with at least one other bone; the exception is the hyoid in the neck, to which the tongue is attached. Joints may be distinguished into three classes on the basis of the amount of movement possible at the point of connection (Fig. 47). The *fixed* joints, such as those found between the bones of the cranium, allow no movement. There are a few *slightly moveable* joints, such as the one between the radius and ulna in the

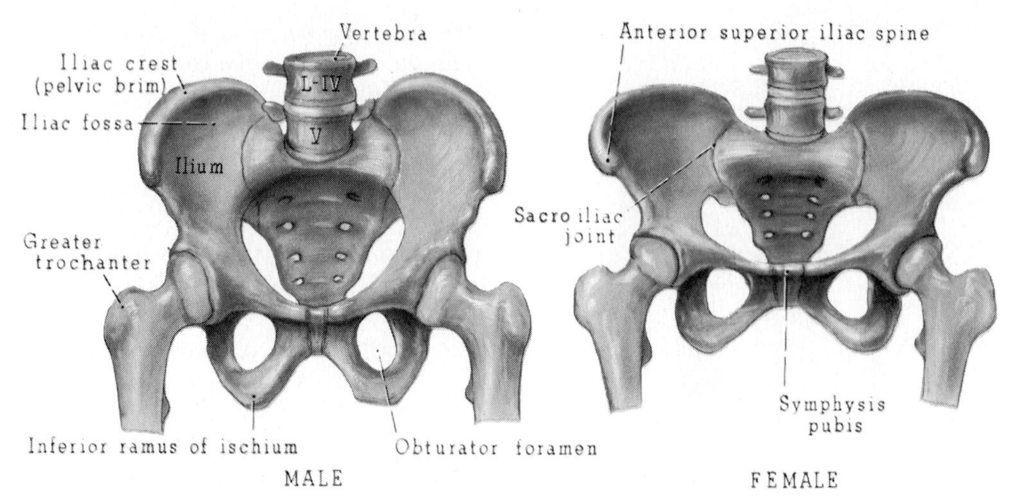

Figure 44. Comparison in proportions of the male and female pelves.

forearm. The vast majority of joints in the body allow considerable movement—sometimes in many directions and sometimes in only one or two directions; these are the *freely moveable* joints. Figure 48 shows the types of movements permitted by some freely moveable joints.

What Is the Structure of a Typical Joint?

All freely moveable joints are bound together by a *capsule,* which fits like a tight sleeve. This capsule is made from strong, fibrous cartilage and is lined with smooth, slippery synovial (si-no′ve-al) membrane. The capsule sleeve holds the bones securely together, and yet permits free movement of the joint. Closely associated with the capsule are the *ligaments,* which are tough fibrous cords or bands that bind and reinforce the capsule (Figs. 49 and 50).

The ends of the bones in all joints are covered by a smooth layer of cartilage. This layer of articular cartilage acts like the rubber heel of a shoe, softening the impact of jolts. The joint space or cavity is filled with synovial fluid, which lubricates the joint.

What Is a Bursa?

A bursa (bur′sah) is a closed sac containing a small amount of synovial fluid and with a synovial membrane lining similar to that of a joint. A bursa can be found between tendons, ligaments and bones, or generally where friction would otherwise develop. Bursae facilitate the gliding of muscles or tendons over bony or ligamentous surfaces (Figs. 51 and 52).

CLINICAL CONSIDERATIONS

What Are Two Common Disorders of Bone Tissue?

Two common diseases of bone tissue are rickets and osteoporosis (*osteo*-bone, *porosis*-porous). *Rickets* is a disorder that occurs when there is too little vitamin D in a child's system, with the consequent inhibition of absorption and deposition of calcium in the bones. As a result, the bones fail to harden and the child may develop curved legs and spine.

Osteoporosis (os″te-o-po-ro′sis), a loss of calcium in the bones, making them porous and brittle (easily broken), occurs commonly in older persons. Its cause has not been clearly determined, although lack of exercise and a progressive loss of sex hormone output are considered to be contributing factors.

What Are the Two Types of Fractures?

The breaking of a bone or cartilage is known as a *fracture.* A fracture is usually

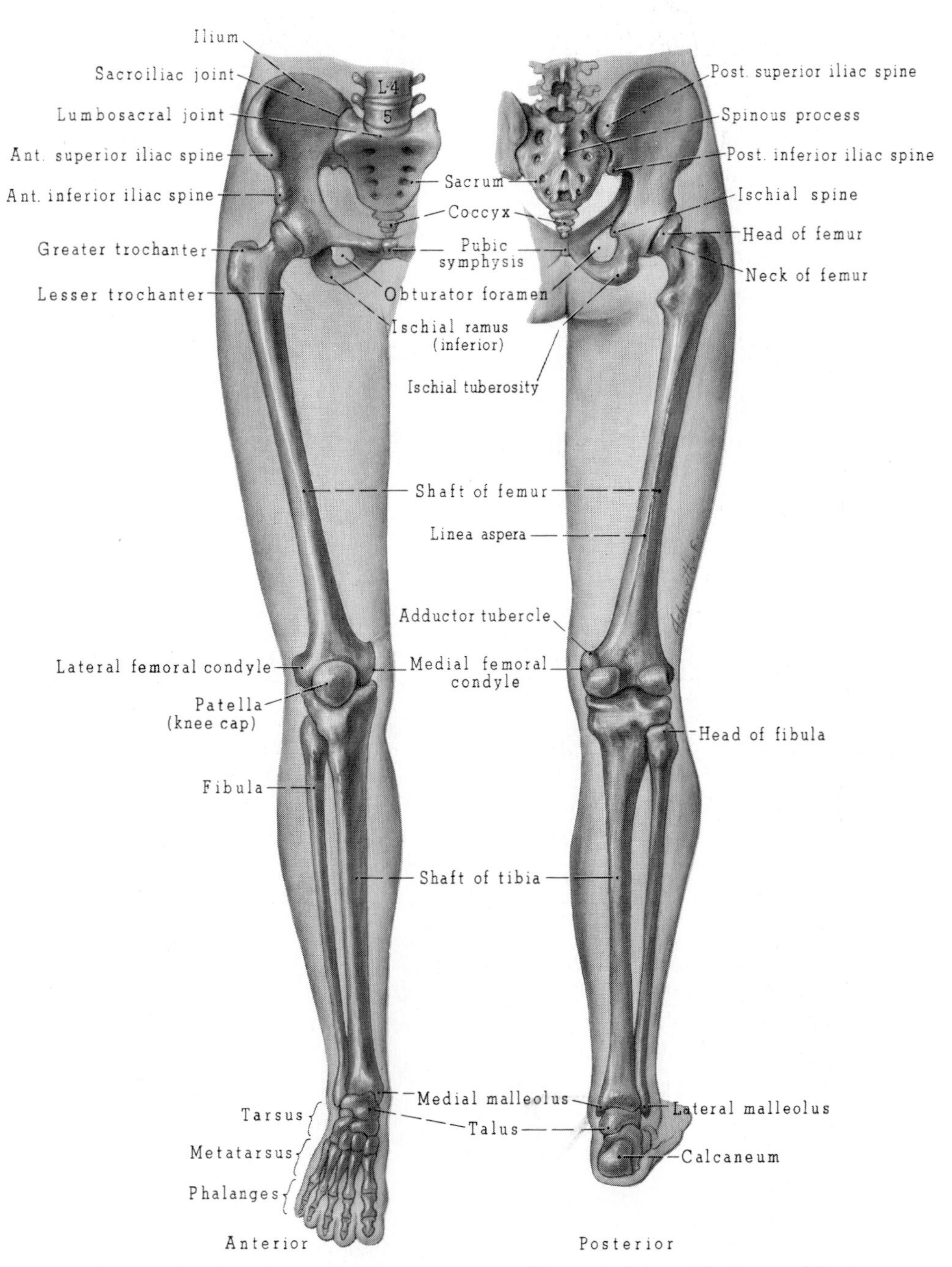

Figure 45. Anterior and posterior views of bones of the right leg and foot.

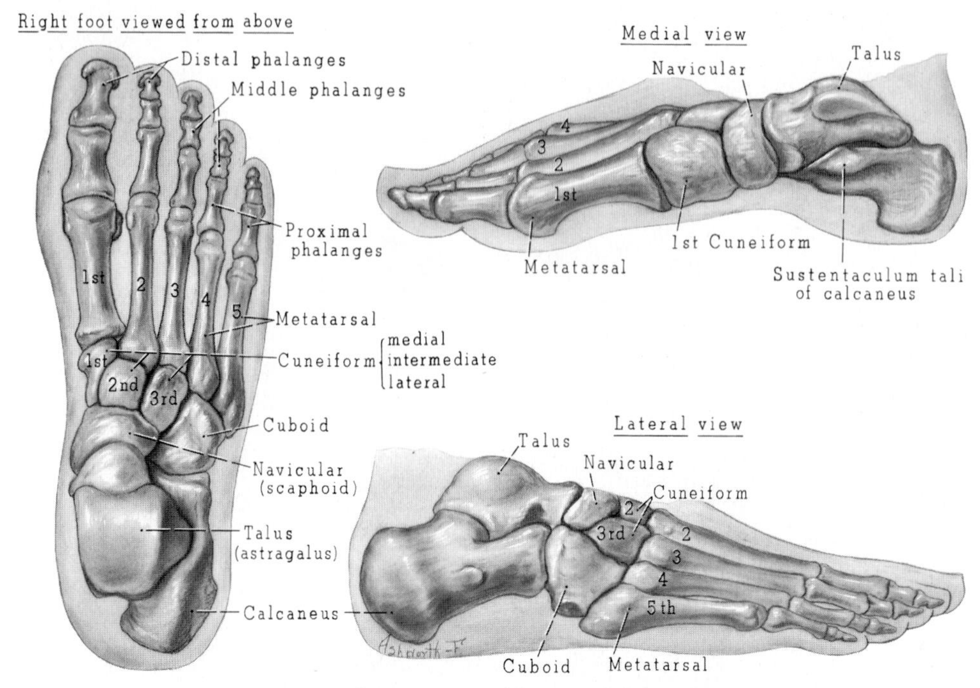

Figure 46. Three views of bones of the right foot.

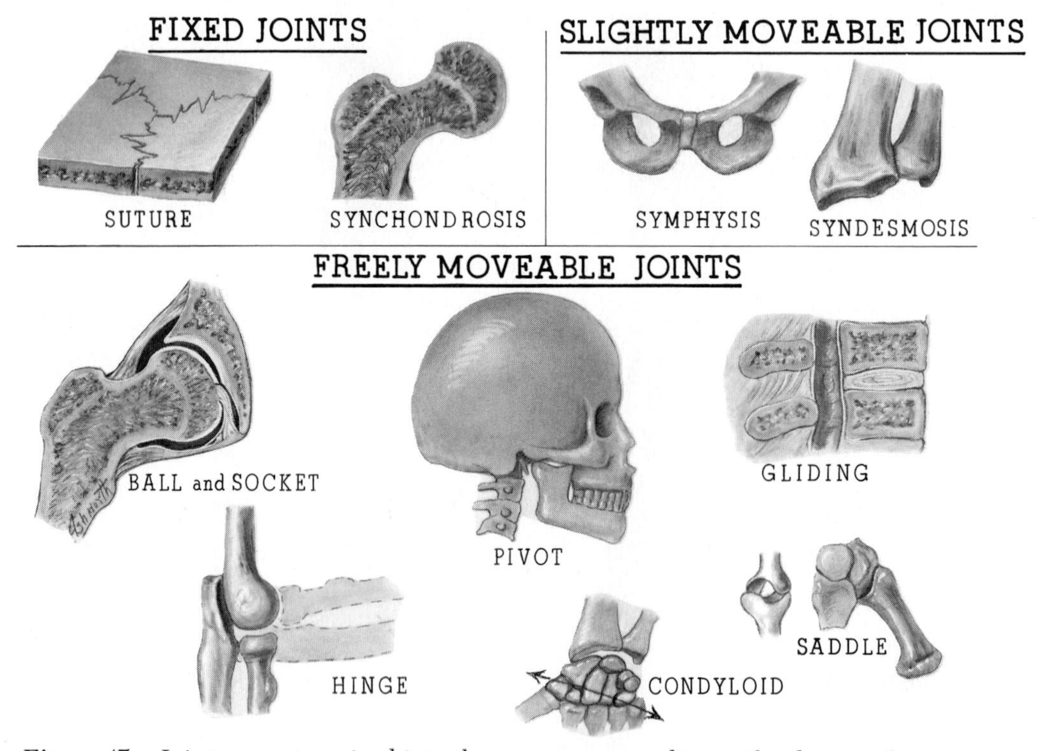

Figure 47. Joints are categorized into three groups, according to the degree of movement permitted. Each of these groups is in turn subdivided with respect to the structural components of individual joints.

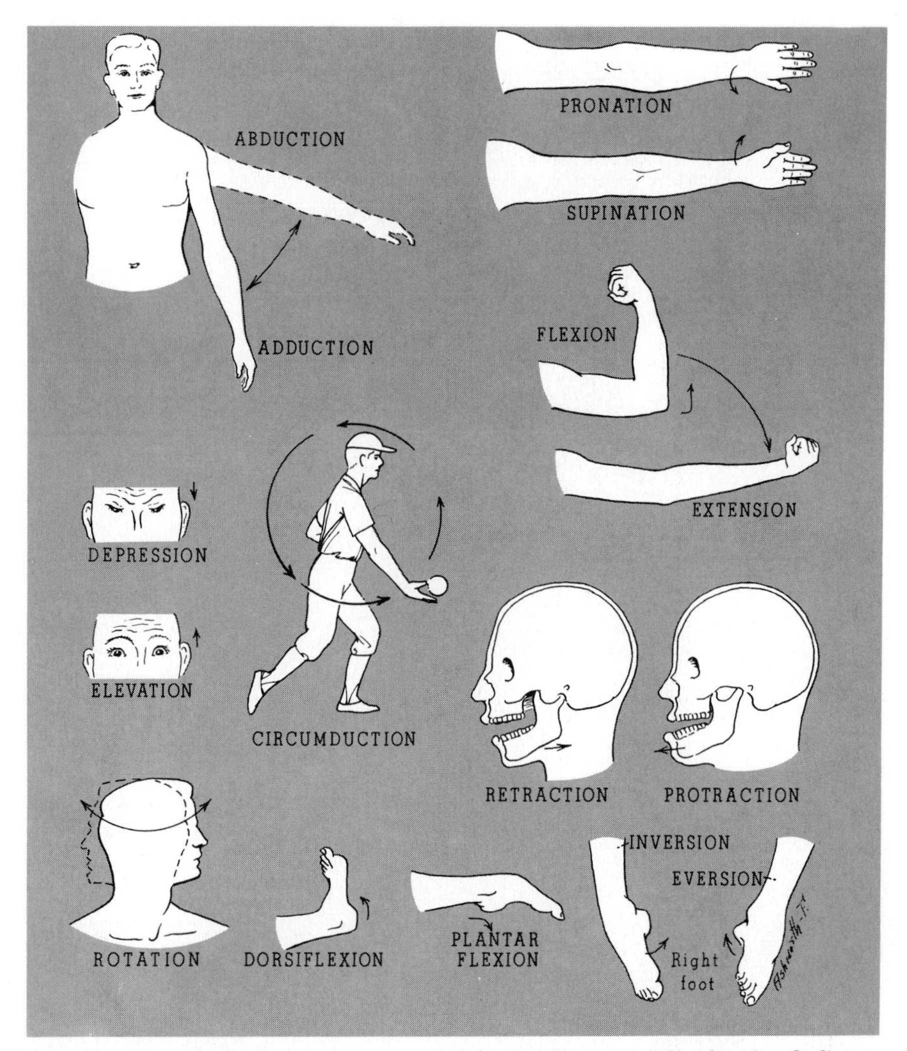

Figure 48. Types of movement permitted by freely moveable (diarthrodial) joints.

accompanied by an injury to the surrounding soft tissue. A fracture is called *compound* if the broken bone protrudes through the skin, or *simple* if it does not. Because of the greater possibility of infection, a compound fracture is the more dangerous of the two. Bone healing occurs best when the fracture ends are repositioned accurately and tightly.

What Are Two Common Disorders of the Joints?

Arthritis (ar-thri′tis), or joint inflammation, significantly immobilizes thousands of people every year. Nearly everyone suffers from some kind of arthritis in some part of his body sooner or later. *Rheumatoid* (roo′mah-toid) *arthritis* is a systemic disease with widespread involvement of connective tissue. Inflammation of the synovial membrane leads to damage and degeneration of cartilage; finally calcification of the damaged cartilage results in joint immobility. *Osteoarthritis* involves a slow wearing away of weakened cartilage, with the underlying bone around the joints developing ivory-like calcium growths called "spurs" or "marginal lippings." These result in a joint stiffness and eventual immobility.

Gout is a metabolic disorder that leads to the deposition of uric acid crystals in and

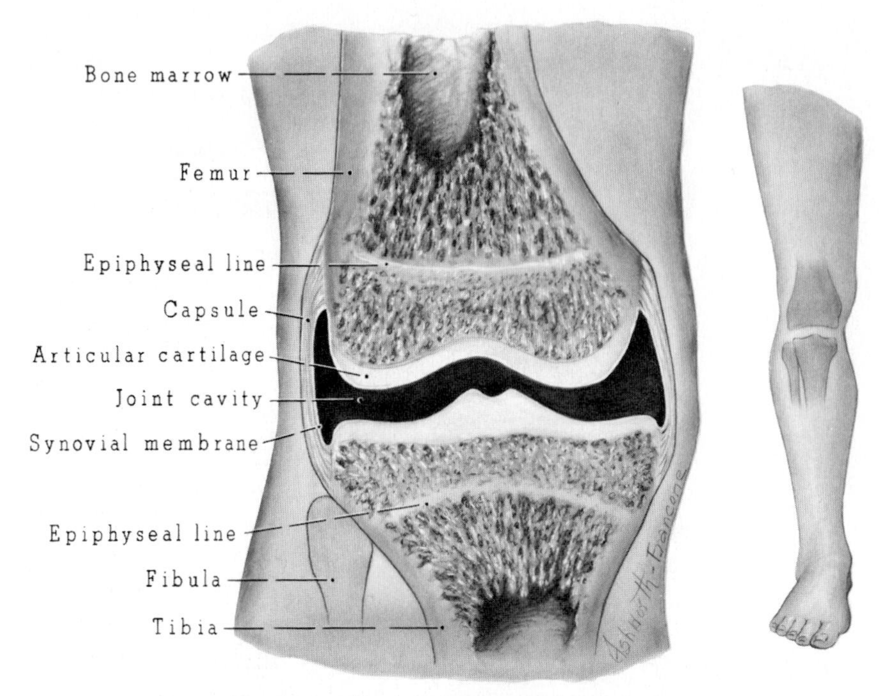

Figure 49. Frontal section through the right knee joint.

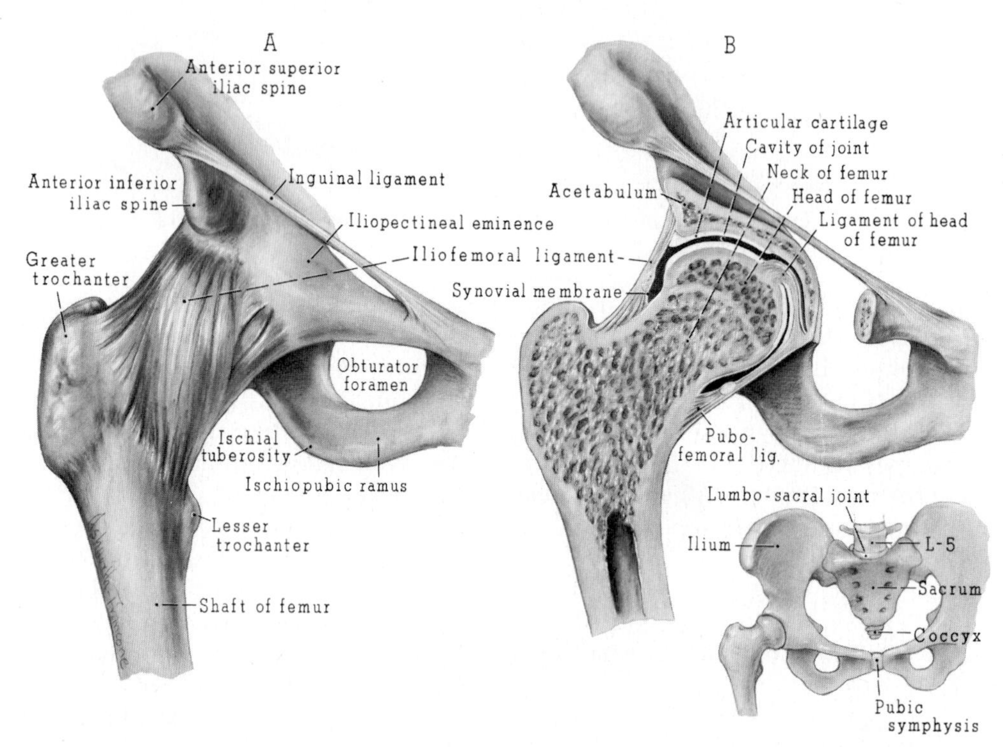

Figure 50. Hip joint, showing ligaments between femur and pelvic bone; intact (*A*) and sectioned (*B*) to show attachments.

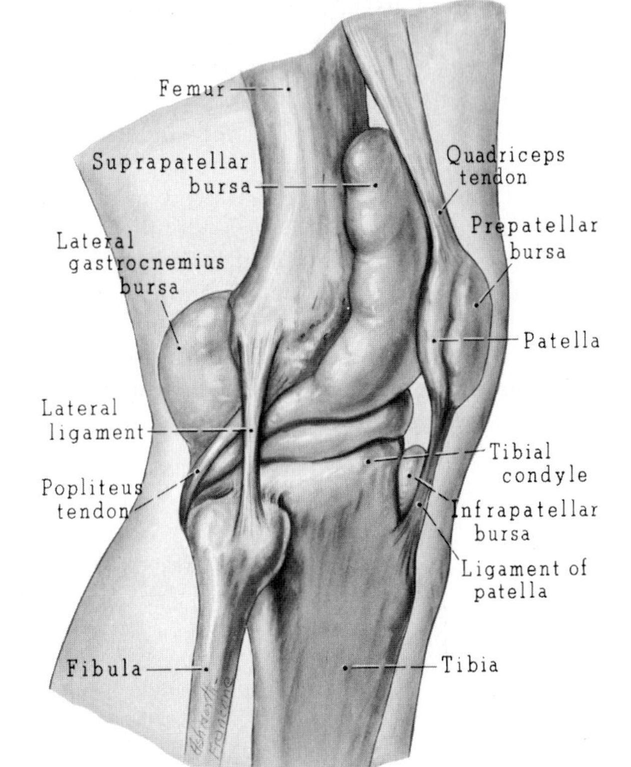

Figure 51. Lateral view of the right knee joint. The bursae have been expanded for clarity.

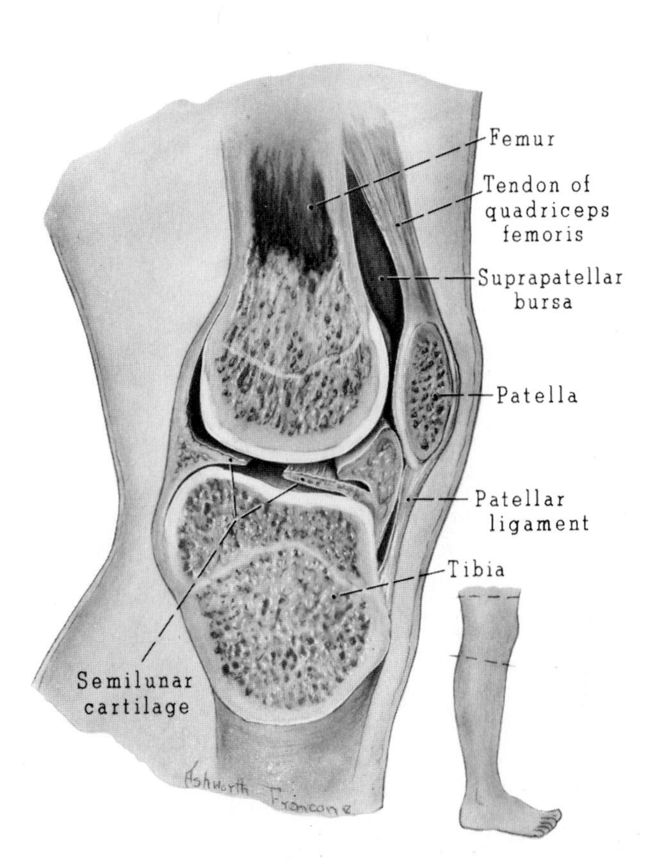

Figure 52. Lateral view of the right knee joint in sagittal section.

about the joint tissues and, together with inflammation of the synovium, results in damage to the articular cartilage and eventual gouty arthritis.

Bursitis (bur-si′tis) is an inflammation of the bursa that may result from excess stress or tension having been placed on the bursa or from some local or systemic inflammatory process. The most frequent location is in the shoulder, where movement of the joint becomes limited and painful. Eventually, with the inflammation, abnormal deposits of calcium occur and further interfere with joint movement.

LEARNING EXERCISES

1. Draw a typical long bone and label as many features as you can.
2. Construct a diagram of the skeleton and label the major bones; check your diagram with Figures 36 and 37.
3. Construct a diagram of the axial skeleton; label the major bones of the skull and the areas of the vertebral column.
4. Based on what you have learned about joints, draw the joint of a typical, freely moveable joint, including and labeling the major parts.

CHAPTER 5

THE MUSCULAR SYSTEM

THE MUSCLES: WHAT THEY DO, HOW THEY DIFFER

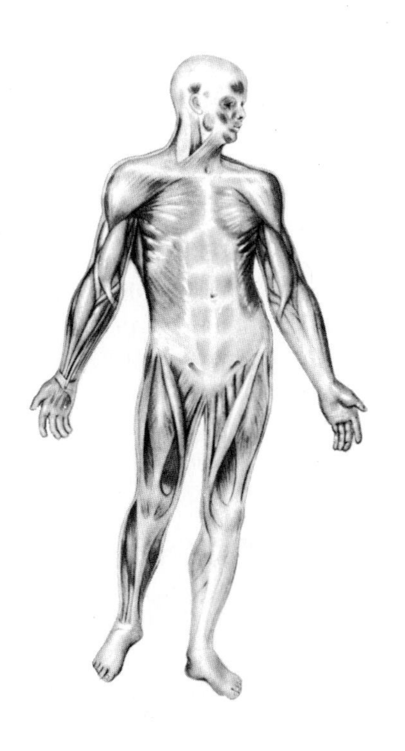

What Is the Function of the Muscular System?

The muscular system is responsible for producing all movements of the body. Every movement we make is the result of the contraction or relaxation of some muscle or group of muscles. For instance, all external or "action" movements, such as walking or turning one's head or swinging a golf club, are caused by coordinated contractions of our skeletal muscles. Of equal importance are the internal or "life supportive" movements brought about by the activity of muscles, such as the circulation of the blood, the passage of food along the digestive tract, and the movement of the chest, diaphragm, and abdomen during respiration.

How Do the Three Types of Muscle Differ in Appearance, Location and Function?

There are three main types of muscle cells: the *striated* (stri′at-ed), the *smooth*, and the *cardiac* (Fig. 53). They make up three corresponding types of muscle tissue—striated muscle tissue, smooth muscle tissue, and cardiac muscle tissue. These three types of muscle tissue, together with some special connective tissue, go into the construction of the three basically different types of *muscle* (Fig. 54). This muscle may be formed into a single muscular organ, such as the biceps of the upper arm (striated muscle), or may compose only one of several layers of a more complex organ, such as the kidney or intestines (smooth muscle).

Striated Muscle. The cells of striated muscle are long and slender, their length being from 20 to 1000 times their width. Because of their exaggerated oblong shape, striated muscle cells are sometimes referred to as *muscle fibers.* Microscopic examination of these fibers reveals that they have many crosswise stripes or striations. The name *stri*-ated is derived from this striped appearance of the fibers. By using special staining techniques, it can also be

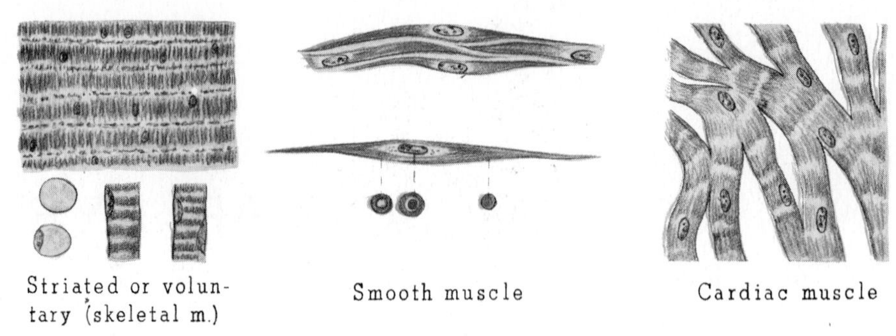

Figure 53. Types of muscle cells.

shown that striated muscle cells (unlike the other two types) have several nuclei.

A *striated* muscle appears to be made up of bundles of fine threads (the fiber-like cells) held together with connective tissue. The striated muscles of the body are all attached to the skeleton and are sometimes called *skeletal muscles.* These muscles or muscular organs are all under conscious voluntary control and, with few exceptions, are the only muscles we can control consciously.

Smooth Muscle. The cells of smooth muscle are of various shapes, but are usually several times longer than they are wide or thick, and therefore are also often referred to as fibers. Unlike the striated muscle cells, the smooth muscle cells con-

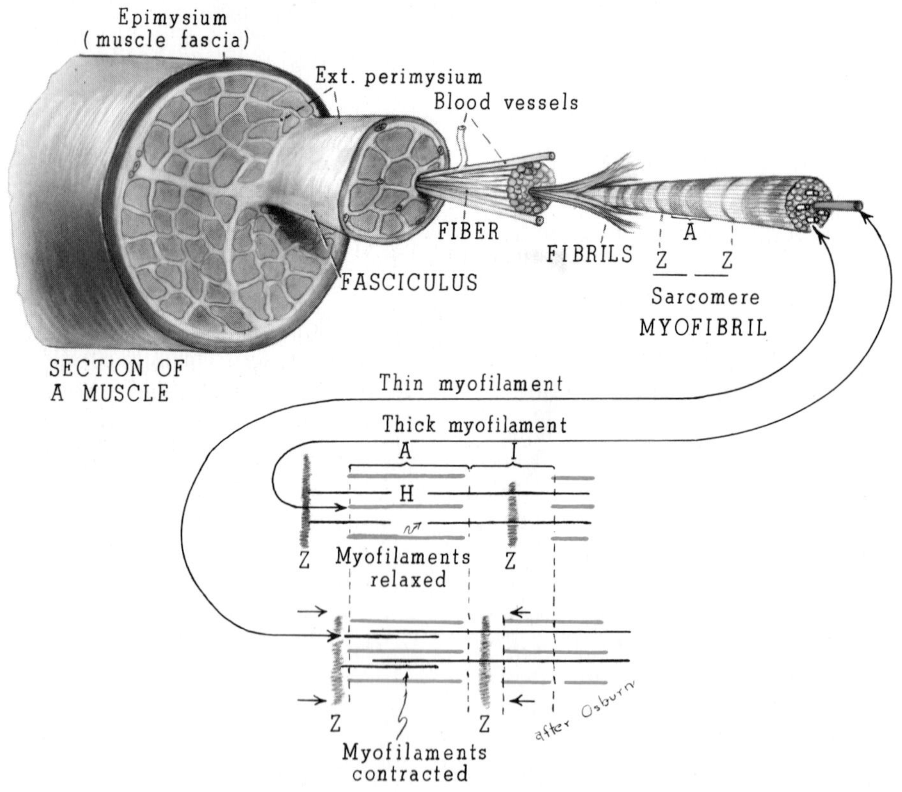

Figure 54. Detail of muscle showing structure and mechanics of muscular contraction.

tain only one nucleus, and they are not striped or banded but appear smooth.

Smooth muscle tissue is found most typically in rolled layers, making up the muscular layers of the digestive tract and other internal (visceral) hollow structures such as the bladder and blood vessels. Smooth muscle is also a major component of the skin. These muscles normally cannot be influenced by the will. They are controlled by the unconscious or automatic parts of the nervous system.

Cardiac Muscle. Cardiac (heart) muscle is involuntary, like smooth muscle, but possesses the striated appearance of skeletal muscle. Interconnecting fibers of cardiac muscle are not single cells but are built up in a chainlike manner from cardiac muscle cells. These cardiac muscle cells are the most evenly proportioned of the three types of muscle cells.

The rapid rhythm of cardiac muscle is possible because this type of muscle tissue has a special ability to receive an impulse, contract, immediately relax, and receive another impulse. During normal daily activity all these events occur about 75 times each minute.

For simplicity, we will use the skeletal muscle as an example in explaining how muscles work. Other aspects of smooth and cardiac muscles will be discussed again in appropriate later chapters.

THE MUSCLES AT WORK

What Are the Three Types of Muscle Contraction?

There are three types of muscle contraction: two active, known as *isotonic* and *isometric contraction,* and one passive, called *tonic contraction.* The terms "isotonic" and "isometric" have been formed from Greek derivatives: "iso-" meaning equal; "tonz" meaning put under tension, and "metr-" meaning measure.

Isotonic (i″so-ton′ik) contraction occurs in *movements* of any part of the body, and always involves a change in the length of the muscle; as for example, in lifting an object or doing exercises, such as jumping jacks. Figure 55 shows the latent period from stimulation to the beginning of contraction; the period it takes to contract and the period for relaxation total less than 1/10 second.

Isometric (i″so-met′rik) contraction does not produce any change in the overall length of the muscle, and therefore does not lead to any movement; examples are holding an object in the air without moving it, or pushing on an object that does not move.

The student may recognize the term "isometric" from recent popular exercise programs which recommend a group of ex-

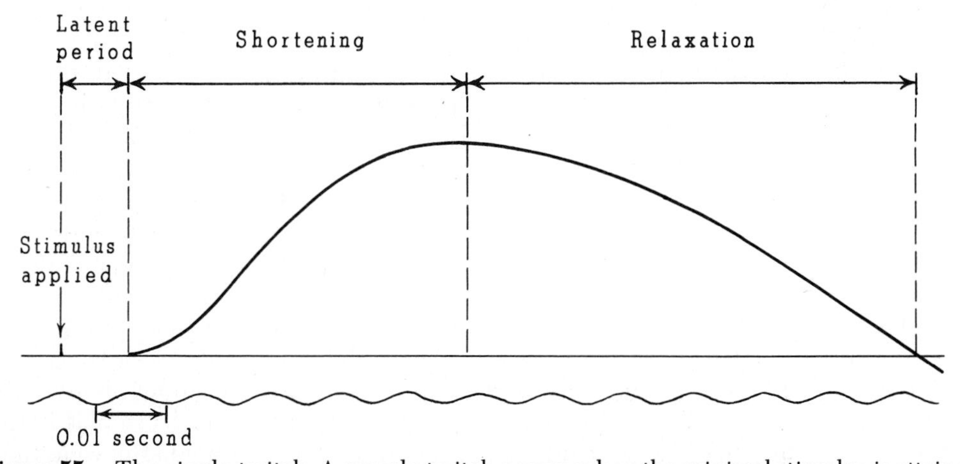

Figure 55. The single twitch. A muscle twitch occurs when the minimal stimulus is attained. All the muscle fibers associated with the stimulated nerve, after a period of latency, contract and then relax. The duration of this twitch is less than 0.1 second. Note that the period for relaxation is longer than the time needed for contraction. (After Carlson and Johnson.)

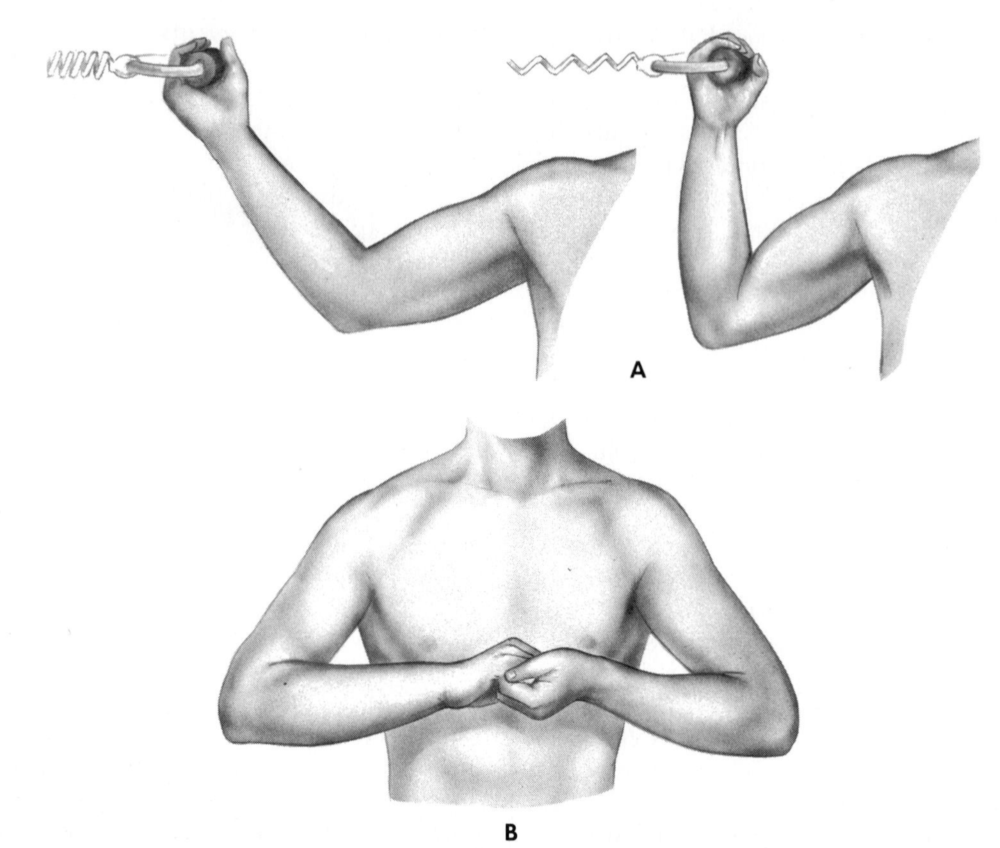

Figure 56. *A* shows isotonic contraction; here there is movement and the shape of the muscle alters to become shorter and wider or thicker. *B* shows isometric contraction; here there is no movement of the skeletal framework but there is exertion of the muscle. In isometric contraction the muscle will become slightly more compact as it is exerted, and later may become slightly larger as it fills with blood.

ercises in relation to fixed objects; that is, where one tightens or contracts one's muscles without producing any actual movement. Figure 56 illustrates isotonic and isometric contraction.

What Is Muscle Tone?

Muscle tone refers to the tone or tension of the muscular system. It is a measure of the degree of constant contraction of the muscles. We usually are not aware of this tonic contraction, since it does not require any conscious exertion; it is, in this sense, passive. The constant tonic contraction that maintains muscle tone is due to a continuous flow of nonconscious stimuli from the brain and spinal cord to each muscle. This constant stimulation increases or decreases, depending on the level of activity of the nervous system. In times of excitement there is an increase in tone; during periods of restfulness a decrease in tone occurs.

Besides serving to maintain a state of readiness of the muscular system, tonic contraction is the basis of good posture. As muscle tone decreases, posture deteriorates. This deterioration of posture causes abnormal pull on ligaments, joints, and bones, which in time may lead to permanent skeletal deformities. Persons with low muscle tone also tire much more easily.

Of primary clinical interest is the fact that individuals with good muscle tone will both look and feel better, and almost invariably have an excellent response to disease or injury.

What Are the Benefits of Exercise?

The benefits of exercise have been well reported in both professional and popular literature. Properly planned and regularly practiced exercise greatly *increases muscle tone* and, as a consequence, improves health and the general ability of the individual to recover from disease or injury.

The general immobility of sick persons can often be as damaging as the primary illness or injury itself. The bedfast patient seems to recover more quickly when a program of regular exercise is maintained during convalescence. The nurse can help and encourage the bedridden to exercise regularly, and by doing so have an enormously important effect on the recovery of the patient.

How Do Skeletal Muscles Produce Movements?

Most skeletal muscles are attached to two bones, with the muscle spanning the joint between the bones. One of the bones always moves more easily than the other in isotonic contraction (contractions producing movement) of any particular muscle. The muscle's attachment to the more stationary bone is called its *origin*. The attachment to the more easily moveable bone is called its *insertion*.

Most voluntary muscles are not inserted directly into bone, but rather through the medium of a strong, tough, nonelastic cord called a *tendon*. Tendons vary in length from a fraction of an inch to more than a foot. In all cases, however, the arrangement is such as to allow the muscle the greatest leverage; that is, so that the muscle will get the most work or power from its activity. Skeletal muscles are sometimes categorized according to the type of action or work they perform. Muscles that bend a limb at a joint are called *flexors* (flek′sors); those that straighten a limb at a joint are called *extensors*. If the limb is moved away from the midline of the body, an *abductor* (ab-duk′tor) is at work; if the limb is brought toward the midline, *adductors* are responsible. There are also other muscles that rotate the involved limb.

LOCATION AND ACTION OF THE IMPORTANT SKELETAL MUSCLES

Figures 57 to 59 and Table 5 include most of the important muscles in the human body. Other organs that contain muscular tissue will be described in appropriate later chapters.

You will find it easier to remember the name, location, and action of each muscle if you keep in mind that muscles are named according to *action* (for example, adductor or extensor); according to *shape* (for example, quadrates); according to *origin* and *insertion* (for example, sternocleidomastoid); according to number of *divisions* (for example, quadriceps or triceps); according to *location* (for example, tibialis or radialis); and according to *direction of fibers* (for example, transversus).

In addition to listing the muscles, Table 5 describes the origin and insertion of each muscle and its principal action and nerve supply.

CLINICAL CONSIDERATIONS

Describe the Symptoms of Several Muscle Disorders

The major symptoms of muscular disorders are paralysis, weakness, pain, atrophy (deterioration) and cramps.

The condition in which a muscle shortens its normal length in the resting state is known as *contracture* (kon-trak′tur). Contractures occur when an individual remains in bed for prolonged periods and the muscles are not properly exercised. Eventually, the muscles readjust to the resting length of a flexed arm or leg. Contractures are treated by the painful and slow procedure of exercising and relengthening the muscle. Contractures can be prevented by

Text continued on page 72.

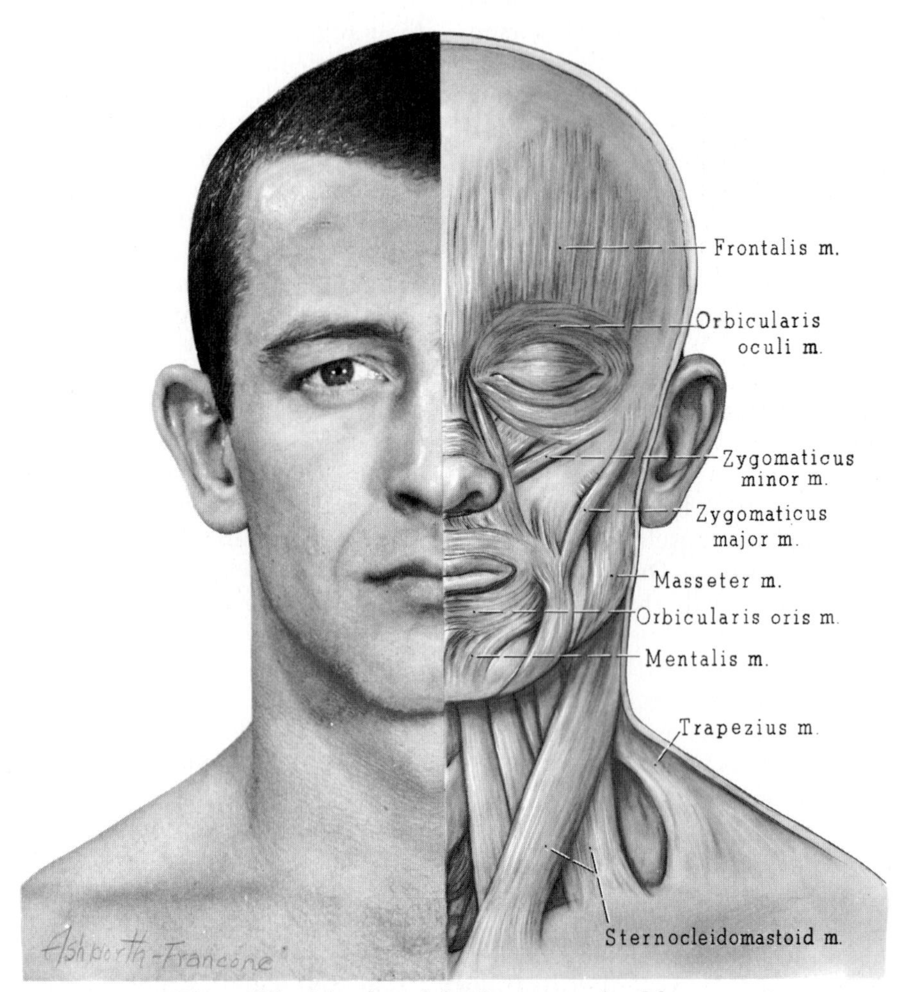

Figure 57. Muscles of the face, superficial layer.

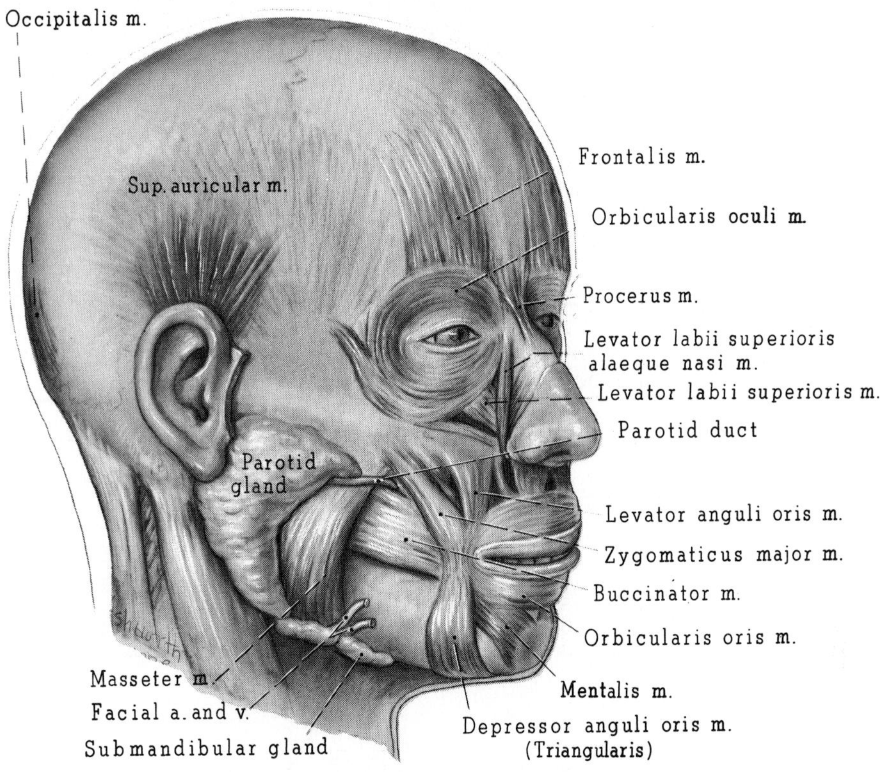

Figure 58. Muscles of the face, deep layer.

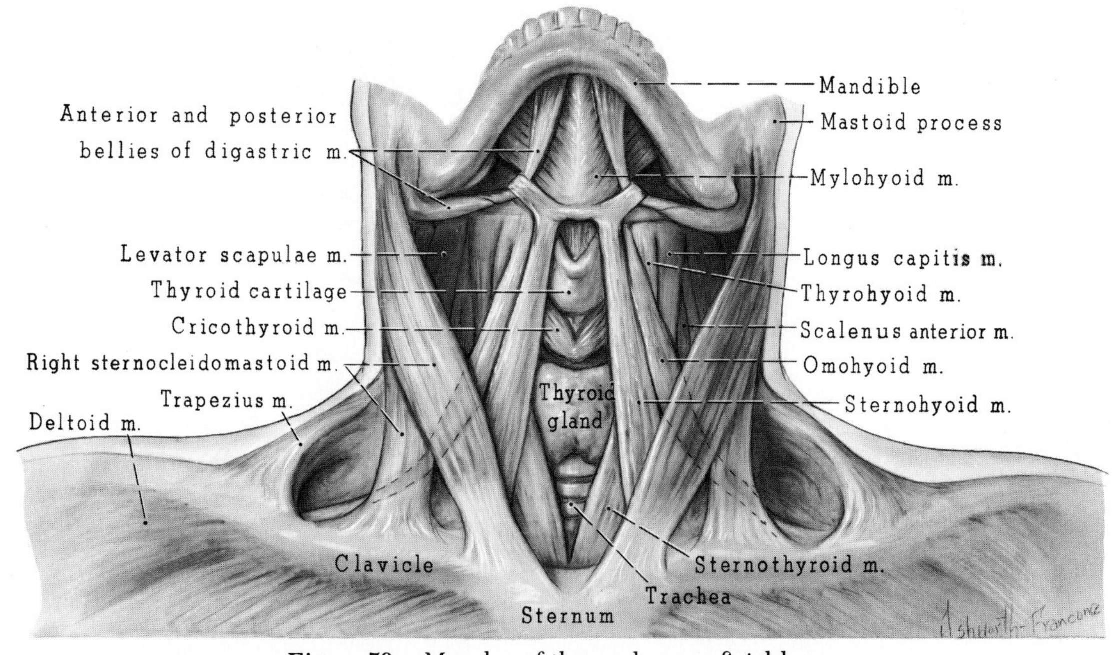

Figure 59. Muscles of the neck, superficial layer.

Table 5. MUSCLES

MUSCLE	ORIGIN	INSERTION	FUNCTION	INNERVATION
Frontalis	Galea aponeurotica	Frontal bone above supra-orbital line	Elevates eyebrows and wrinkles skin of forehead	Facial
Zygomaticus major	Zygomatic bone	Orbicularis oris	Pulls angle of mouth upward and backward when laughing	Facial
Orbicularis oris	Muscle fibers surrounding the opening of the mouth	Angle of mouth	Closes lips	Facial
Masseter	Zygomatic process and adjacent portions of maxilla	Angle and lateral surface of ramus of mandible	Closes jaw	Trigeminal
Sternocleidomastoid (right and left)	Two heads from sternum and clavicle	Tendon into mastoid portion of temporal bone	Flexes vertebral column; rotates head	Spine accessory
Trapezius	Occipital bone, seventh to twelfth thoracic vertebrae	Acromial process of clavicle and spine	Draws head to one side; rotates scapula	Spinal accessory
Pectoralis major	Anterior surface of sternal half of clavicle; sternum; six upper ribs	Crest and greater tubercle of humerus	Flexes, adducts, rotates arm	Anterior thoracic
Deltoid	Clavicle; scapula	Lateral surface of body of humerus	Abducts arm	Axillary
Latissimus dorsi	Vertebrae; ilium	Humerus	Extends, adducts, rotates arm medially; draws shoulder downward and backward	Thoracodorsal
Biceps brachii	Scapula	Radius	Flexes lower arm; rotates hand	Musculocutaneous
Triceps brachii	Scapula humerus	Ulna	Extends and adducts forearm	Radial
Brachioradialis	Humerus	Lower end of radius	Flexes forearm	Radial
Flexor carpi radialis	Humerus	Second and third metacarpals	Flexes, abducts wrist	Median

Flexor pollicis longus	Radius	Base of distal phalanx of thumb	Flexes second phalanx of thumb	Posterior interosseus
Rectus abdominis	Crest of pubis and ligaments covering symphysis	Cartilages of fifth, sixth, and seventh ribs	Flexes vertebral column; assists in compressing abdominal wall	Branches of seventh to twelfth intercostal costal
External oblique	Lower eight ribs	Anterior half of outer lip of iliac crest; anterior rectus sheath	Compresses abdominal contents	Branches of eighth to twelfth intercostal; iliohypogastric
Diaphragm	Xiphoid process; costal cartilages; lumbar vertebrae	Central tendon	Pulls central tendon downward to increase vertical diameter of thorax	Phrenic
Iliopsoas; psoas major; psoas minor	Ilium; vertebrae; femur	Femur; ilium; vertebrae	Flexes and rotates thigh Flexes trunk	Second and third lumbar First lumbar
Gluteus maximus	Ilium; sacrum; coccyx	Femur	Extends thigh	Inferior gluteal
Gluteus medius	Ilium	Strong tendons that run into lateral surface	Abducts, rotates thigh medially	Superior gluteal
Quadriceps femoris group (rectus femoris, vastus lateralis, vastus medialis, vastus intermedius)	Ilium and femur	Patella and tibia	Extends leg, flexes thigh	Femoral
Gastrocnemius	Femur and capsule of knee	Tendon calcaneus (Achilles tendon or hamstring)	Points toes; flexes leg; supinates foot	Tibial
Tibialis anterior	Upper tibia	Undersurface of medial cuneiform and base of first metatarsal	Dorsally flexes foot	Deep peroneal
Abductor hallucis	Calcaneus	Proximal phalanx of great toe	Abducts, flexes great toe	Medial plantar

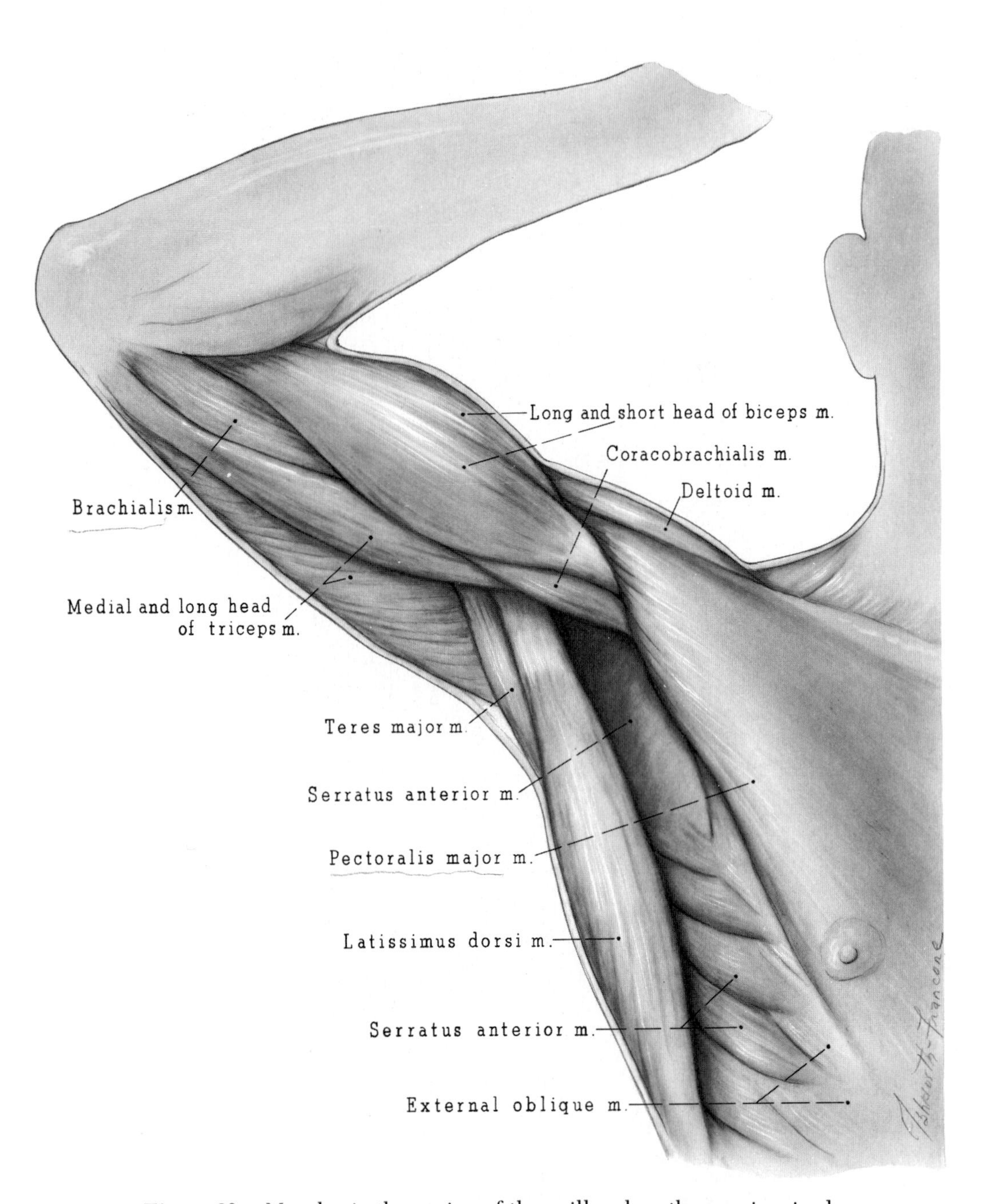

Figure 60. Muscles in the region of the axilla when the arm is raised.

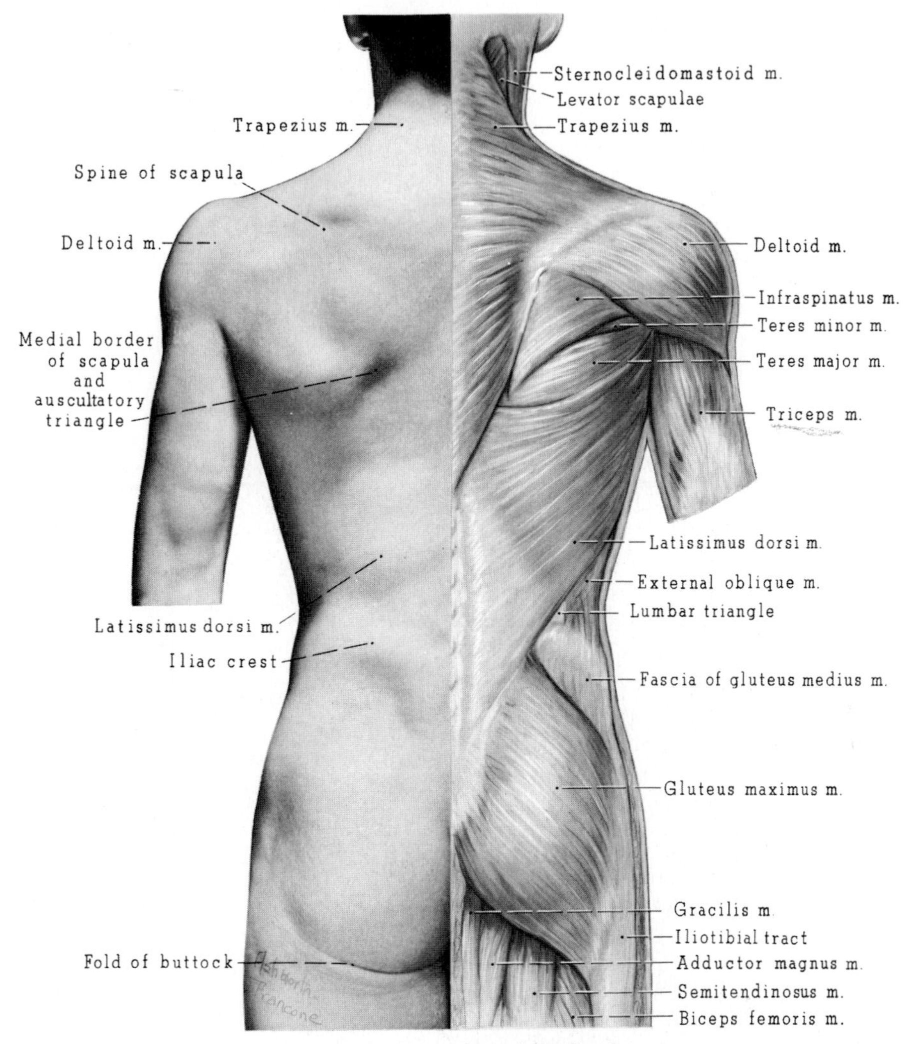

Figure 61. Muscles of the back.

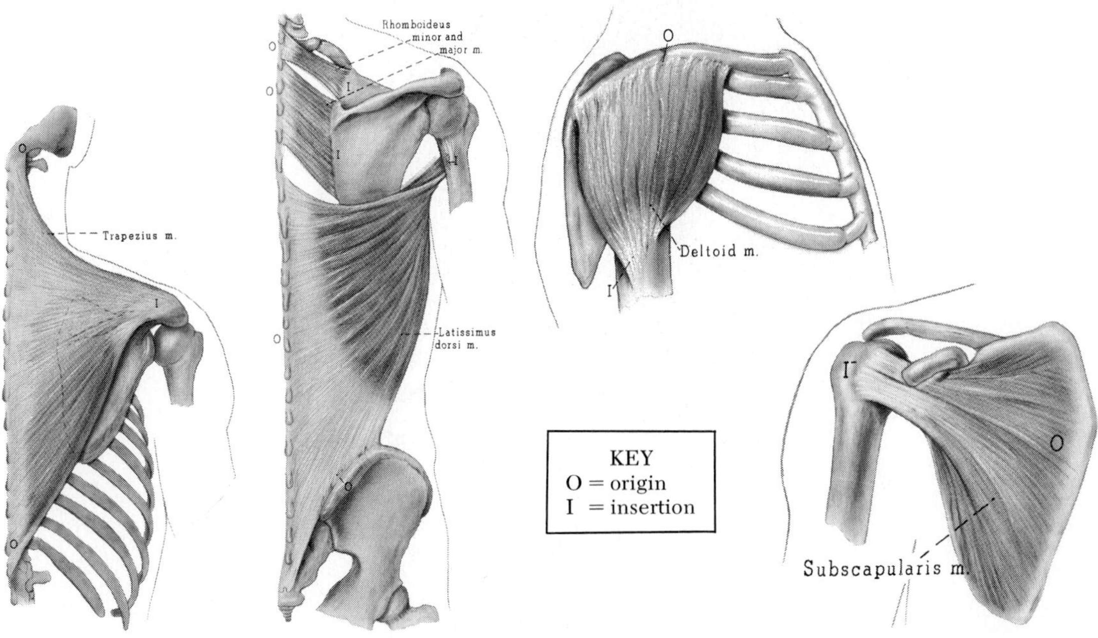

Figure 62.

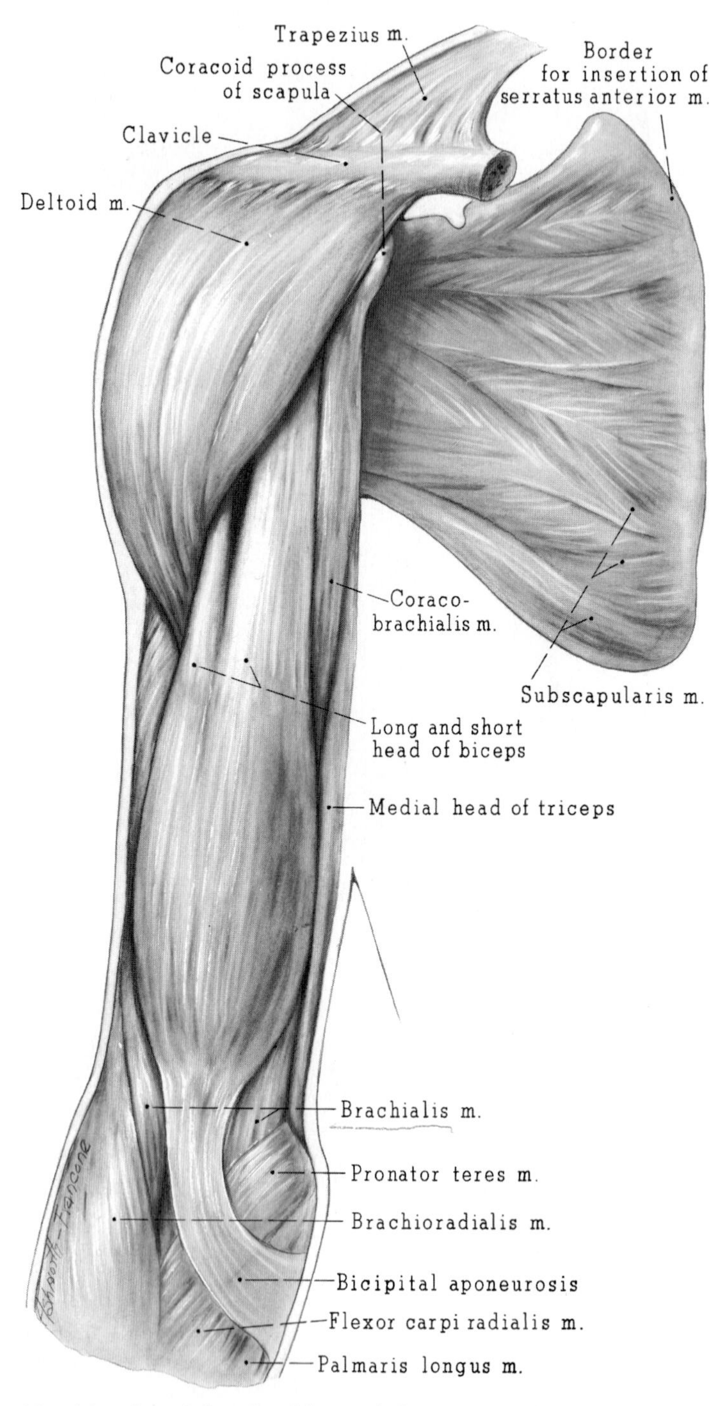

Figure 63. Muscles of the shoulder and the upper right arm, anterior view.

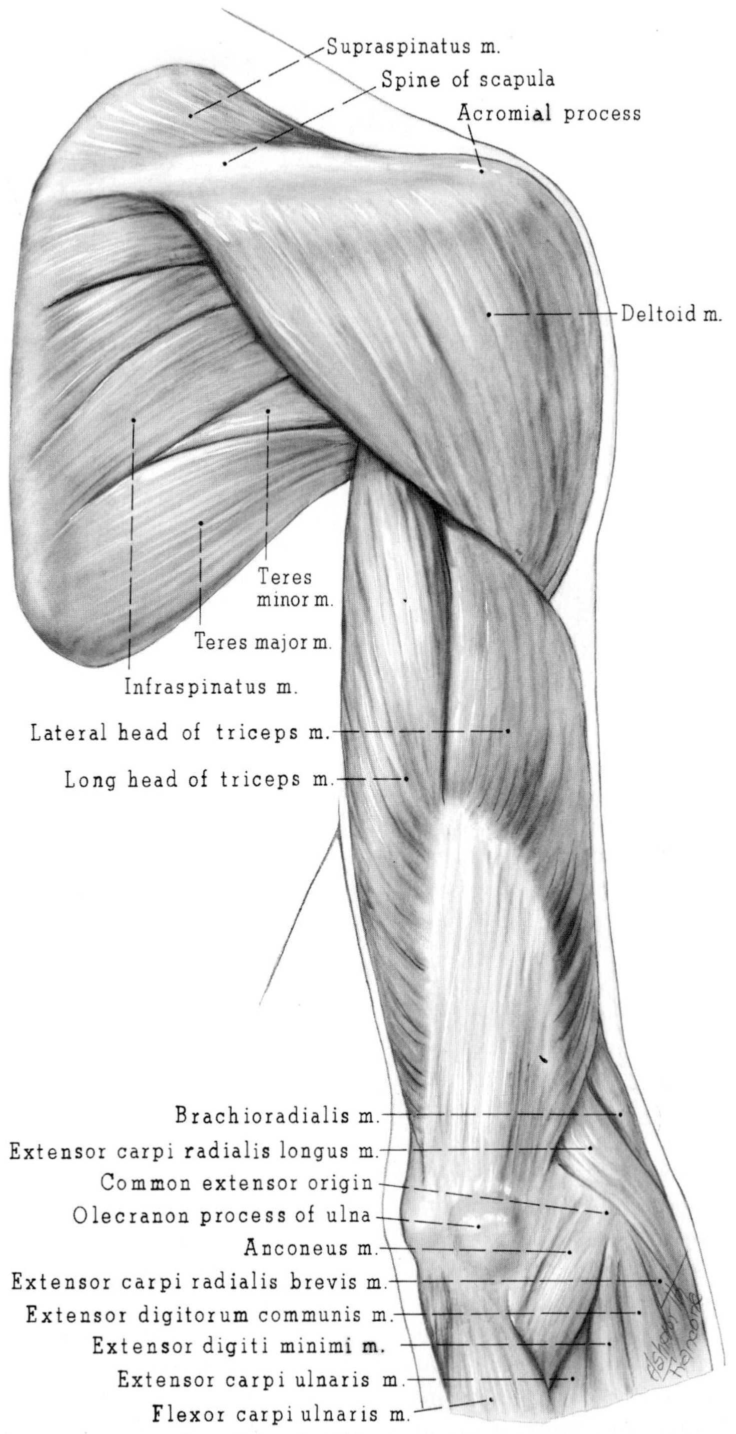

Figure 64. Muscles of the shoulder and upper arm, posterior view.

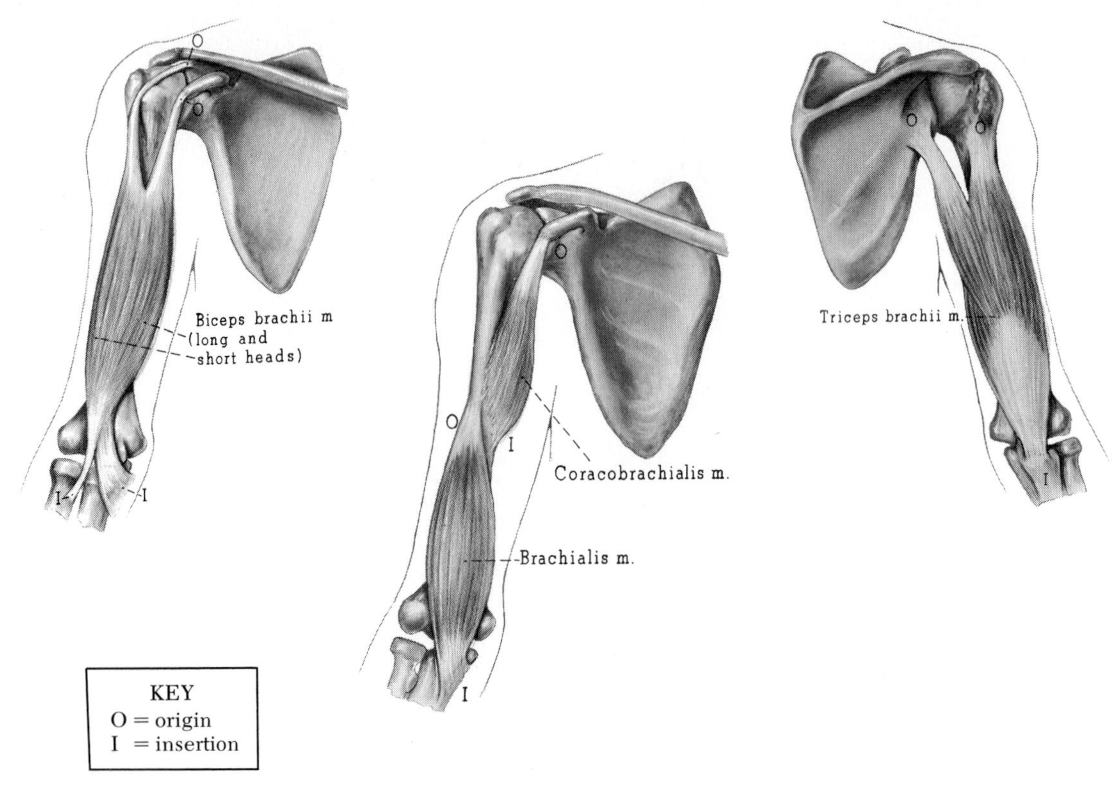

Figure 65.

keeping the body in correct alignment when resting, and by periodically exercising the muscles. Muscular exercise can be either active (by the patient himself) or passive (by someone else).

Paralysis is a loss of responsiveness of muscles; this condition invariably results from disease or injury to the nervous system, usually damaging the brain or spinal cord. In cases of paralysis or lack of exercise, muscles tend to *atrophy* (at′ro-fe), meaning to deteriorate, shrink, and waste away. Muscle *hypertrophy* is the opposite of atrophy. Hypertrophy is an abnormal building up or increase in the size of the muscle; this may result from overexercise.

Two other disorders of the muscular system are muscular dystrophy and myasthenia gravis. The exact cause and treatment of both are unknown. *Muscular dystrophy* (dys- means difficult, and trophy means to nourish) occurs most often in males, and is a slowly progressive disorder that ends in complete helplessness. *Myasthenia gravis* (*my-* means muscle, *asthenia* means weakness, and *gravis* means heavy) is characterized by easy fatigability of muscles. It involves an impairment of conduction of signals between the muscles and the nerves that normally serve them.

Where Should Intramuscular Injections Be Given? Why?

Figures 80 and 81 show the best locations for intramuscular injections. These areas are considered to be relatively safe because they are away from major blood vessels and nerves. When the needle has been inserted into the muscle, one should always *pull back* on the syringe plunger before injecting; this provides an immediate test of whether the needle is inside a vessel—if blood is drawn in, the needle is in a vessel and should be reinserted. If an injection goes directly into a major vessel or nerve, it can cause serious problems.

Text continued on page 87.

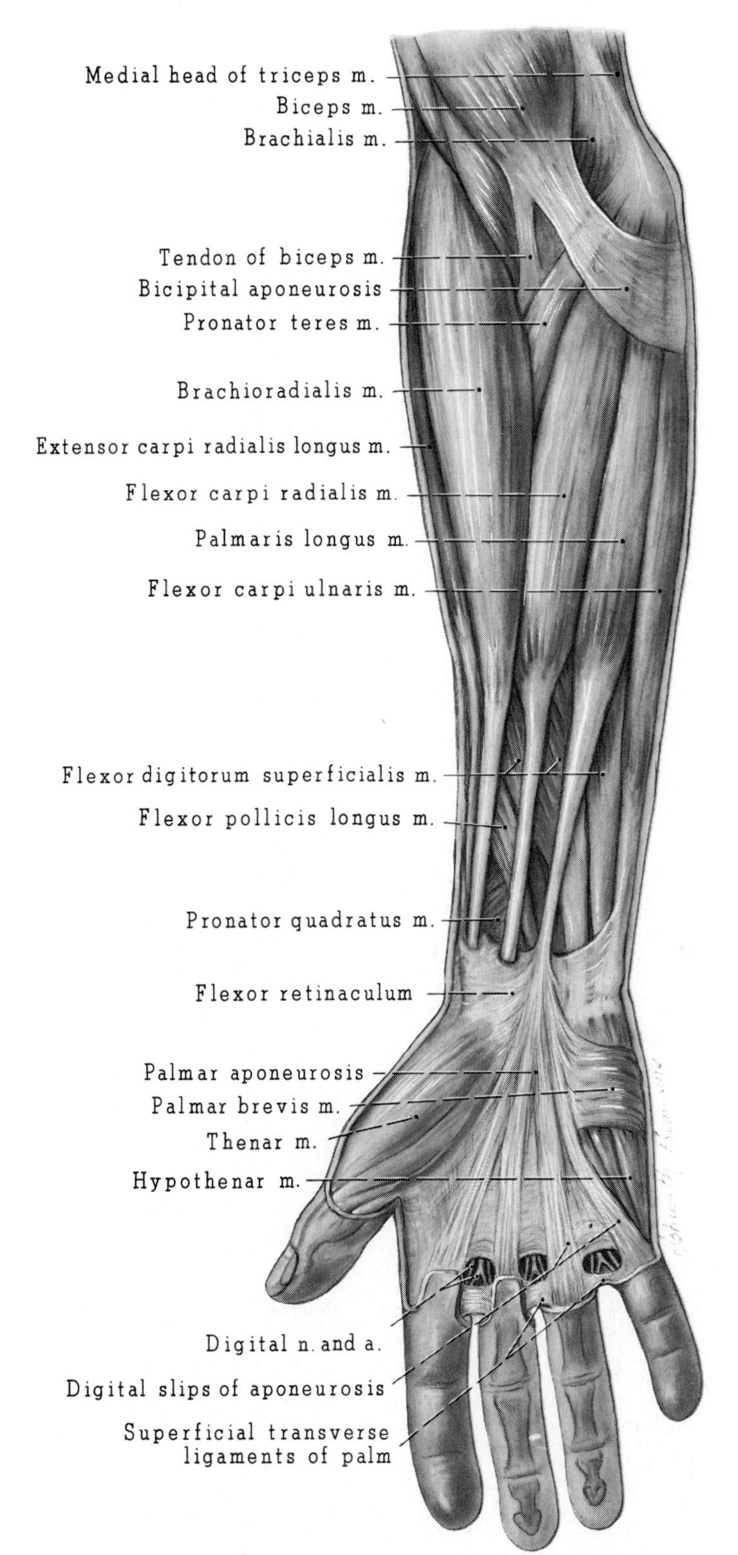

Figure 66. Muscles of the palmar aspect of the right hand and forearm.

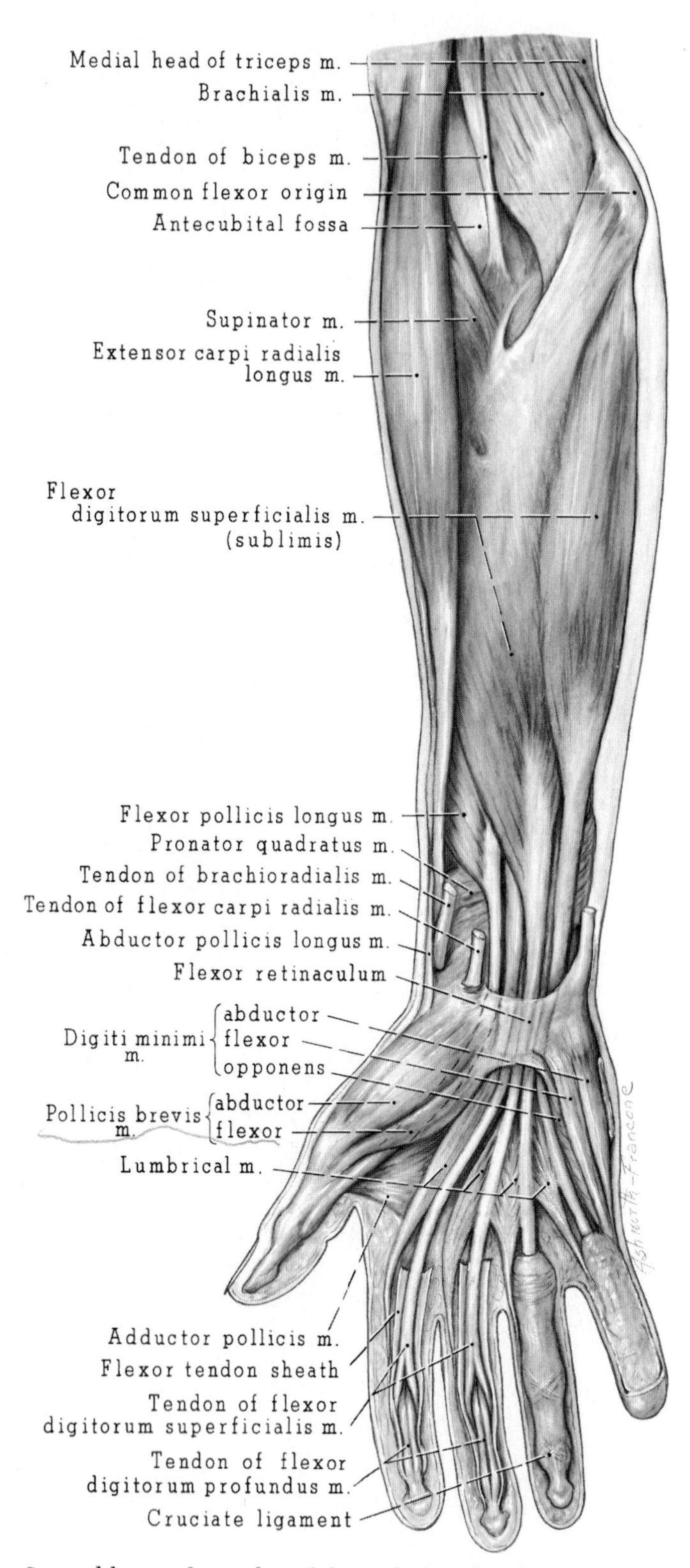

Figure 67. Second layer of muscles of the right hand and forearm, palmar aspect.

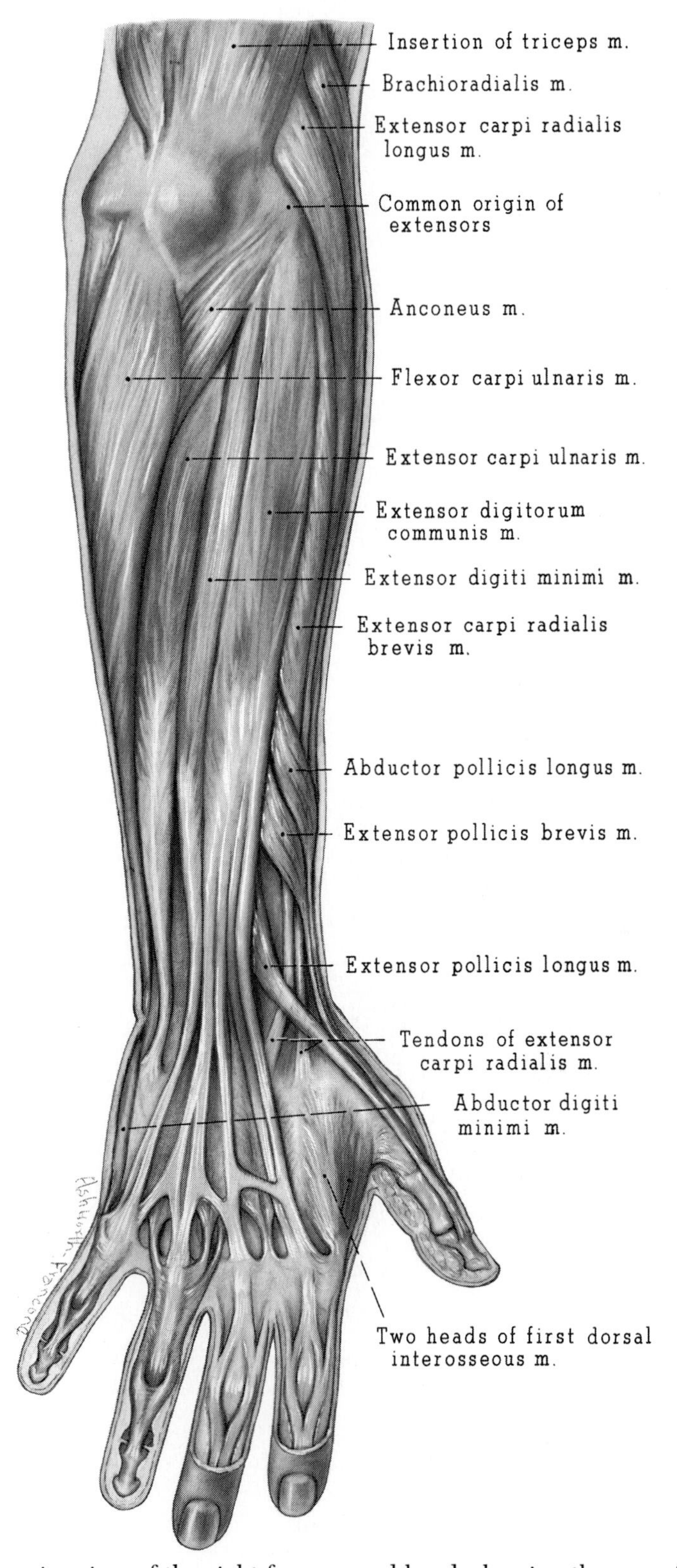

Figure 68. Posterior view of the right forearm and hand, showing the superficial muscles.

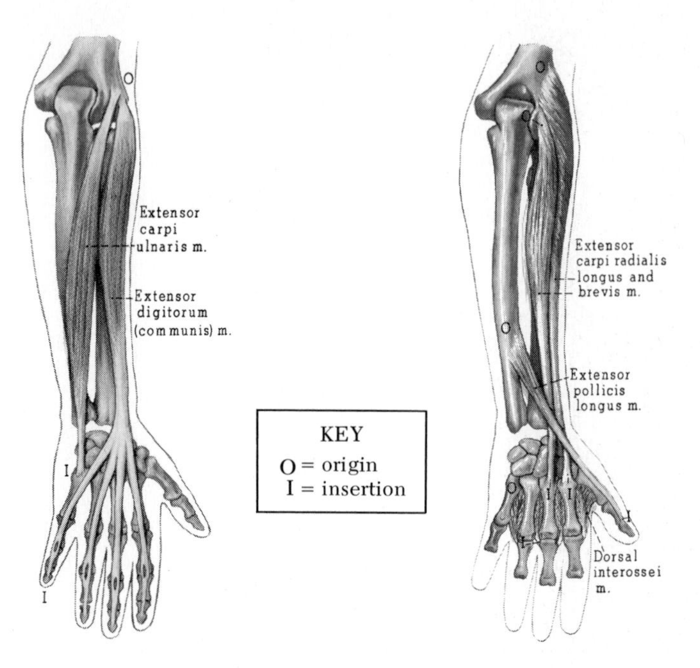
Extensor
carpi
ulnaris m.
Extensor
digitorum
(communis) m.
KEY
O = origin
I = insertion
Extensor
carpi radialis
longus and
brevis m.
Extensor
pollicis
longus m.
Dorsal
interossei
m.

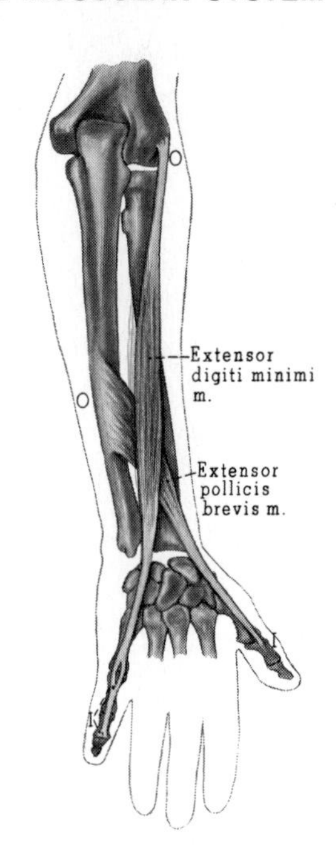
Extensor
digiti minimi
m.
Extensor
pollicis
brevis m.

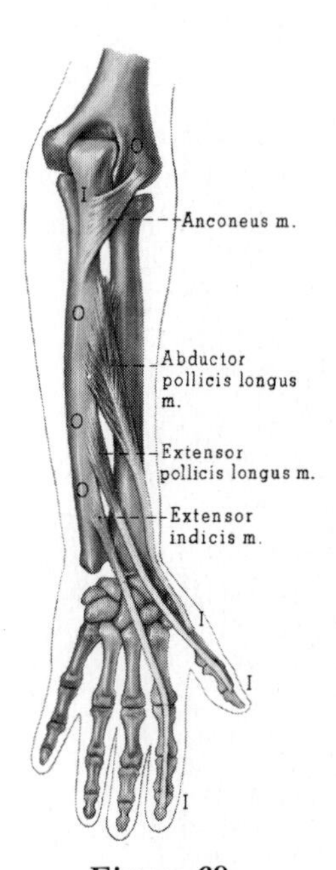
Anconeus m.
Abductor
pollicis longus
m.
Extensor
pollicis longus m.
Extensor
indicis m.

Figure 69.

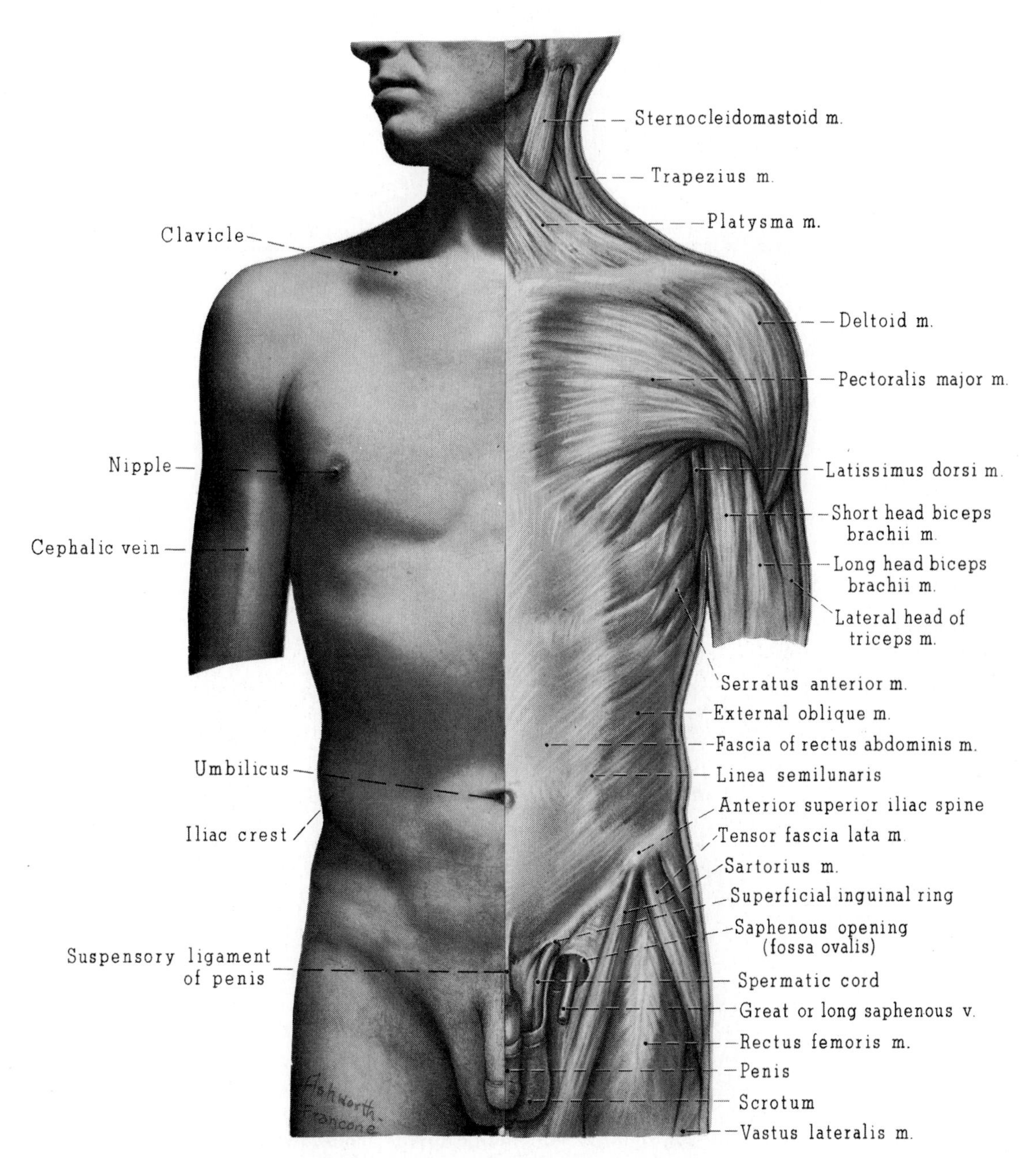

Figure 70. Muscles of the anterior surface of the male.

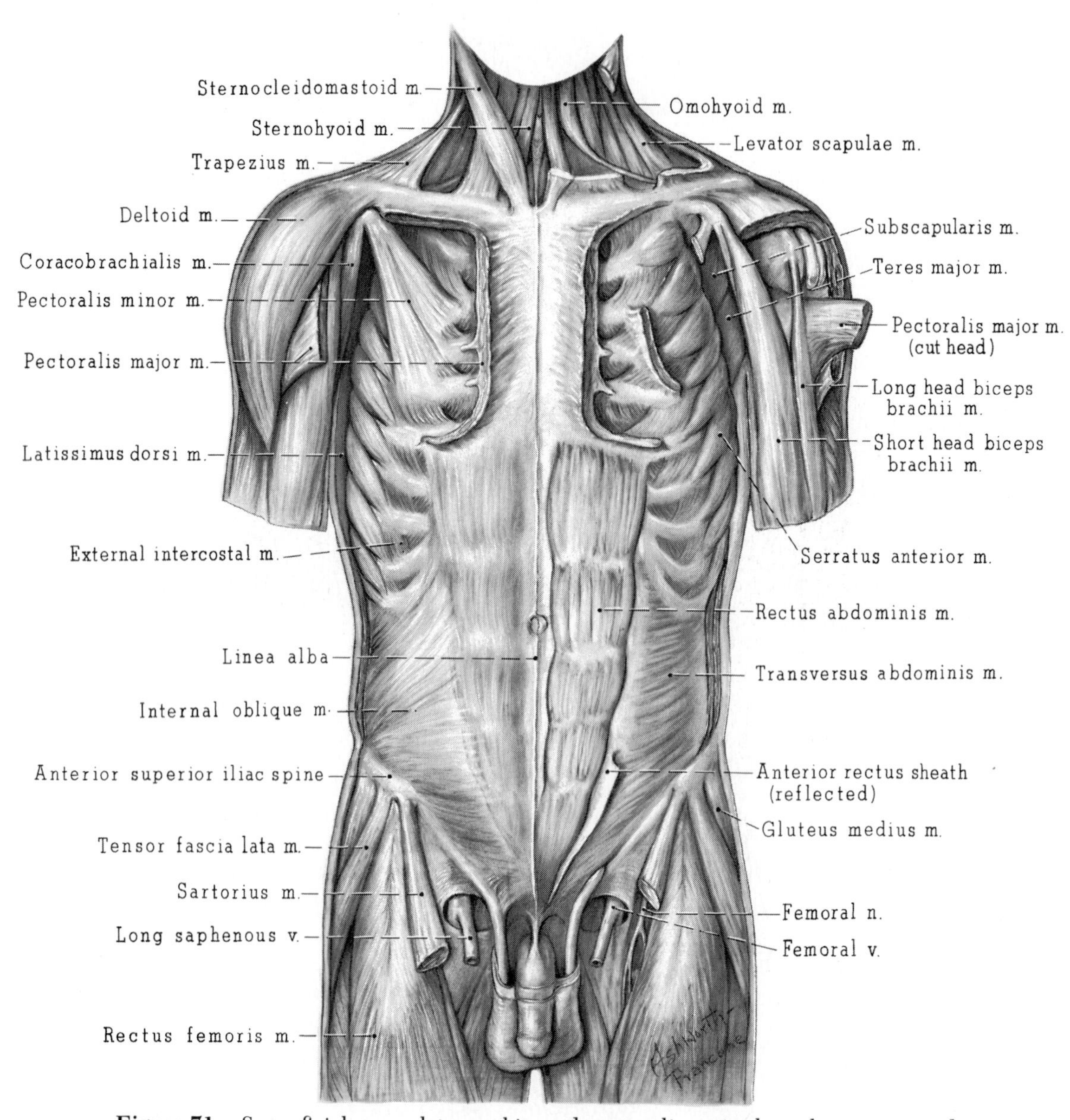

Figure 71. Superficial musculature; skin and pectoralis major have been removed.

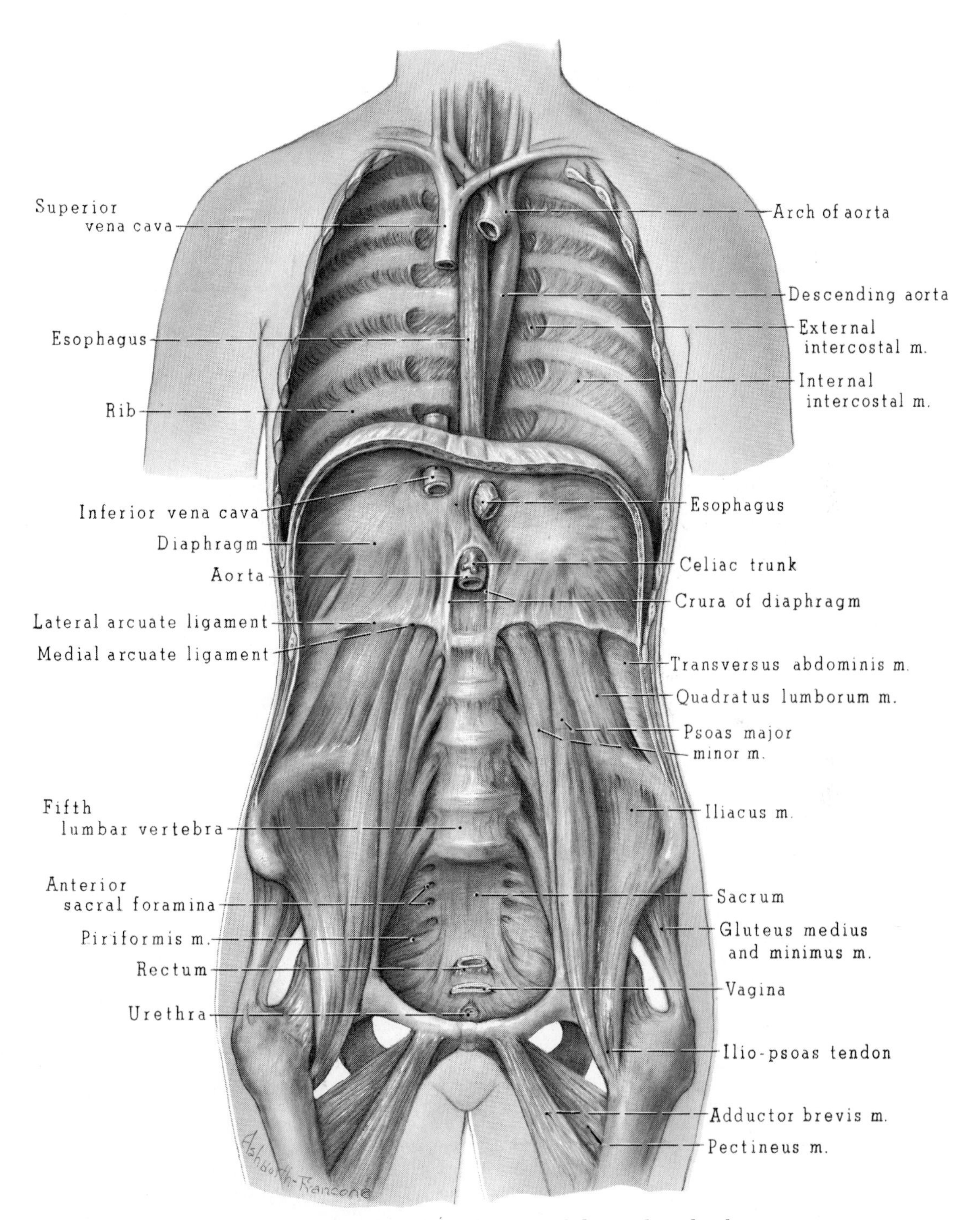

Figure 72. Deep muscles of the thoracic, abdominal, and pelvic cavities.

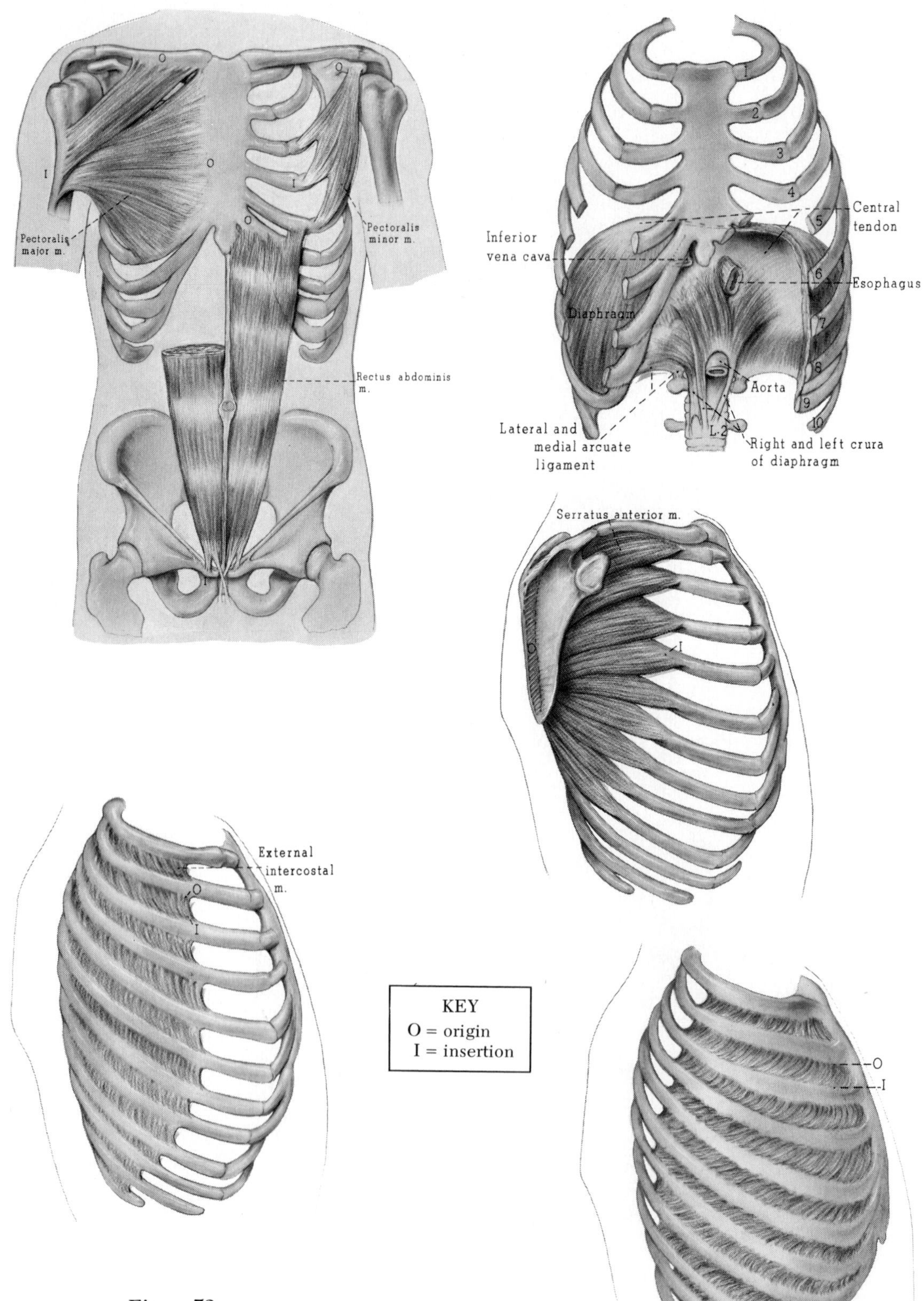

O
O
I
O
I
O
I
Pectoralis major m.
Pectoralis minor m.
Rectus abdominis m.
1
2
3
4
5
6
7
8
9
10
Central tendon
Inferior vena cava
Esophagus
Diaphragm
Aorta
L.2
Lateral and medial arcuate ligament
Right and left crura of diaphragm
Serratus anterior m.
O
I
External intercostal m.
O
I
KEY
O = origin
I = insertion
O
I
Internal intercostal m.

Figure 73.

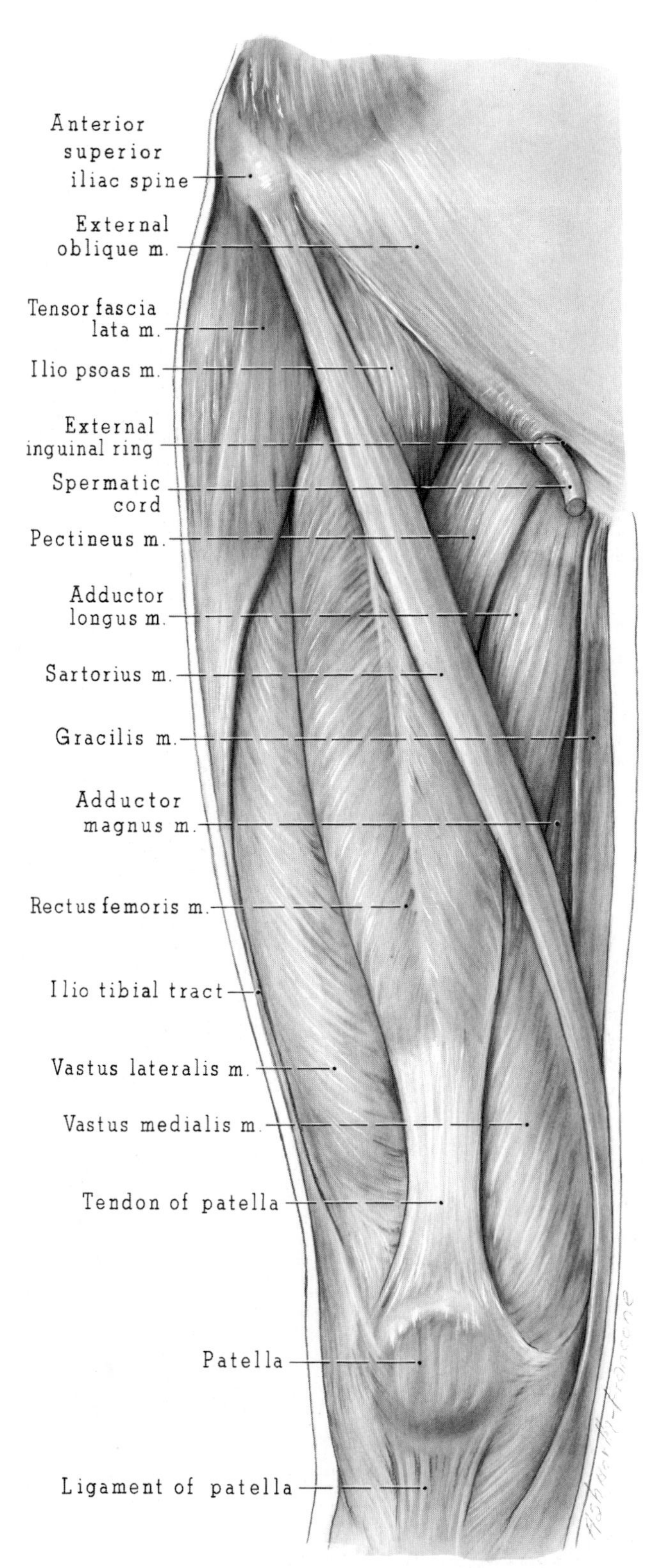

Figure 74. Superficial muscles of the right upper leg, anterior surface.

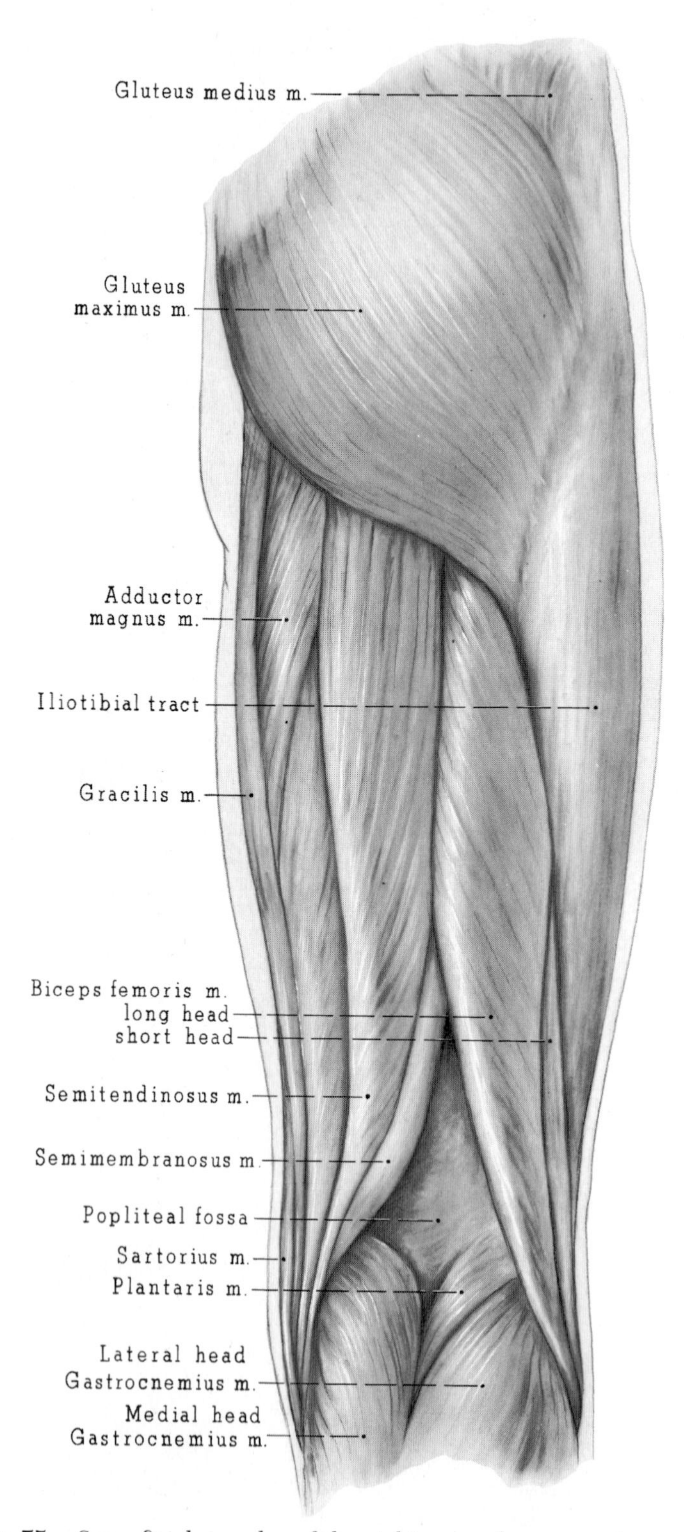

Figure 75. Superficial muscles of the right upper leg, posterior surface.

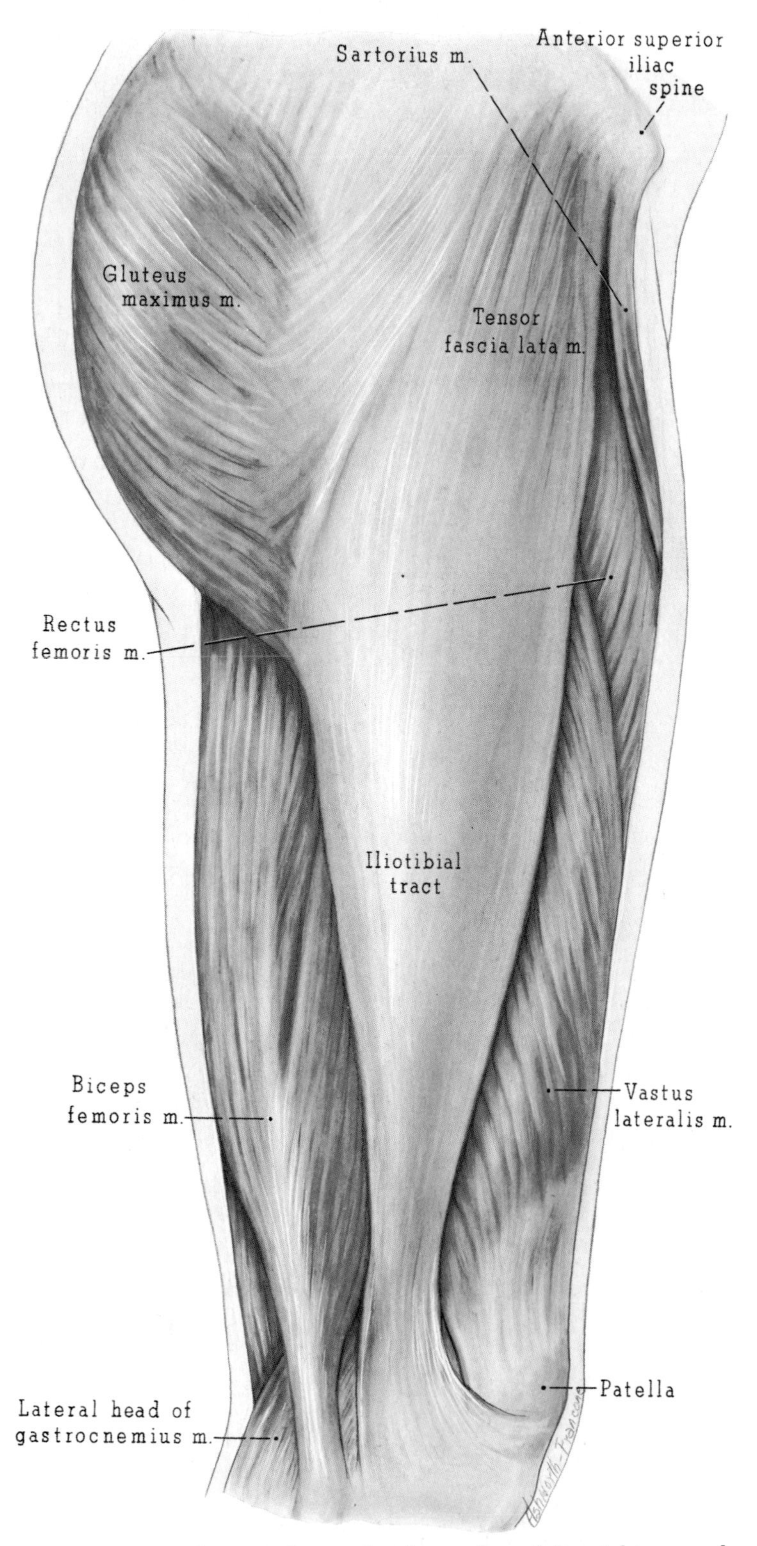

Figure 76. Lateral view of superficial muscles of the right upper leg.

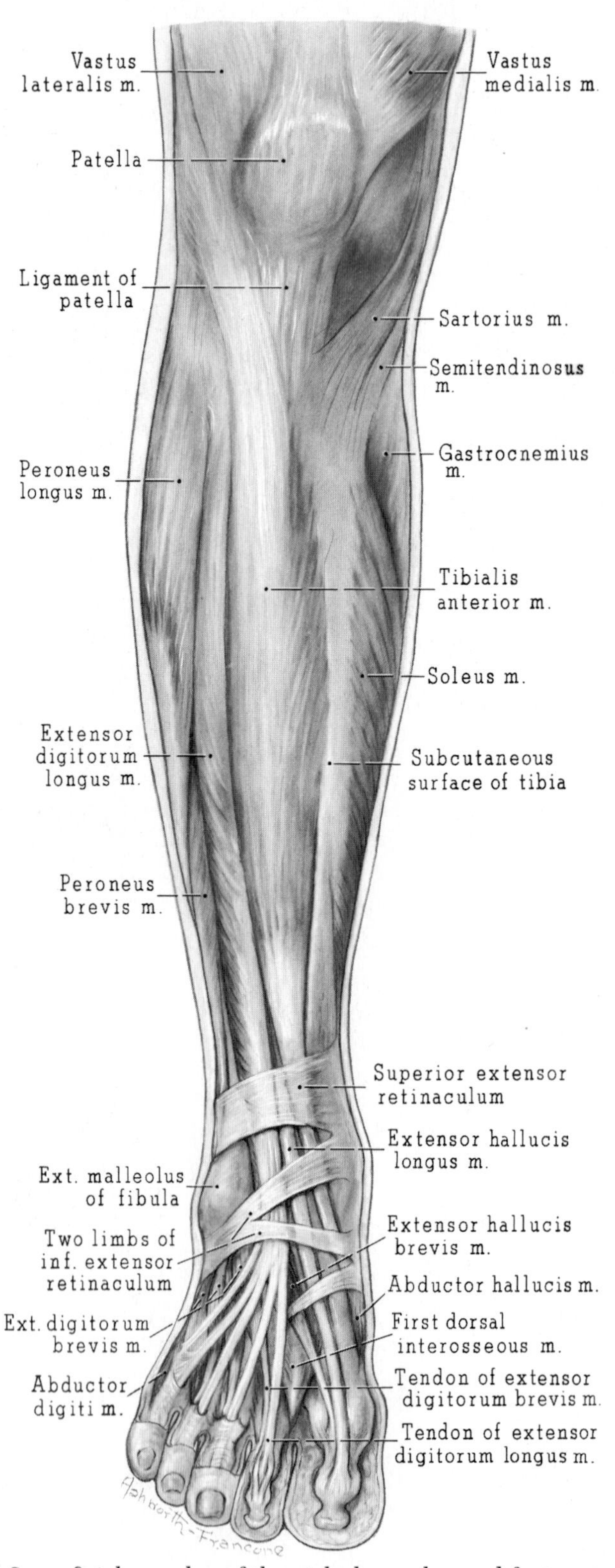

Figure 77. Superficial muscles of the right lower leg and foot, anterior surface.

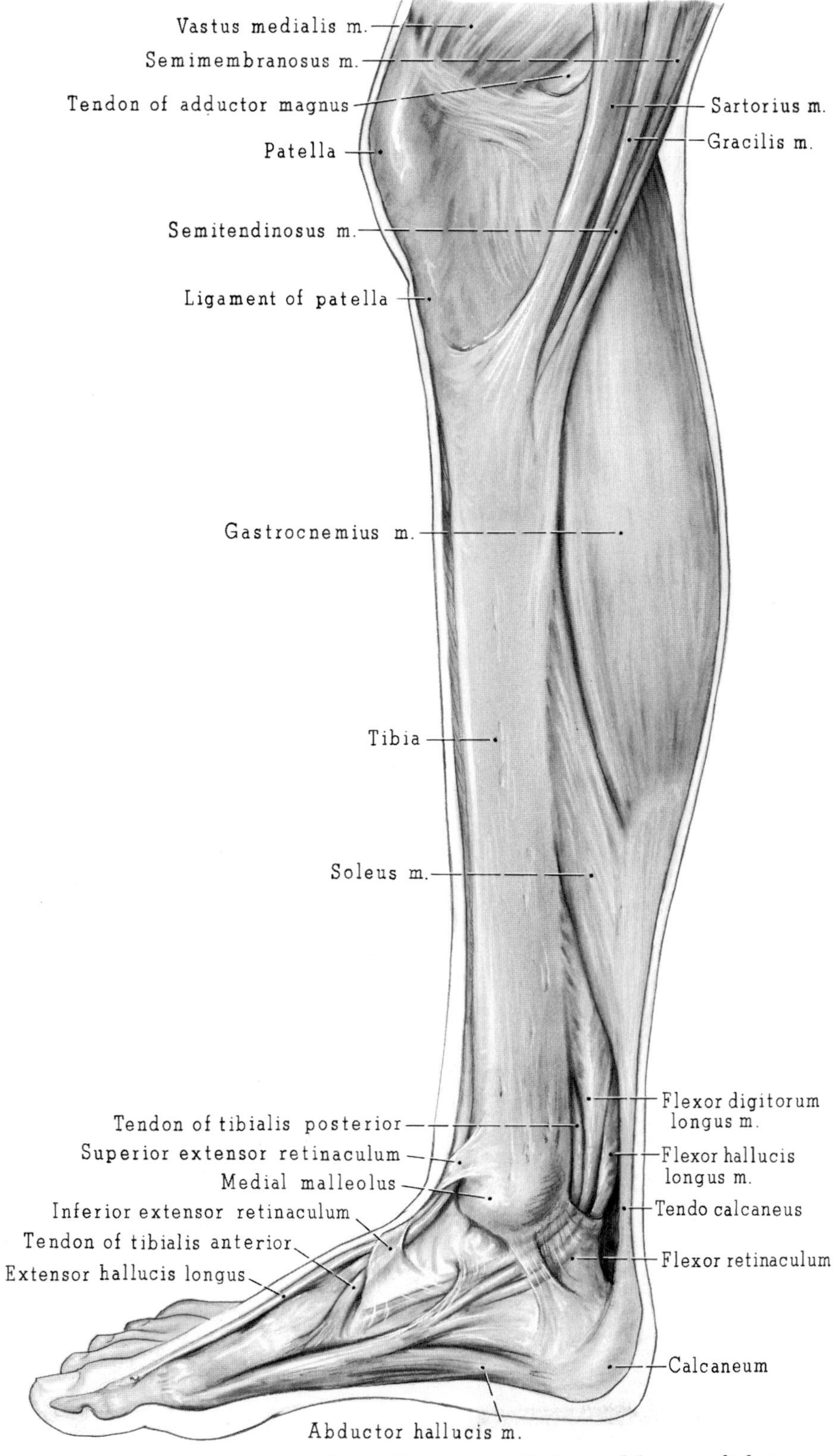

Figure 78. Superficial muscles of the lower right leg and foot, medial view.

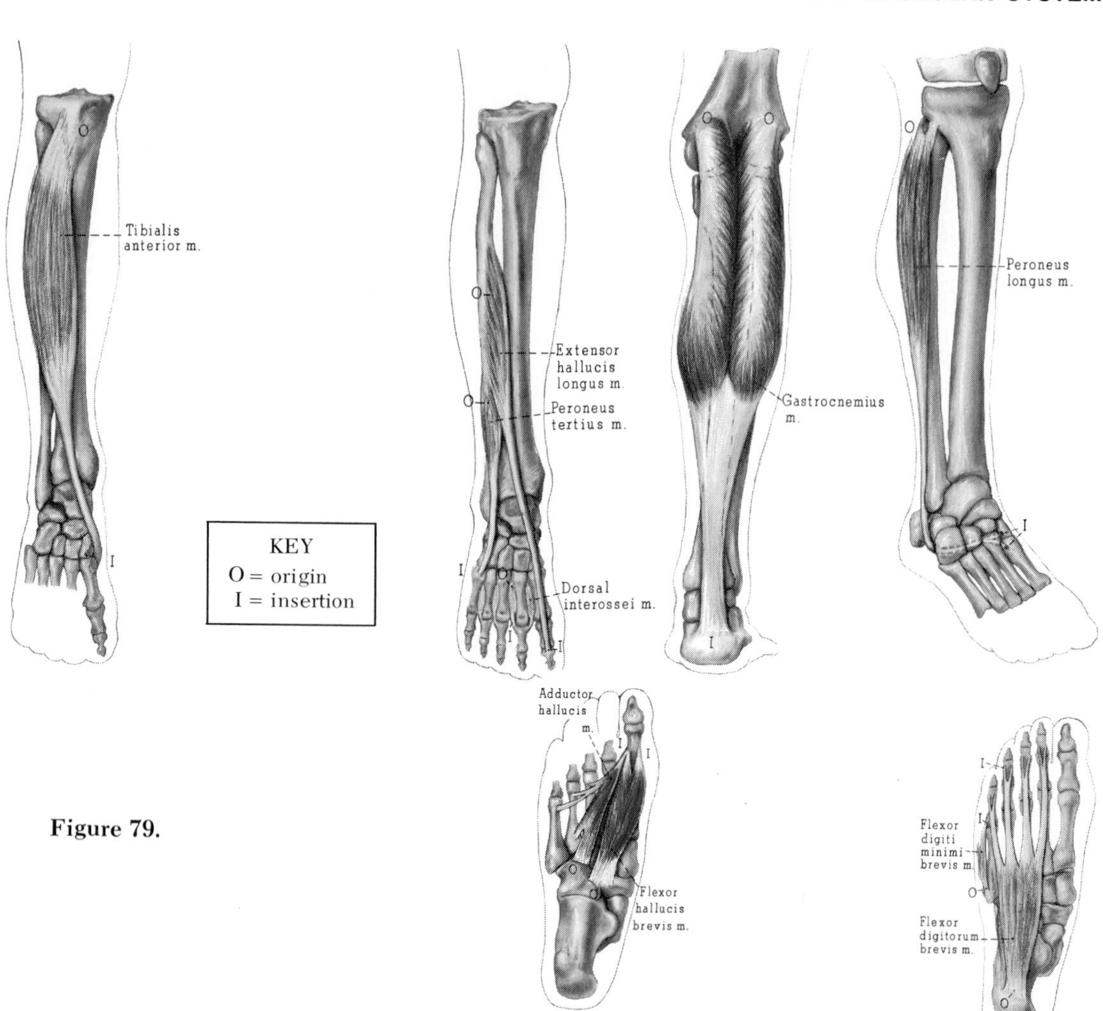

Figure 79.

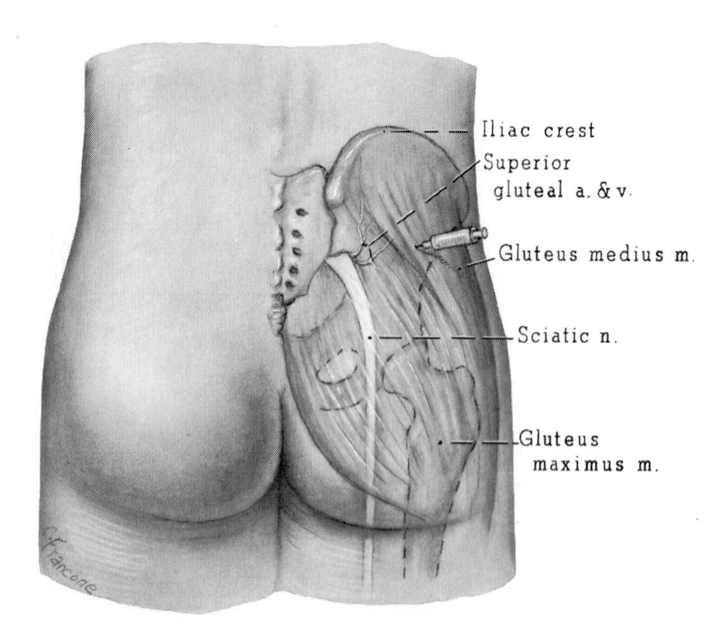

Figure 80. Intramuscular injection in the gluteal region should be in the upper outer quadrant.

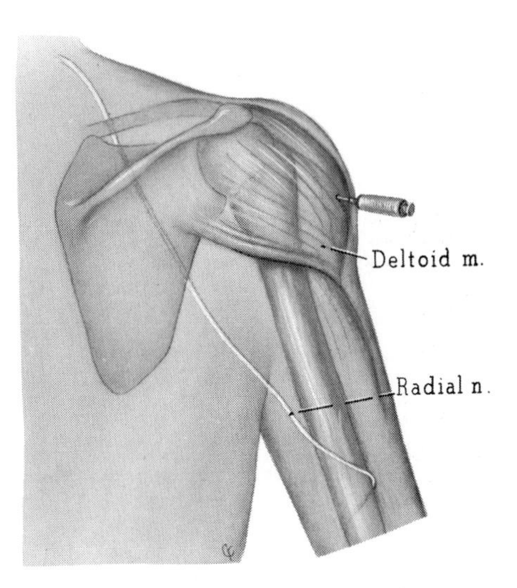

Figure 81. Intramuscular injection into the deltoid, two to three fingerbreadths below the acromion.

LEARNING EXERCISES

1. Using Table 5, find the listed muscles in the diagrams.
2. Copy the left column (names of muscles) on a sheet of paper, and then, with the book closed, write down the location and function of each muscle.

CHAPTER 6

THE BLOOD

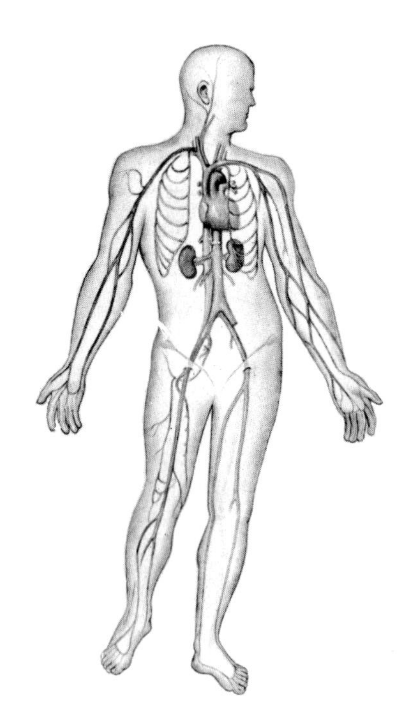

BLOOD: WHAT IT DOES, WHAT IT CONTAINS

What is the Function of the Blood?

The primary function of the blood is the internal transportation of cellular nutrients and wastes. Digested nutrients absorbed from the intestines and oxygen collected in the lungs are carried by the blood to all the cells in the body. The major waste product, carbon dioxide, is carried to the lungs for expulsion, whereas other wastes and excesses of cellular activity are carried to the kidneys for selective elimination.

Another function of the blood is accomplished principally by the white blood cells; they play a central role in the prevention of infection.

The blood also transports *hormones*, which are chemical messengers or stimulants secreted directly into the blood by special glandular cells. The function of hormones will be discussed in more detail in Chapter 12, The Endocrine System.

What Are the Major Constituents of the Blood?

The general appearance of blood might lead one to suppose that it is one simple substance. However, when blood is examined under a microscope or analyzed chemically, it is found to be composed of many distinct components (Figs. 82 and 83).

One simple way to separate blood is by use of the centrifuge (sen′tri-fuj). A few minutes of low-speed spinning yields, in the lower portion, the *formed elements* (cells, etc) and in the upper portion, the straw-colored *plasma* (plaz′mah). This centrifuge procedure is used for a standard type of blood test—to determine the percentage of formed elements in a person's blood. The term *hematocrit* (he-mat′o-krit) refers to the percentage of formed elements in the blood by volume (Fig. 84). Thus, if the percentage of formed elements in the centrifuge tube turns out to be 45, then the hematocrit is reported as 45. The

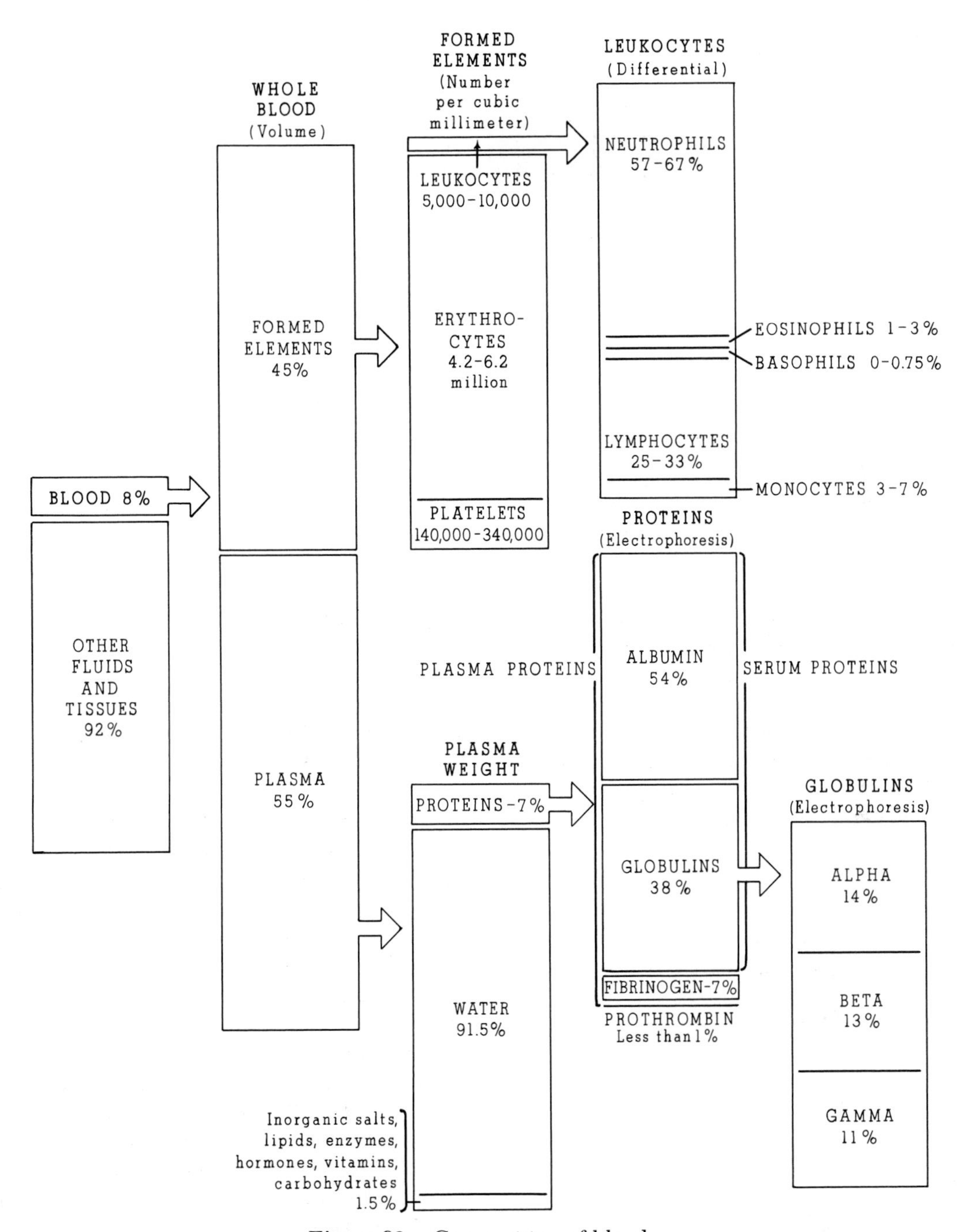

Figure 82. Composition of blood.

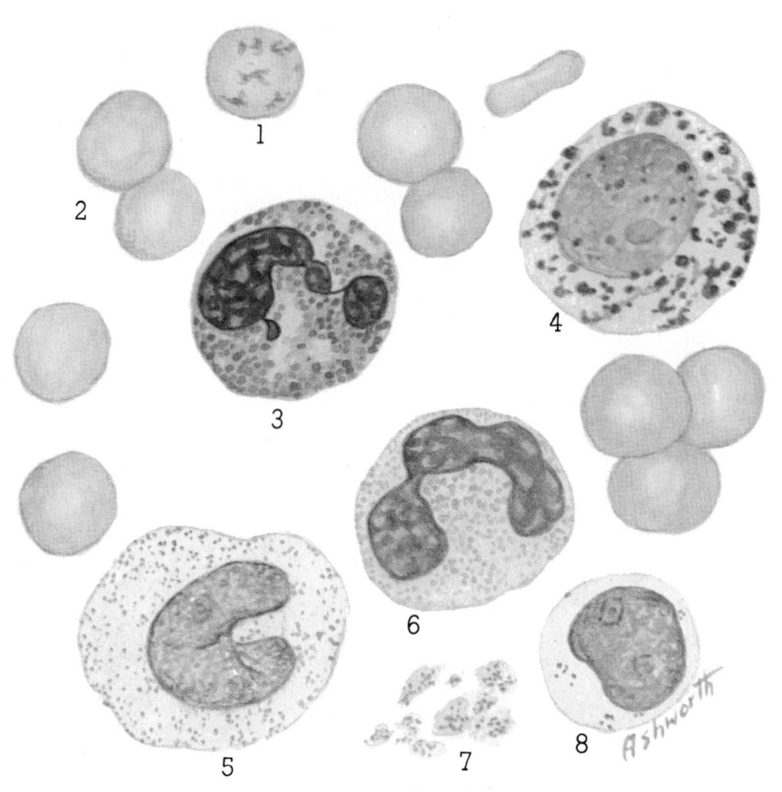

Figure 83. Blood cells: *1*, reticulocyte; *2*, erythrocyte; *3*, eosinophil; *4*, basophil; *5*, monocyte; *6*, neutrophil; *7*, platelets; and *8*, lymphocyte.

word hematocrit is derived from the Greek words *haimat*, meaning blood, and *crino*, meaning to separate; literally, then, it means to separate the blood.

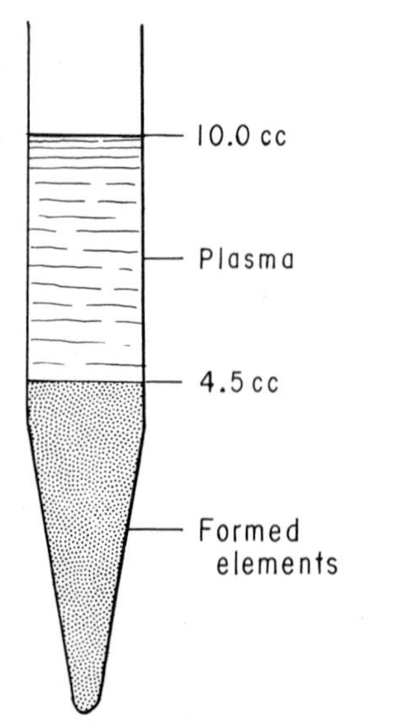

Figure 84. Demonstration of hematocrit.

What Are the Formed Elements? What Functions Do They Perform?

The formed elements consist of the red blood cells, the white blood cells, and the platelets (Fig. 83). All the formed elements develop and mature from the same basic type of cell, known as a hemocytoblast (cell from which blood grows).

The *red blood cells*, or *erythrocytes* (e-rith′ro-sits), function to transport oxygen from the lungs to the cells throughout the body. The name erythrocyte is derived from the Greek words, *erythros*, which means red, and *kytos*, which means hollow or cell. This oxygen-carrying task is accomplished through the action of a special protein, *hemoglobin* (he″mo-glo′bin), which is a major constituent of erythrocytes.

There are several types of *white blood cells*, or *leukocytes (leukos* means white and *kytos* means cell). The general function of leukocytes (lu′ko-sits) is to provide a defense against any virus, bacteria, or other foreign protein substance that enters into the blood stream or tissues. Two types of leukocytes, *neutrophils* (nu′tro-fils) and *monocytes*, operate by devouring any microorganisms or foreign protein. For example, the pus found around an infected wound is made up primarily of leukocytes that have died in combating the infection. Neutrophils comprise about 56 to 67 per cent of the white blood cells. Two other types of white blood cells are eosinophils and basophils.

Lymphocytes (lim′fo-sits) make up about 25 to 33 per cent of the leukocytes, and operate to protect us from infection by a process known as immune response. A great deal of research is being done in an attempt to fully understand the immune response system. Very generally, one can say that when a virus, bacteria, or foreign protein, known generally as an antigen (an′ti-jen) enters the body, the immune system produces a special substance (antibody) that either destroys or alters the antigen to make it harmless. Understanding the immune response is of major concern in current cancer research. The term "antigen" derives from anti body + the Greek word *gennan*, meaning "to produce"; in other words, an antigen is anything that will lead to the production of antibodies.

Platelets are the tiny cell-like parts that originate with the breakup or partial disintegration of the hemocytoblasts. As we will see below, *platelets* function in the mechanism of blood clotting (coagulation).

What Is Plasma? Serum?

Blood *plasma* is a straw-colored liquid composed of a solution of water (91 per cent) and various chemical substances, (9 per cent) – primarily special blood proteins. Plasma is obtained by removing the formed elements from *whole* (complete) *blood*, usually by means of spinning whole blood in a centrifuge. Plasma can only be stored if an anticoagulant is added to prevent coagulation. (Coagulation, or clotting, is the process by which blood or plasma changes from a liquid state to that of a soft, jelly-like solid).

Blood *serum* is obtained from plasma by first causing or allowing the plasma to coagulate and then removing the fibrous clotting material. This last step which removes the clotted fibrous material from the solution, leaving serum, is commonly achieved as above, by further use of a centrifuge. Serum then is identical to plasma except that the clotting factor has been removed. Serum can be stored almost indefinitely and there is, of course, no need to add an anticoagulant.

What Are the Functions of the Four Major Plasma Proteins?

The plasma proteins are dissolved proteins, not directly associated with formed elements. The plasma proteins are so called because they will always be found in blood plasma in solution; that is, after the formed elements are removed. Proteins dissolve in water in the same way that sugar dissolves. The four major plasma proteins are albumin, globulin, fibrinogen, and prothrombin.

Albumin (al-bu′min) is the most plentiful plasma protein (54 per cent). It is crucial in maintaining the proper osmotic balance between the blood and the interstitial (between cells) fluids. If albumin concentration decreases for any reason, the amount of water in the blood also decreases, making the blood thicker. In injury, such as a severe burn, albumin leaks from the damaged capillaries to the interstitial fluids. As a result water cannot be kept in the blood compartment, and blood volume drops, leading to thickening of the blood. If the loss is severe, shock results. Treatment to counteract this disorder includes intravenous infusion of *serum albumin*, (serum containing a high concentration of albumin).

Globulin (glob′u-lin) is important since it contains antibodies involved in the body's immune (infection fighting) mechanisms. When examined chemically, globulin can be separated into three groups: alpha, beta, and gamma. The *gamma globulin* is the most active antibody factor. Gamma globulin injections from previously affected persons have been used in the control and prevention of epidemics of some infectious disease such as infectious hepatitis (liver inflammation).

Fibrinogen (fi-brin′o-jen) and *prothrombin* (pro-throm′bin) are present in lesser amounts than albumin and globulin. These plasma proteins function in the coagulation process discussed below. Since these two proteins are used up in the coagulation process, would you expect to find them in serum? Albumin and globulin are both found in serum.

How Are Blood Cells Formed?

Development of blood cells within the bones begins during the fifth month of fetal life. Blood-forming elements initially appear in the centers of the bone marrow cavities; the blood-forming centers later expand to occupy the entire marrow space. This widely dispersed blood cell formation continues until puberty, when the marrow in the ends of all the long bones becomes less cellular and more fatty. In the adult, only the skull, vertebrae, ribs, sternum, pelvis, humerus, and femur retain active red marrow formation. In elderly individuals, areas of bone marrow, once occupied by active cell production, become fat laden. This helps to explain the difficulty elderly individuals experience in regenerating lost blood; it is because they have very little active red bone marrow remaining.

Two types of blood cells are not formed in the bone marrow. Lymphocytes and monocytes are formed by lymphatic tissue, chiefly in the lymph nodes and the spleen.

All blood cells originate from undifferentiated cells called hemocytoblasts (literally "blood cells"). As these primitive cells mature they undergo alterations in nuclear and cytoplasmic characteristics, producing the red blood cells, white blood cells, and platelets (Fig. 83).

What Factors Stimulate an Increase in the Rate of Blood Cell Formation?

Red blood cell production is stimulated by any factor that lowers the oxygen available to bone marrow or body tissue. Thus, high altitude, hemorrhage (blood loss), nutritional deficiencies, and respiratory disturbances may stimulate increased red blood cell formation. The life span of erythrocytes is approximately 80 to 120 days. When their usefulness is impaired with age, the red cells are destroyed by the spleen, liver, and bone marrow. Two to ten million red cells are destroyed each second, yet, because of replacement, the number of circulating cells remains remarkably constant.

Many diseases are characterized by a change in the number of circulating white cells (leukocytes). An increase in white cell count usually indicates an acute infection. More generally, anything that stimulates an immune response will increase the rate of white blood cell formation.

PROTECTION AGAINST BLOOD LOSS

What Are the Three Mechanisms in Stopping Blood Loss?

There are three separate mechanisms involved in checking the flow of blood from an injury: formation of the platelet block, contraction of the related blood vessels, and formation of a fibrin clot (Fig. 85).

First, when a vessel is cut or damaged, platelets rapidly accumulate at the site of blood loss. This collection of platelets forms a temporary plug that is capable of stopping blood loss in small arteries and veins.

Second, soon after the first step begins there is a contraction of the blood vessels around the immediate area of the injury. This reduces the amount and rate of blood

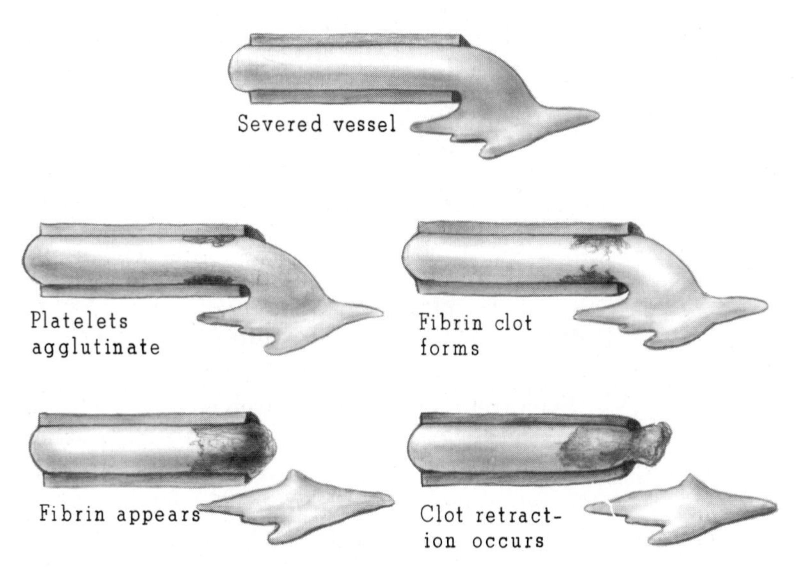

Figure 85. Formation of clot following injury.

flow around the wound. Later a more generalized constriction of vessels occurs, resulting in an overall decrease in the flow of blood to the area.

Third, next begins the actual process of blood clotting, or coagulation. This takes place with the formation of the insoluble fibrin clot through a conversion of the normally soluble plasma protein fibrinogen into insoluble *fibrin.* This occurs at the point where the platelets have built up earlier. Notice the Greek suffix *-gen,* meaning "to produce," in the word fibrino*gen*; this means, "that which produces fibrin," which is precisely the outcome of the coagulation process—soluble fibrinogen turns into insoluble, fibrous, or fiber-like, fibrin.

The three essential steps in this third stage, the coagulation (or clotting) stage, are (a) thromboplastin is formed by platelet-plasma interaction; (b) this stimulates conversion of prothrombin to thrombin; and, finally, (c) the conversion of fibinogen to fibrin, which is dependent on the presence of thrombin (Fig. 86).

What Are Two Common Abnormalities of the Clotting Mechanism?

Two common abnormalities of the clotting mechanism are internal clotting, also called thrombosis (throm-bo′sis), and hemophilia (he″mo-fil′e-ah).

Thrombosis. The term thrombosis, derived from the Greek word *thrombos,* meaning clot, describes clotting inside of a "normal" blood vessel. A clot, or thrombus, forming in the blood vessels of the leg or arm may cause some minor local damage. However, if an internal clot should block the blood supply to the heart or brain, it can be fatal. An *embolus* (Greek for plug) is an internal clot that has become dislodged from its place of origin and has lodged elsewhere in the body. The condition is spoken of as *embolism.*

Hemophilia. This is a hereditary bleeding disease that is characterized by delayed coagulation of the blood. It results from a diminished coagulation factor in the plasma. There are actually several types of hemophilia, classified according to which coagulation factor is deficient. Treatment consists of transfusions with fresh blood or administration of the deficient factor.

BLOOD TYPES AND TRANSFUSIONS

How Are Blood Groupings Related to Transfusions?

The transfusion of whole blood from one person to another can be a lifesaving

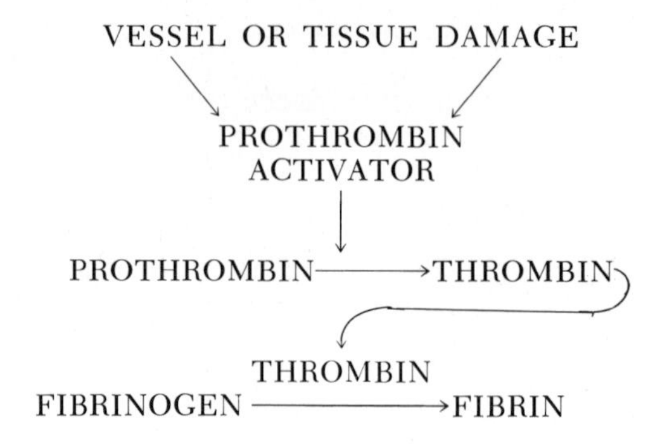

Figure 86. Phases of coagulation.

procedure. However, the safe administration of whole blood from donor to recipient requires typing and crossmatching. These procedures are necessary, since a patient receiving blood of a type that reacts with his own will experience a serious, and sometimes fatal, reaction. The potential reaction is due to the presence of the *immune response system,* which attacks and alters foreign proteins that enter the blood. Whether one person's blood will be foreign to another person (or to his immune response system) depends on the types of proteins that make up the blood cells and other blood factors. If the proteins of the donor blood are foreign to the recipient's, then this will trigger an immune response and already existing or newly produced antibodies will attack the transfused blood. Since the amount of blood transfused is usually large (one pint or more), a very serious or fatal condition can result from this reaction.

The systems of blood classification are based, therefore, on the pesence of specific, known antigens (proteins foreign to some people in the population) in the red blood cells and on the presence of specific antibodies in the plasma or serum. The primary classification system identifies antigens A and B, antibodies anti-A and anti-B, and a factor known as Rh (see Table 6 and Fig. 87).

How Are the Different Blood Types Related?

Blood groups are named for the protein antigens contained in the red blood cells. Blood type A contains A-type protein (also

Table 6. BLOOD GROUPING

TYPE	PERCENTAGE OF POPULATION	RED CELL ANTIGENS	PLASMA ANTIBODIES
ABO			
A	41%	A	Anti-B
B	10%	B	Anti-A
AB	4%	A, B	
O	45%	*	Anti-A, anti-B
Rh (D)			
Positive	85%	Rh	
Negative	15%		†

* Type O blood is sometimes called the "universal donor," since it does not contain agglutinogens A and B.
† Anti-Rh does not occur naturally in blood, but will result if an Rh negative individual is given Rh positive blood.

RECIPIENT'S BLOOD GROUP
(Plasma or serum tested)

Donor's Blood Group — Red Cells Tested		AB	A	B	O
	AB				
	A				
	B				
	O				

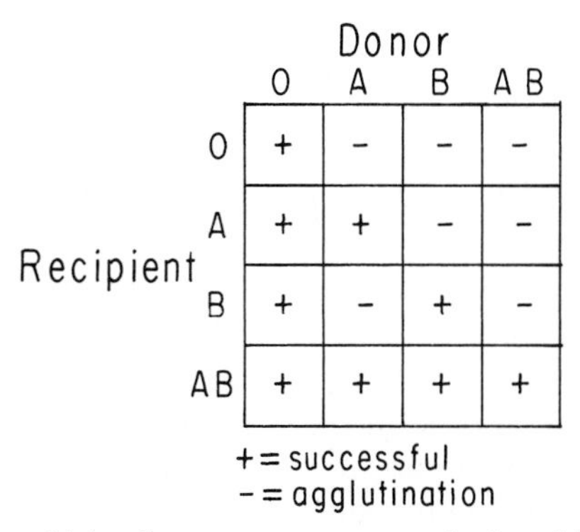

Recipient \ Donor	O	A	B	AB
O	+	–	–	–
A	+	+	–	–
B	+	–	+	–
AB	+	+	+	+

+ = successful
– = agglutination

Figure 87. Cross matching of blood types is necessary before blood from donor is administered to recipient. If the two types are not compatible agglutination, a serious or fatal reaction, may occur.

called A antigen), and type B blood contains B-type protein (also called B antigen). Type AB blood has both these protein antigens, and type O blood has neither.

The potential problem of a transfusion reaction arises because of the additional presence of antibodies to these antigens (A and B). If A antigen is contacted by antibody A (also called anti-A), a serious problem arises; the antagonism between these substances leads to a clumping or *agglutination* (ah-gloo-ti-na′shun) of the red blood cells—a very dangerous condition (Fig. 87).

The antibodies for A and B antigens are distributed on a sort of opposite or reciprocal basis. That is, type A blood contains type B antibody. And type B blood contains type A antibody. As a result, people with type A blood cannot receive type B blood, nor is the reverse possible.

Type AB blood, which contains both A and B antigen, does not contain either A or B antibodies. That is, it has no antibodies, relevant to the ABO classification system for transfusions. As a consequence, any of the other blood types can be given to a person with type AB blood. It is referred to as the *universal recipient.*

Type O blood has *both* A and B antibodies, but contains neither A nor B antigen. Type O blood cannot receive type A, which contains A antigen, type B, which has B antigen, or type AB, which has both antigens but neither A nor B antibody. Type O blood can receive only type O blood. However, since it has no antigens, type O blood can be given to any of the other types. Type O is called the *universal donor.*

The antibodies contained in the blood that is actually transfused seem to have a negligible effect outside their host body. So what is important is the donor's antigens and the recipient's (host) antibodies.

What Is The Rh Factor?

The Rh factor, so named because it was first found in the blood of the Rhesus monkey, is a system consisting of 12 antigens. Of these, "D" is the most antigenic; the term Rh positive, as it is generally employed, refers to the presence of agglutinogen "D" The Rh negative individual does not naturally possess this D antigen, and consequently will form D antibodies (by means of the immune response system) when he receives blood cells containing D.

The anti-D antibody does not occur naturally in the blood (recall that A and B antibodies occur naturally). As a result, the initial transfusion of Rh positive blood into an Rh negative individual may merely sensitize the recipient and cause development of antibodies (somewhat like a vaccination), but without the occurrence of severe symptoms. However, once sensitized, the recipient will probably experience a severe reaction to subsequent infusions of Rh positive blood. Rh negative blood can always be given to an Rh positive person. Why? Because Rh negative blood doesn't have the antigen (D).

The same reaction often occurs during pregnancy when the fetus of an Rh negative mother is Rh positive. Some positive red cells in the fetus leak across the placenta into the Rh negative blood of the mother, which produces antibodies in response to this invasion. The maternal antibodies then cross the placenta to the fetus and cause a reaction. When a woman is sensitized by this process during the first pregnancy, the first child is often normal and only in subsequent pregnancies does any severe reaction occur.

Many other antigenic factors have been discovered in the blood of various individuals, but their occurrence is relatively rare. Nonetheless, in any transfusion there remains a substantial danger of an immune reaction. Because of the sensitizing process of the immune response, people receiving transfusions again after several weeks, months or years are in greater danger.

CLINICAL CONSIDERATIONS

What Can Blood Counts Tell Us?

A *blood count* is the number of either white or red blood cells per cubic millimeter of whole blood, and is arrived at by counting under a microscope the number of cells in a sample of blood in a given area.

As noted earlier, many diseases are characterized by an increase in the white (cell) blood count. Normally, the total white cell count ranges from 5000 to 10,000 per cubic millimeter; however, it may be as high as 500,000 per cubic millimeter in leukemia.

Leukemia. Derived from the Greek words meaning white blood, this disease is characterized by a rapid and abnormal growth of white blood cells, leukocytes, in the peripheral blood. It is generally thought of as a blood cancer (malignant disease). The word "benign" is derived from the Latin word meaning "to be kind," and is opposed to the word "malignant," which is derived from the Latin word meaning "to be malicious (unkind)."

Infectious Mononucleosis (mon″o-nu″kle-o′sis). This is a benign disease associated with an increase in mononuclear leukocytes (the type of leukocytes with a *single* nucleus—*mono* means one). It usually occurs in children and young adults, and is believed to be caused by a virus. The patient with infectious mononucleosis evidences a slightly elevated temperature, enlarged lymph nodes, fatigue, and a sore throat.

Normal red blood cell count is approximately 5,400,000 per cubic millimeter in males and 4,900,000 per cubic millimeter in females. In other words, there are about 500 to 1000 red cells for every white cell in the blood. A reduction in the total number of red cells or in the amount of hemoglobin in the body results in anemia. *Anemia* (from the Latin words meaning "lack of blood") is defined as a decrease in the capacity of the blood to transport oxygen to the tissues. Two common types are pernicious anemia and iron deficiency anemia.

Anemias. In *pernicious* (per-nish′us) *anemia,* there is a decrease in the total number of erythrocytes, and those that are produced are large, oddly shaped, and fragile. *Iron deficiency anemia* usually follows a loss of blood or occurs during other periods when the demand for iron is unusually great. Without iron, hemoglobin formation is retarded.

Sickle cell anemia is a condition in which the red blood cells are abnormally shaped and do not function properly. It is genetically based, affecting primarily persons with a negroid heritage. Its prevalence has recently been explained by the fact that persons carrying one gene for sickle cell are more resistant (and thus better adapted) to malaria, a virus that is common in areas of predominantly negroid population. Persons with only one gene for sickle cell who are resistant to malaria do not develop sickle cells. However, persons with two sickle cell genes will probably die young because they will develop sickle cell anemia.

LEARNING EXPERIENCES

1. Following Figure 82, outline the blocks on a separate sheet of paper, and then, with the book closed, fill in the blocks with the percentage of each blood constituent.
2. Make a table, as in Figure 87, and then, with the book closed, indicate with pluses and minuses, which transfusions will be successful.

CHAPTER 7

THE CIRCULATORY SYSTEM, LYMPHATIC SYSTEM, AND ACCESSORY ORGANS

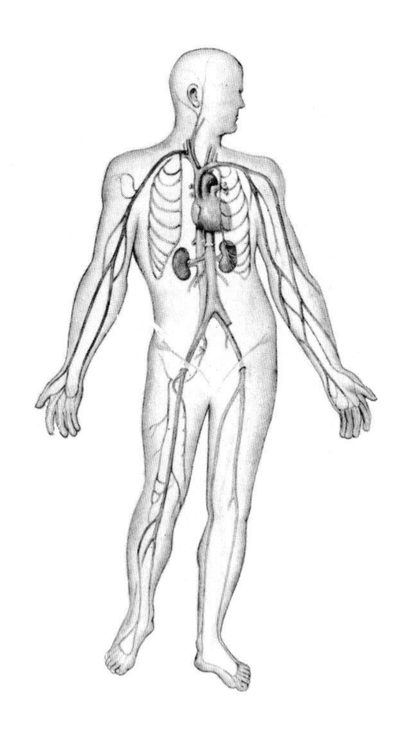

THE CIRCULATORY SYSTEM

What Is the Function of the Circulatory System?

The circulatory system is the *transportation* system of the body. It provides a link—either direct or indirect—between each individual cell and the major homeostatic organs. By way of the circulatory system each cell in the body has fairly rapid communicative access to the lungs for oxygen (or for elimination of carbon dioxide), to the digestive tract for nutrients, and to the kidneys for elimination of cellular wastes.

The circulatory system itself should be considered as primarily a homeostatic organ system. The successful operation of each of the other homeostatic organ systems (respiratory, digestive, urinary) is inseparable from the effective operation of the circulatory system.

The circulatory system also performs two other "helping" functions. First, it helps the endocrine system by transporting the substances secreted by the endocrine glands. These substances help to regulate bodily processes; for example, one substance regulates growth, others stimulate sexual development at puberty. These endocrine substances (hormones) are all delivered to their site of action by the circulatory system. The second, additional "helping" function performed by the circulatory system is in the bodily process that protects against infection. The circulatory system transports antibodies and specialized bacteria-fighting white blood cells to the area of possible infection.

What Structures Make Up the Circulatory System?

Strictly speaking the expression "circulatory system" should refer to both the *blood circulating system* and the *lymph circulating system.* In actual practice the expression "circulatory system" usually refers to the blood circulating system alone; this is particularly true in medicine, since the traditional concern and most immediate medical problems are with the blood system. The lymph circulating system is usually referred to separately as the *lymphatic* (lim-fat'ik) *system.*

The blood circulatory system is made up of the blood, the heart, and the blood vessels and capillaries. The lymphatic system is made up of lymph (the plasma-like tissue fluid), the lymph nodes, and the lymph vessels. The lymphatic system has no pump corresponding to the heart in the blood system.

In the first part of this chapter we will consider the blood circulating system, which we shall refer to as the circulatory system unless otherwise specified. In the latter part of the chapter the lymphatic system will be considered.

THE HEART

What Is the Heart?

The heart is the pump of the circulatory system. It is a four-chambered muscular organ lying in the thoracic cavity between the two lungs (Fig. 88). The structures of the heart include the *pericardium* (*peri* means "around", and *cardium* means "heart"), the wall surrounding the heart; the *valves* associated with the chambers; and the *cardiac arteries* and *veins* that supply blood to the heart tissue.

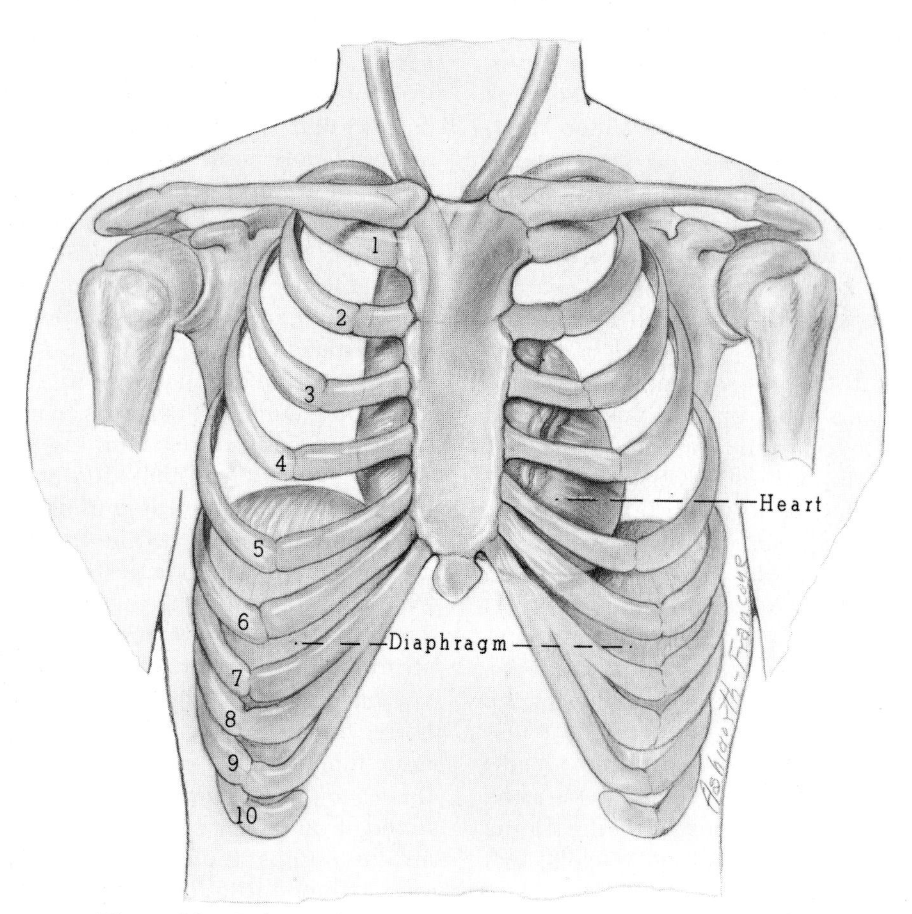

Figure 88. Relationship of the heart and diaphragm to rib cage.

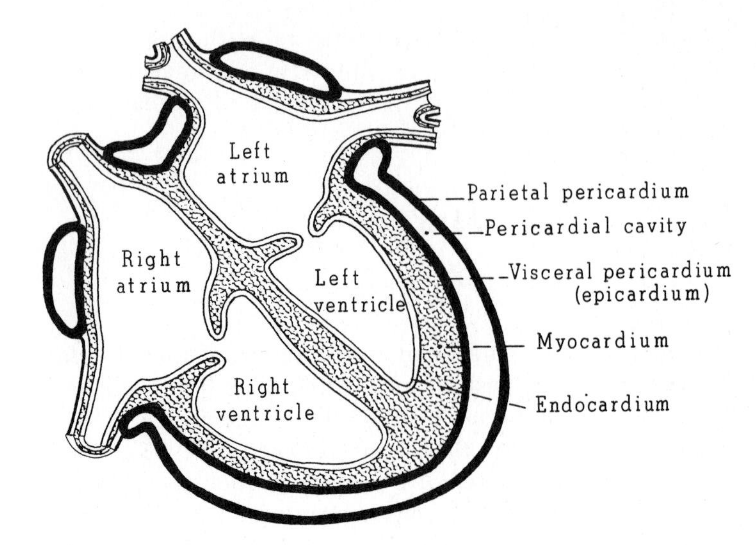

Figure 89. Heart wall and pericardium. (Note the thickened left ventricular wall.)

The pericardium is a saclike structure surrounding and supporting the heart. It consists of two layers, an external fibrous layer called the *parietal* (pah-ri'e-tal) *pericardium,* and an inner serous (from Latin, meaning water) layer, which adheres to the heart and becomes its outer layer, called the *visceral* (vis'er-al) *pericardium* (also called the epicardium (*epi* = upon, *cardium* = heart). Serous fluid found between these two layers lubricates the two membranes with every beat of the heart as their surfaces glide over each other (Fig. 89).

The wall of the heart consists of three distinct layers—the epicardium (external layer), the myocardium (middle layer), and the endocardium (inner layer). The *epicardium* (visceral pericardium) is a serous layer containing some connective tissue and stored fats. The *myocardium* (from Greek, *mys* = muscle, *cardium* = heart) is the muscular layer of the heart, and consists of interlacing bundles of cardiac muscle fibers. This layer is responsible for the ability of the heart to contract. The *endocardium* (*endo* = internal, *cardium* = heart) lines the cavities of the heart, covers the valves, and is continuous with the lining membrane of the large blood vessels; that is, it is one continuous internal lining of the heart and major vessels.

What Are the Four Chambers of the Heart?

The heart is divided into right and left halves by a lengthwise *septum* (Latin for wall or partition) or wall of tissue. Each half is subdivided into two chambers. The upper chambers on each side are called the *atria* (a'tre-ah), singular *atrium* (Latin for entrance hall), and the larger, lower chambers are called the *ventricles* (from the Latin for belly or cavity). Figure 90 shows the four chambers and the phases of the cardiac cycle.

The atria serve as receiving chambers for blood from the various parts of the body, and pump blood into the ventricles. Since the distance the atria pump their blood is short (into the adjoining ventricles), they are less muscular. The *right atrium* is a thin-walled chamber that receives blood from all parts of the body except the lungs. Three large veins empty into the right atrium: the *superior vena cava,* bringing venous blood from the upper portion of the body; the *inferior vena cava* bringing venous blood from the lower portion of the body, and the *coronary sinus,* draining blood from the heart itself. The right atrium pumps the deoxygenated (bluish) venous blood into the right ventricle.

The *right ventricle* is a thick-walled,

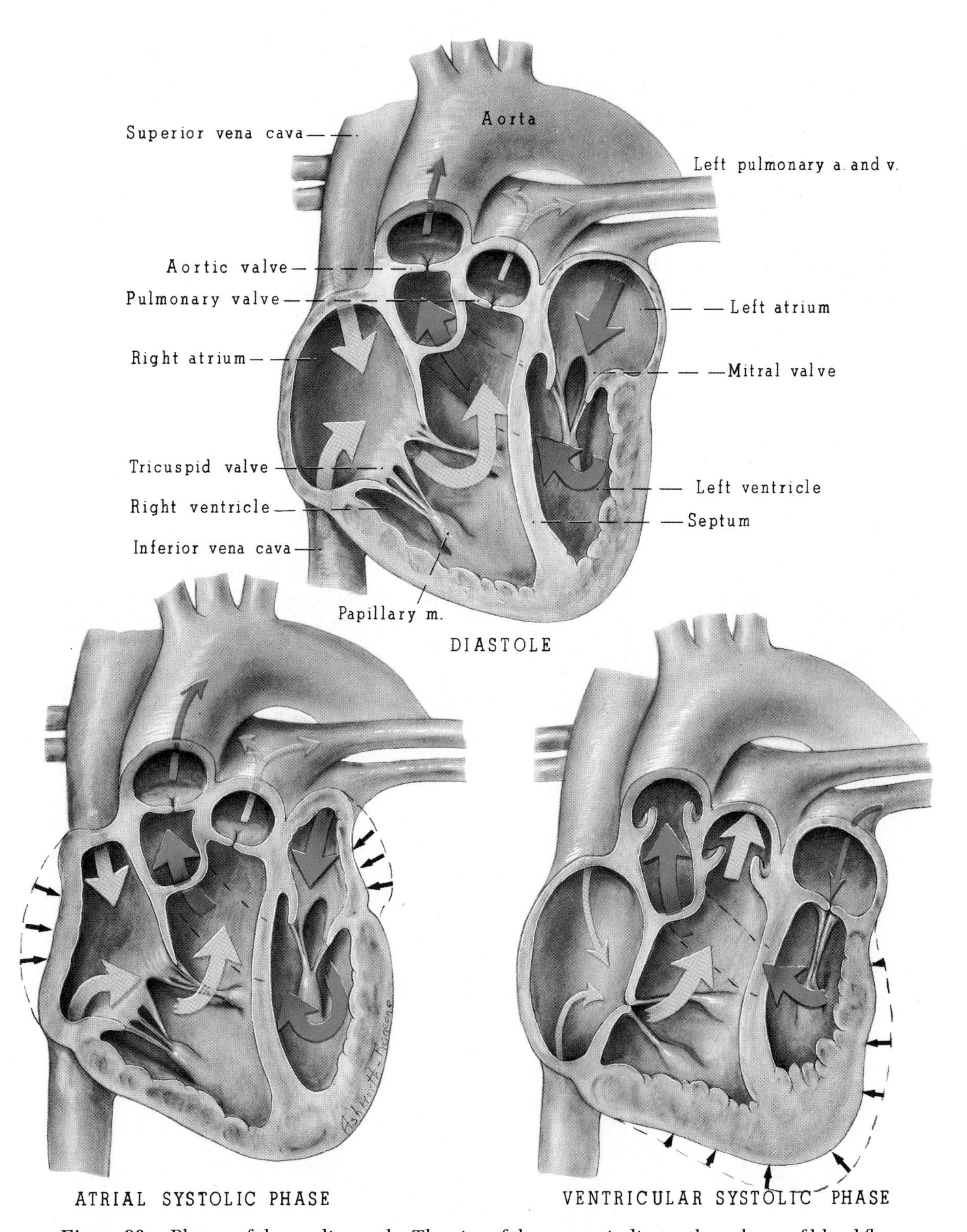

Figure 90. Phases of the cardiac cycle. The size of the arrows indicates the volume of blood flow.

muscular chamber that pumps the blood, received from the right atrium, out of the heart through the *pulmonary* (from Latin for lung) *artery* into the lungs. This chamber must be very powerful to push the blood through the thousands of capillaries in the lungs and back to the left atrial chamber of the heart.

The *left atrium* receives the now oxygenated (bright red) blood from the lungs through the four *pulmonary veins.* From this receiving chamber the blood is then pumped into the left ventricle.

The *left ventricle* is the most muscular chamber. Its walls are three times as thick as those of the right ventricle. This powerful pumping chamber forces the oxygenated blood out through the *aorta* to all parts (upper and lower portions) of the body except the lungs. The blood returns to the heart at the left atrium.

What Are the Valves? What Is Their Function?

The four heart *valves* are membranous structures designed to prevent *backflow* (in the wrong direction) of blood during the heart's pumping cycle. There are two types of valves: the atrioventricular valves and the semilunar valves.

The *atrioventricular* (a″tre-o-ven-trik′u-lar) *valves* are thin, leaflike structures between the atria and ventricles. They prevent backflow from the ventricles to the atria during the period when the ventricles are pumping (contracting). Between the right atrium and the right ventricle is the *tricuspid* (*tri* = three, *cuspid* = cusps) *valve,* so called because it consists of three irregularly shaped flaps (or cusps) formed mainly of fibrous tissue. The opening between the left atrium and the left ventricle is guarded by the *mitral* or *bicuspid valve,* so named because it consists of two flaps. The bicuspid valve is stronger and thicker, since the left ventricle is a more powerful pump (Fig. 91).

Blood is propelled through the tricuspid and bicuspid valves when the atria contract. When the ventricles contract the valves close, resisting any pressure of the blood which might force them open into the atria.

The *semilunar* (*semi* means half; *lunar* means moon) *valves* are pocket-like structures attached at the points at which the pulmonary artery and aorta leave the ventricles. The *pulmonary valve* guards the opening between the right ventricle and the pulmonary artery (which leads to the lung). The *aortic valve* guards the opening between the left ventricle and the aorta. Figure 92 is an x-ray photo showing implanted artificial valves.

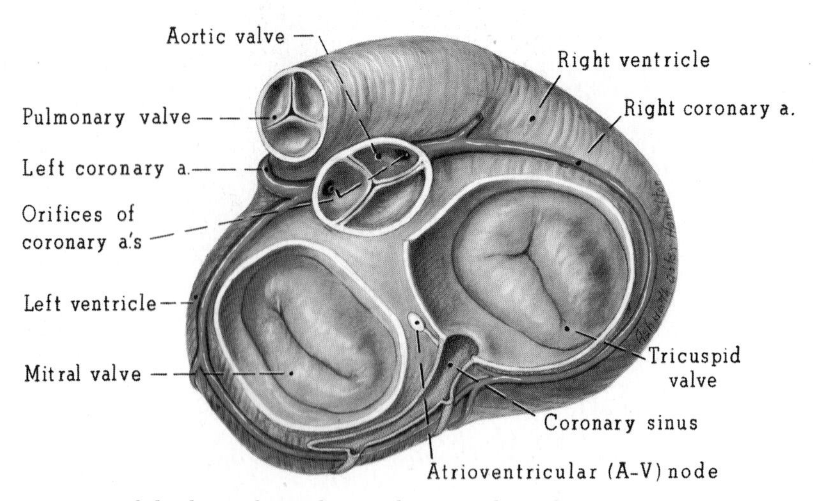

Figure 91. A view of the heart from above, showing the valves, coronary arteries, and sinus.

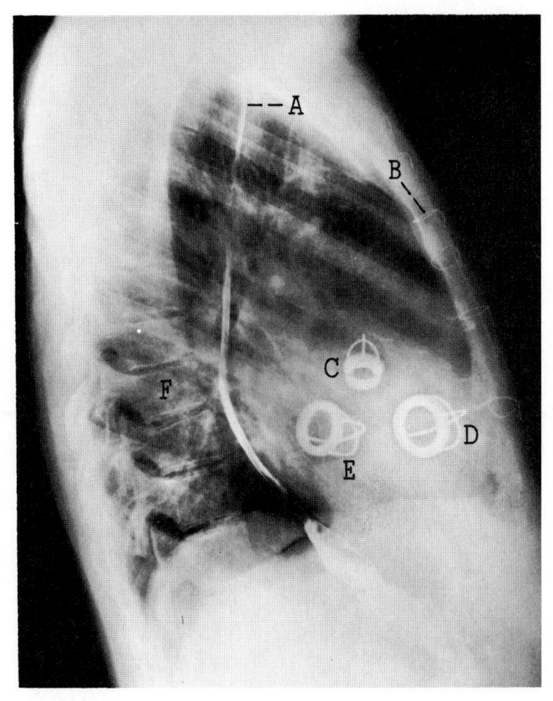

Figure 92. Three artificial valves implanted in a patient's heart by Dr. A. Starr of the University of Oregon Medical School. *A*, esophagus; *B*, wire sutures in sternum; *C*, aortic valve; *D*, tricuspid valve; *E*, mitral valve; and *F*, vertebral column.

What Causes the Valves to Open and Close?

When electrical impulses spread through the ventricular muscle of the lower heart the two ventricles contract at the same time. This contraction of the ventricle is referred to as *ventricular systole* (sis'to-le); the term "systole" is from the Greek word meaning a contracting. This results in a rise in the ventricular pressure—the pressure inside the ventricular chambers. When ventricular pressure exceeds atrial pressure, the atrioventricular valves (tricuspid and bicuspid) are forced shut.

When the ventricular pressure builds up to exceed the pressure in the pulmonary artery and the aorta, the semilunar (pulmonary and aortic) valves open.

The period of ventricular contraction is followed by relaxation of the ventricular muscle and an abrupt fall in the ventricular blood pressure. When ventricular pressure falls below pulmonary (artery) and aortic pressure, the pulmonary and aortic valves are forced shut, preventing backflow. Further relaxation of the ventricles follows, resulting in a drop of ventricular pressure below that of atrial pressure. At this point the atrioventricular valves open and the ventricles begin filling again as the atria contracts.

In short, the heart valves open and close as the pressure of the blood on either side of the valves changes. And this is primarily due to the action of the powerful ventricles.

What Produces the Heart Sounds?

The characteristic heart sounds are caused by the sudden deceleration of the blood when the heart valves close. The first sound occurs when the tricuspid and bicuspid valves close. It has a characteristic dull quality and low pitch, and has been described classically by the syllable "lubb." The second heart sound is produced when the pulmonary and aortic valves close. It is described by the syllable "dupp," and is of a snapping quality. The first heart sound is followed after a short pause by the second. A pause about two times longer comes between the second sound and the beginning of the next cycle. The opening of the valves is silent.

What Are the Elements of the Heart's Electrical Stimulating System?

Specialized sections of the myocardium initiate the sequence of events in the cardiac cycle and control the cycle's regularity. The *sino-atrial* (si″no-a'tre-al) *node* (SA node) is located in the wall of the right atrium. Its regular rate of electrical (electrochemical) discharge sets the rhythm of contraction for the entire heart, and for this reason it is known as the *pacemaker.*

The atrioventricular node (AV node), located in the lower right wall (septum) between the atria, conducts the electrochemi-

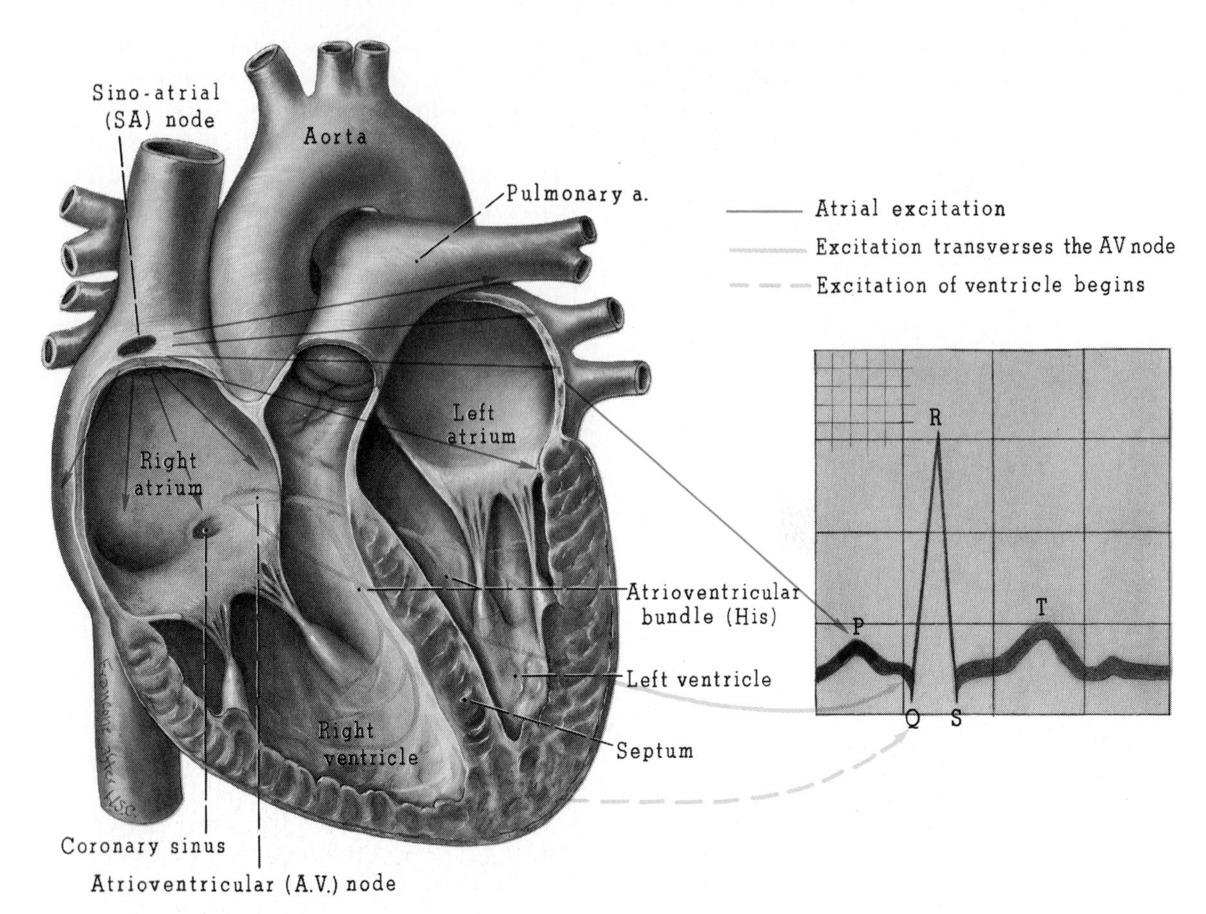

Figure 93. Conducting system of the heart showing source of electrical impulses produced on electrocardiogram.

cal impulses set up by the SA node to the *bundle of His*, named after its discoverer, Wilhelm His, Jr. The bundle of His conducts the electrochemical impulses from the AV node to the ventricles and activates depolarization—an electrochemical process that causes the cardiac muscle fibers to contract (Figs. 93 and 94).

What Is the Cardiac Rate and Rhythm?

The normal heart *rate* is 60 to 100 beats per minute at regular rhythmic intervals. *Tachycardia* (tak″e-kar′de-ah) is a condition characterized by a rhythmic beat at a rate of over 100 per minute; the term tachycardia is from the Greek *tachys* meaning quick, and cardia = heart. This occurs because the sinoatrial (SA) node is stimulated to set the pace at an increased rate. It can follow exercise or emotional disturbances, or it can result from disease. When the sinoatrial node has a rate of less than 60 beats per minute, *bradycardia* (*brady* = slow, *cardia* = heart) is present. This rate may be normal in some individuals, particularly in well-conditioned athletes.

What Factors Determine Cardiac Output?

Cardiac output is the volume of blood pumped by the heart over a given interval, usually calculated for a minute; that is, the number of beats per minute times the average volume, or the sum total of the volumes of all the beats in a minute. Under normal resting conditions, cardiac output approximates *4 to 5 liters per minute*—an amazing fact considering that the total blood volume of an average man is only 5 to 6 liters. The average volume of blood ejected by each beat of the heart is 60 to 70 millimeters; this is referred to as the *stroke volume.*

The cardiac output over a specific period depends on several factors, including the volume of the venous blood returning to the heart, the number of beats per minute, and the force of each contraction.

Venous Return. Cardiac output increases with an increase in venous return. Venous return is influenced by the following factors: contraction of skeletal muscles squeezing the associated veins and forcing the blood to move; and an increase in arterial or capillary blood pressure. With decreased blood, as in hemorrhage, venous return is lowered, and consequently cardiac output tends to decrease.

Heart Rate. Frequency of heart beat influences cardiac output. However, even with a constant heart rate, an increased venous return will increase cardiac output. In order to accomplish this the heart must work harder, pumping a greater volume with each stroke.

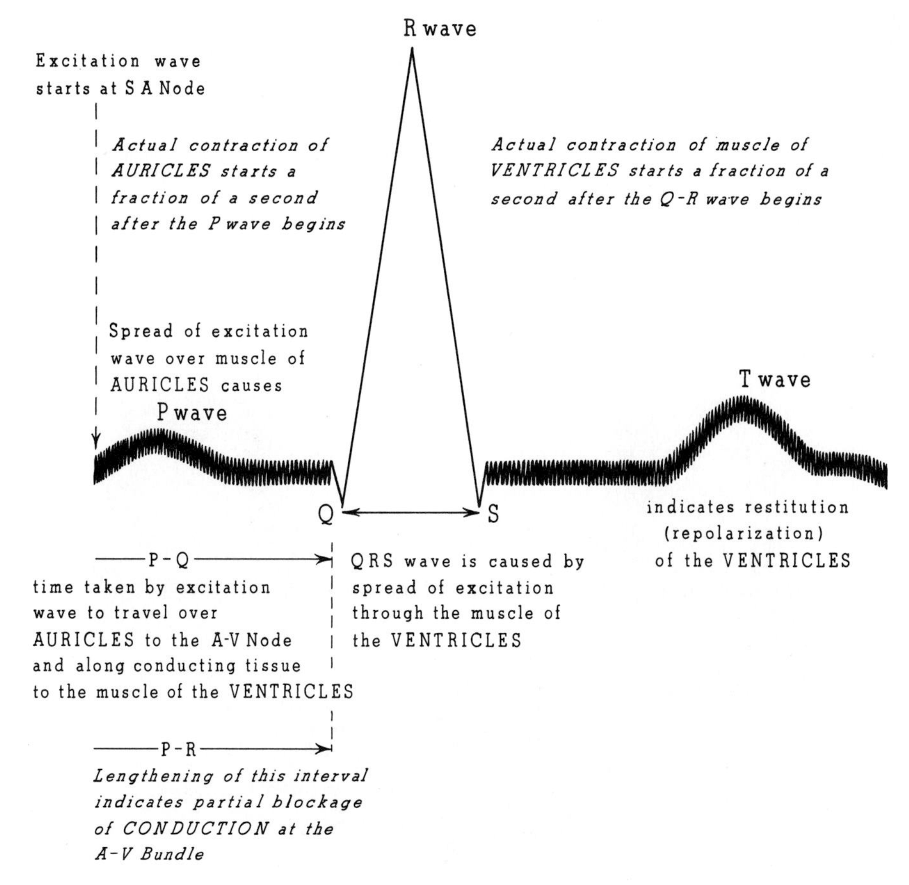

Figure 94. Electrocardiogram.

Force of Contraction. The force or power of each cardiac contraction depends, interestingly, on the initial length of the cardiac muscle fibers. *Starling's Law of the Heart* states that "the energy of contraction is proportional to the initial length of the cardiac muscle fiber." As venous return increases it tends to expand the heart. This increases the strength of contraction, producing an increased output without any change in rate.

The heart could handle increased venous return by speeding up, but it seems to prefer to be stretched and contract harder instead. During exercise and periods of excitement both mechanisms operate; that is, more beats and stronger and bigger (more voluminous) contractions.

THE BLOOD VESSELS

What Are the Three Kinds of Blood Vessels?

There are three kinds of blood vessels: the arteries, the capillaries, and the veins.

Arteries. The arteries (from the Greek word for pipe) carry oxygenated blood away from the heart. They are thicker than the other vessels and are composed of three layers of tissue, the most functional of which is made up of smooth muscle fibers and elastic connective tissue. The inner layer is made up of *epithelial* tissue, which is continuous through all the vessels and the heart (endocardium). The largest artery in the body is the *aorta,* which leaves from the powerful left ventricle of the heart. Slightly smaller arteries branch off the aorta, and still smaller arteries branch off these. The smallest arteries are called *arterioles,* and they pass the blood into the tiny capillaries (Figs. 95 and 96).

Capillaries. There are literally thousands of miles of tiny capillaries supplying the tissues of the body; if all of them were laid end to end, they would form a tube 62,000 miles long. The term "capillary" is derived from the Latin word for hair, indicating their thinness; however, capillaries are actually several times thinner than a human hair, and can only be seen through a microscope. The wall of a capillary is only one layer of *epithelial cells* thick, and most capillaries are just large enough in diameter to allow cells of blood to pass through in single file. The exchange of waste products for oxygen and nutrients is constantly occurring between the blood and the tissue spaces through the thin walls of the capillaries. The capillaries are, in this sense, the secret of the success that the circulatory system has in linking the in-

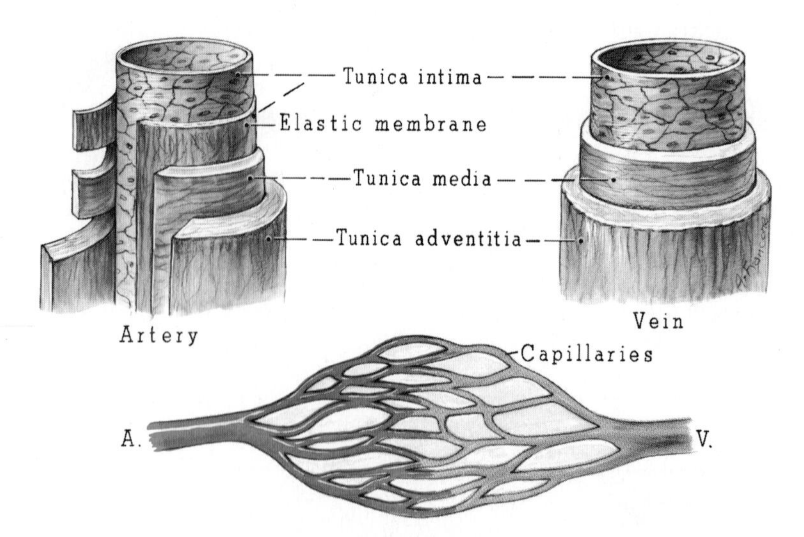

Figure 95. Component parts of arteries and veins.

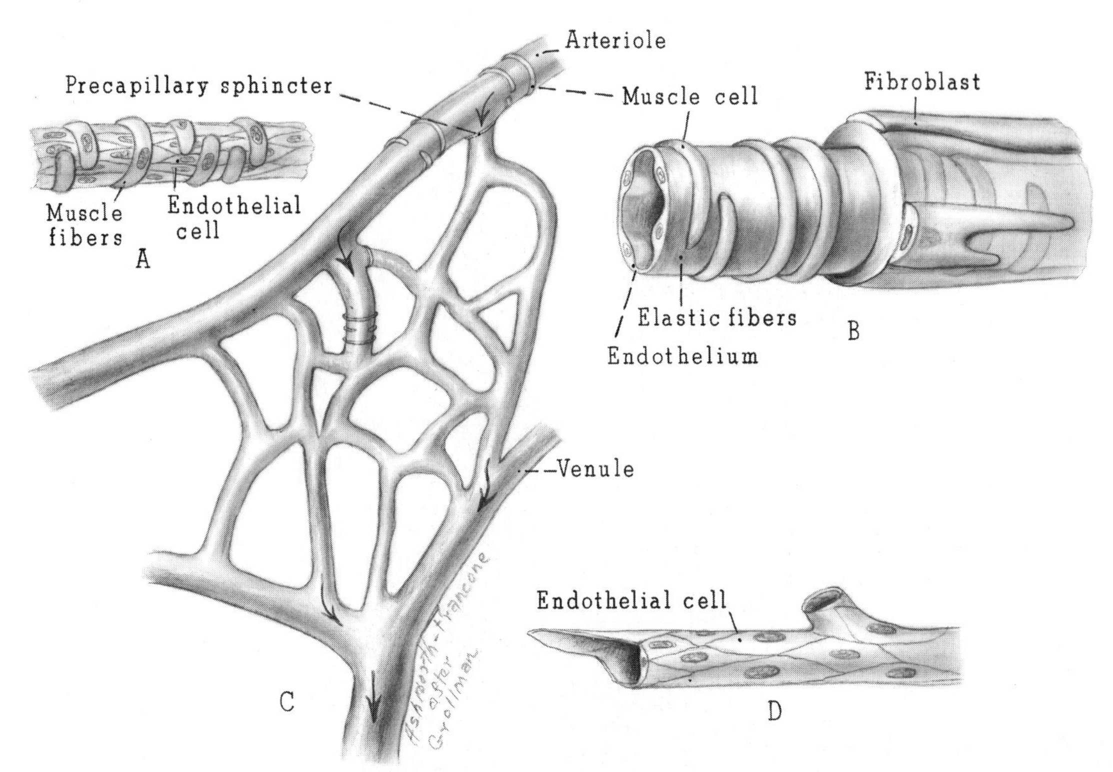

Figure 96. Diagrammatic representation of a capillary bed showing (*A*) a portion of capillary from the arteriole side, (*B*) arteriole, (*C*) capillary bed, and (*D*) true capillary.

dividual cells to the major homeostatic organs (lungs, digestive system, kidneys). At the ends of the capillary beds, the deoxygenated blood flows into the smallest veins, the *venules* (ven′uls).

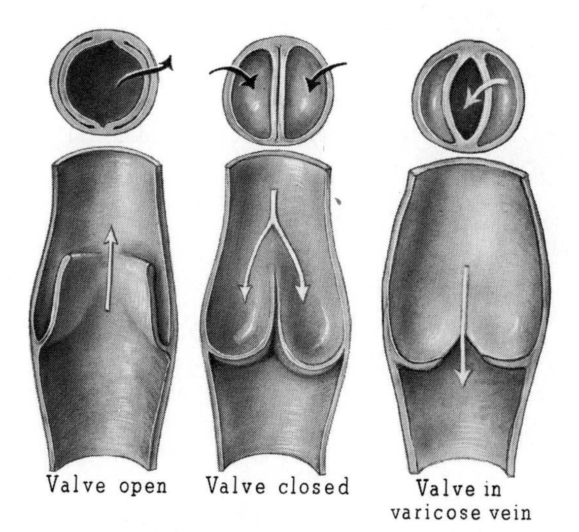

Figure 97. Veins contain bicuspid valves which open in the direction of blood flow, but prevent regurgitation of flow when pockets become filled and distended.

Veins. The venules empty blood into veins, and these in turn drain into still larger veins. The veins (Latin for vines), like the arteries, are composed of three layers, but are thin walled and less muscular than the arteries. The veins also contain small *valves* along their length to prevent *backflow* of blood during periods when the blood pressure changes (Fig. 97). If the valves break down, *varicose* (var′i-kos) *veins* result; "varicose" is Latin for dilated vein.

CIRCULATION

What Are the Three Main Circuits of the Circulatory System?

The term "circulation" comes from the Latin word *circulatio*, referring to movements in a circle or through a circular course. The overall circulatory system can be studied in terms of three smaller, interrelated flow circuits: the pulmonary sys-

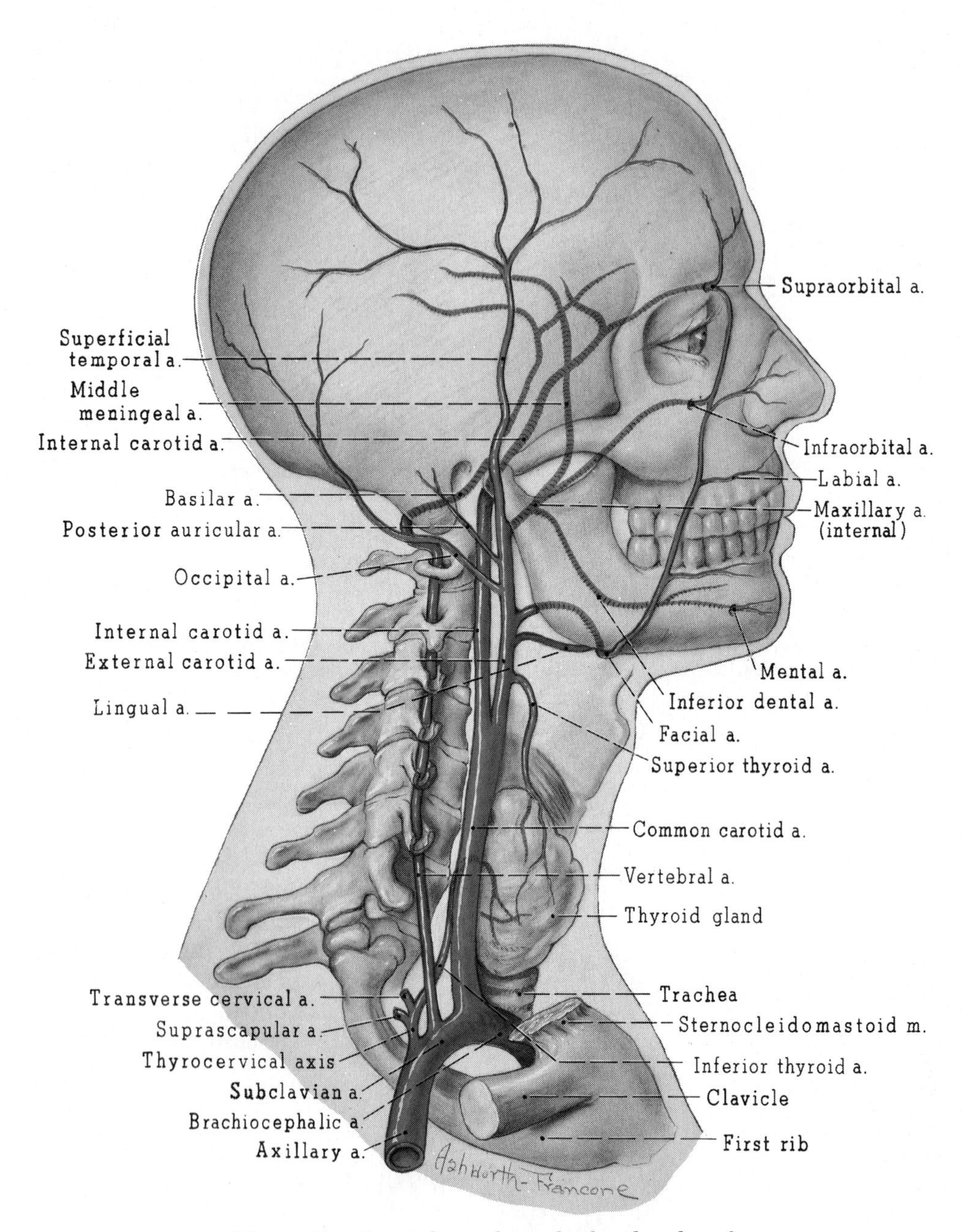

Figure 98. Arterial supply to the head and neck.

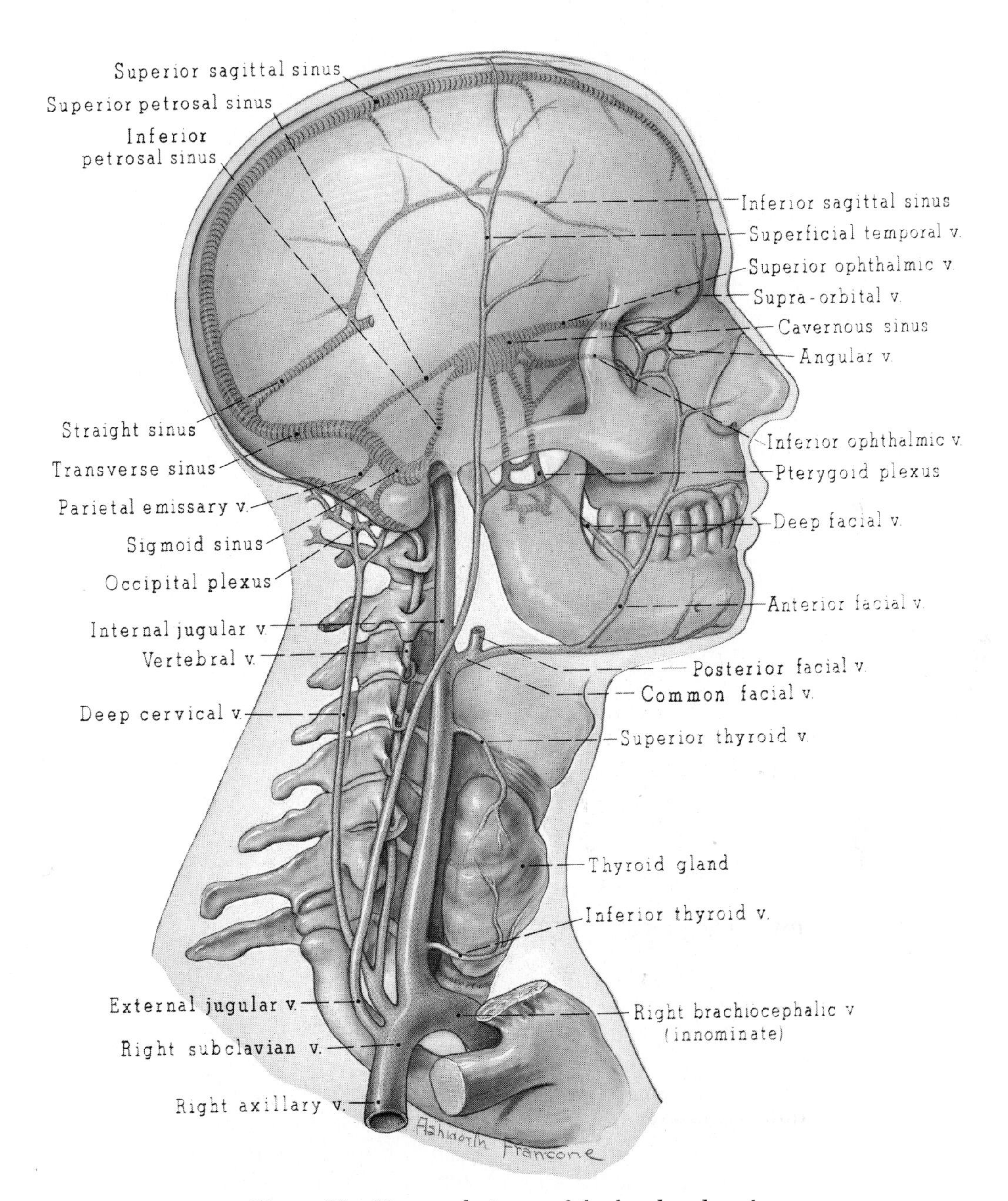

Figure 99. Venous drainage of the head and neck.

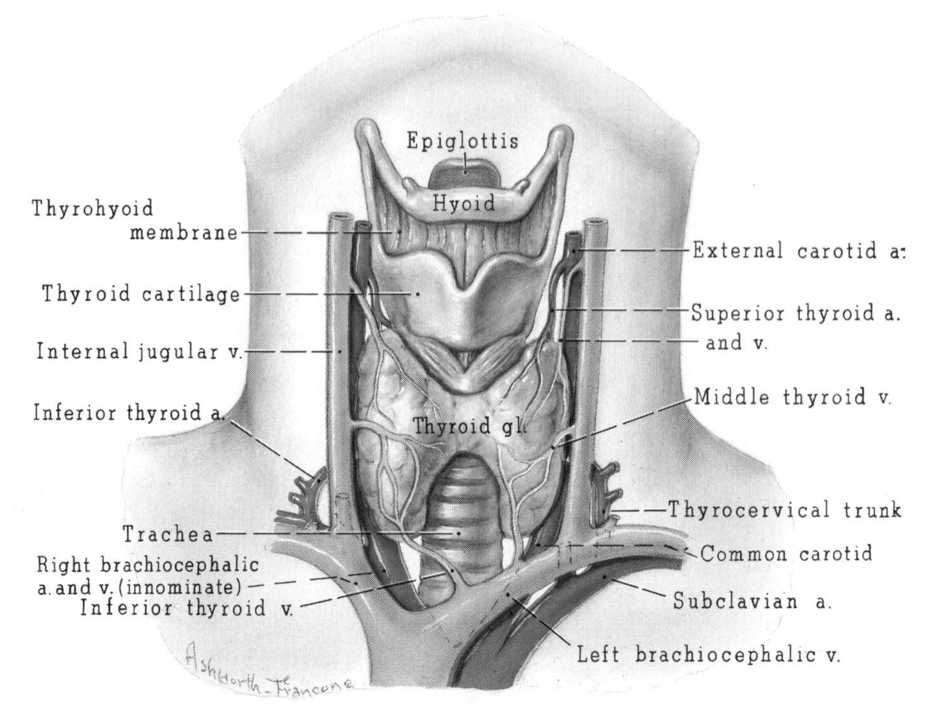

Figure 100. Arterial supply and venous drainage of the thyroid gland.

tem, the systemic system, and the portal system (Figs. 98 through 109).

Pulmonary Flow. The pulmonary system carries blood from the heart to the lungs and back to the heart again; more specifically, the blood travels from the right ventricle through the *pulmonary artery* to the lungs. Upon entering the lungs the pulmonary arteries quickly branch down to capillaries, which surround the air sacs (alveoli), exchanging oxygen and carbon dioxide. Gradually, the capillaries reunite, taking on the characteristics of veins. The veins join to form the *pulmonary veins,* which carry oxygenated blood from the lungs to the left atrium. Figure 110 shows the relationship between the systemic and pulmonary circuits.

Systemic Flow. The systemic (sistem'ik) circulation is the main flow circuit, as its name implies. It carries the oxygenated blood from the heart to all areas of the body except the lungs, and then back again to the heart. All systemic arteries spring from the *aorta.* The veins of the systemic circulation flow into either the inferior vena cava or the superior vena cava, which in turn empty into the right atrium.

Portal Flow. The portal (Latin for gate) system is really part of the systemic circulation, but is distinguished by the fact that blood from the spleen, stomach, pancreas, and intestines first passes through and branches out into the liver before going on to the heart. The liver then receives blood from two major vessels, the *hepatic* (Greek for liver) *artery* (20 per cent) and the *portal vein* (80 per cent). The hepatic (he-pat' ik) artery supplies the liver with the necessary oxygenated blood. The blood from the intestinal tract, which is rich in newly absorbed nutrient substances, passes into the liver through the portal veins. The liver cells are very active in taking up nutrients from the portal blood after a meal. The blood leaving the liver flows through the hepatic vein, which empties into the inferior vena cava. Figure 111 shows the relationship between the portal and systemic circuits.

Text continued on page 120.

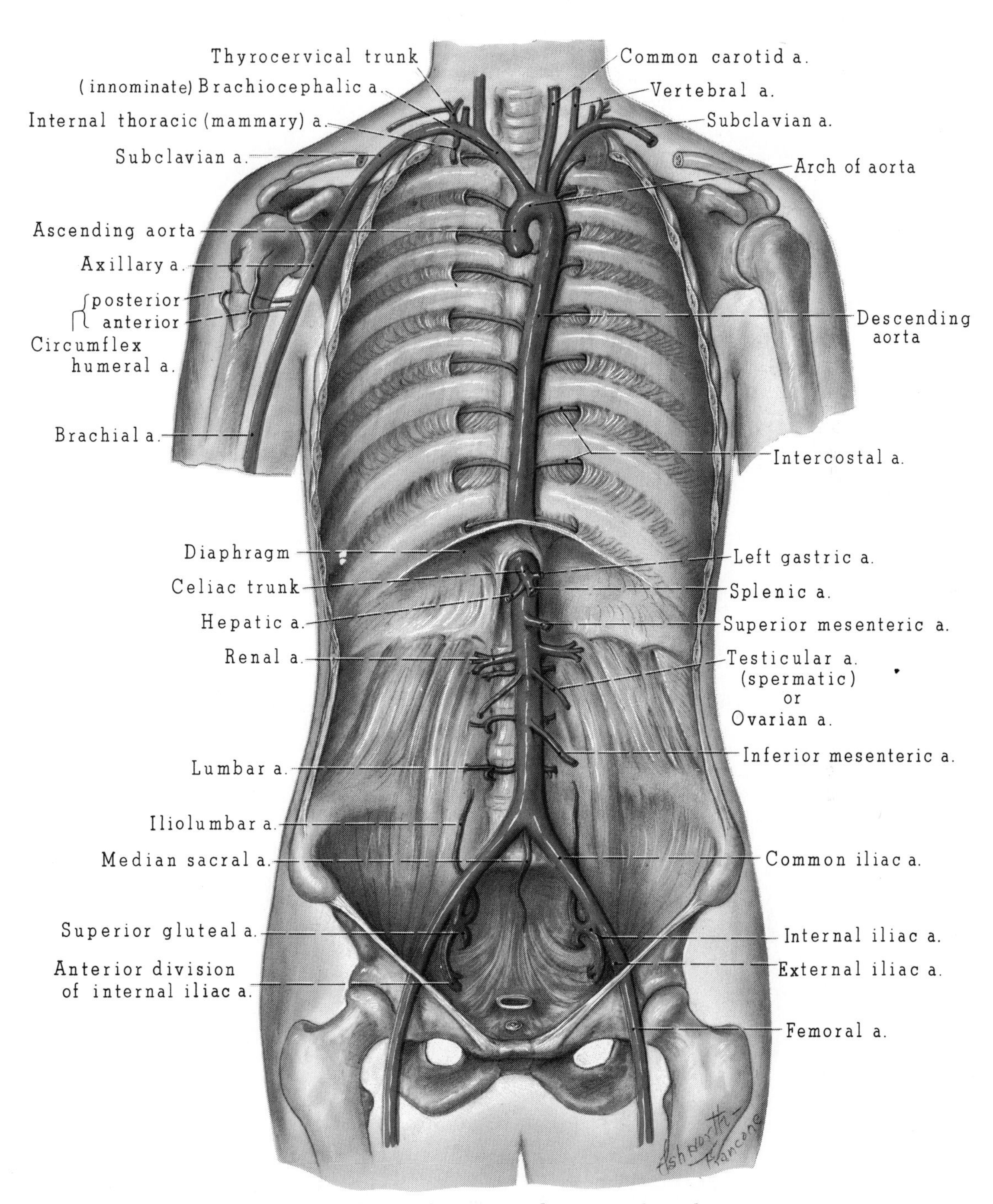

Figure 101. The aorta and its major branches.

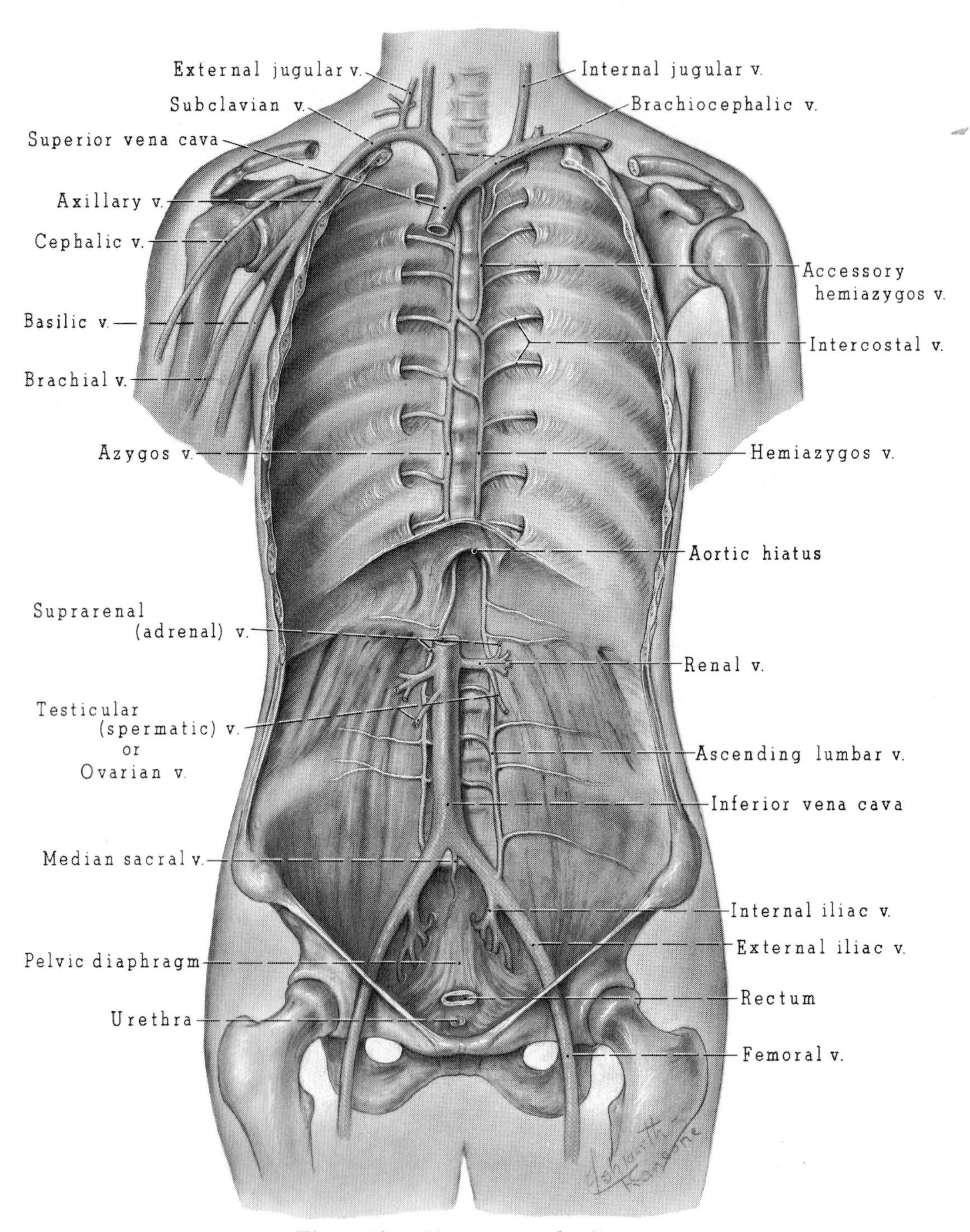

Figure 102. Vena cava and tributaries.

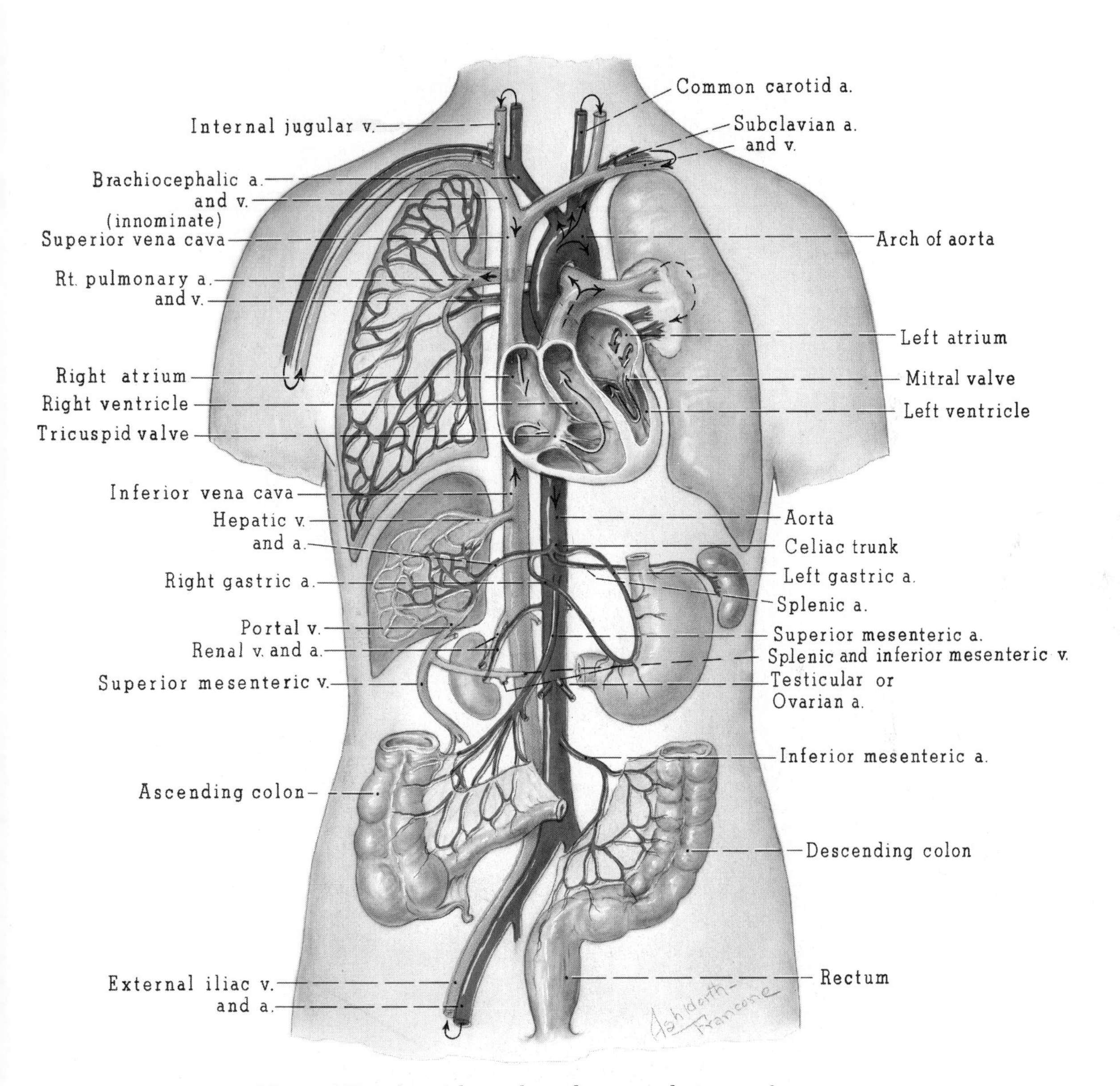

Figure 103. Arterial supply and venous drainage of organs.

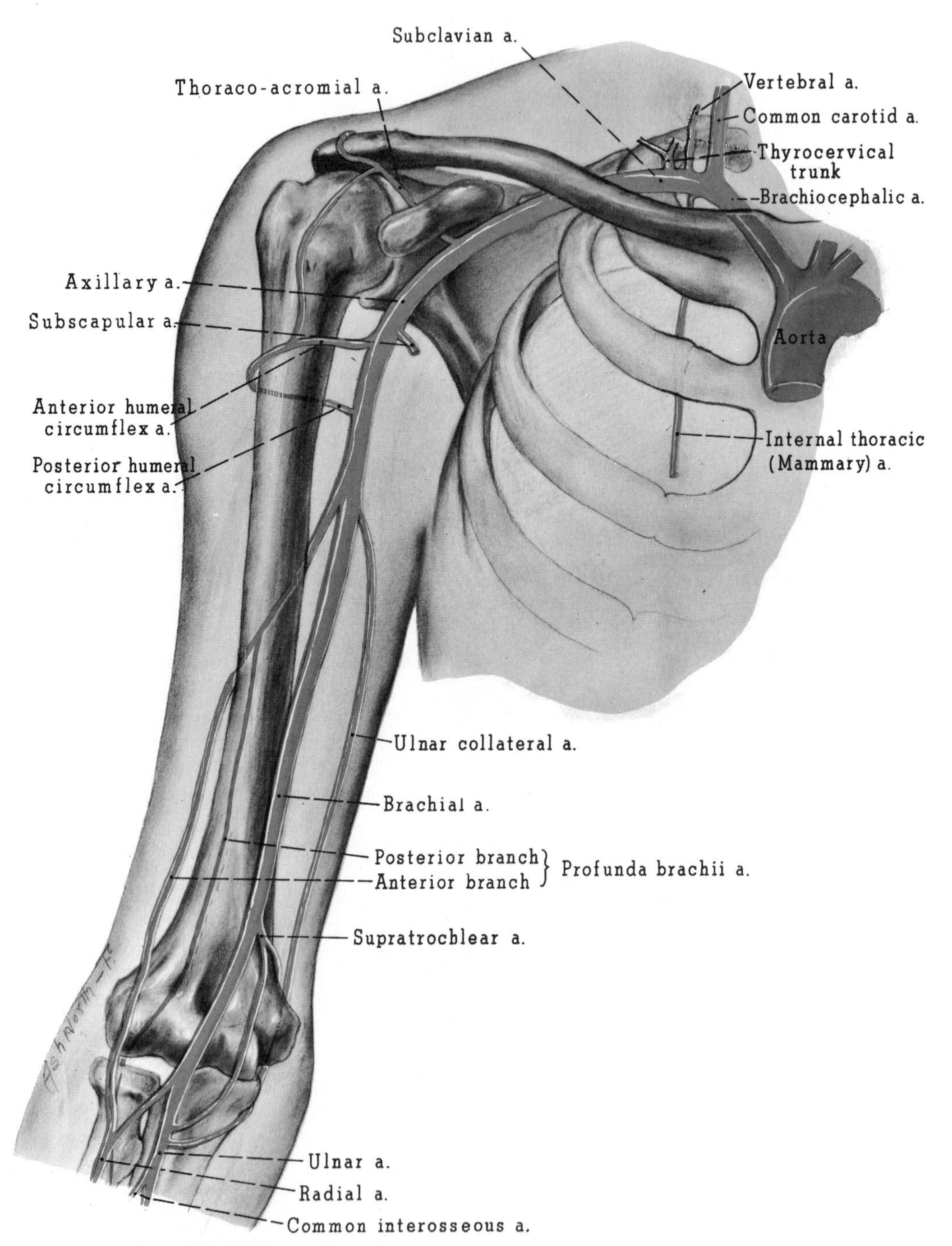

Figure 104. Arteries of the right shoulder and upper arm.

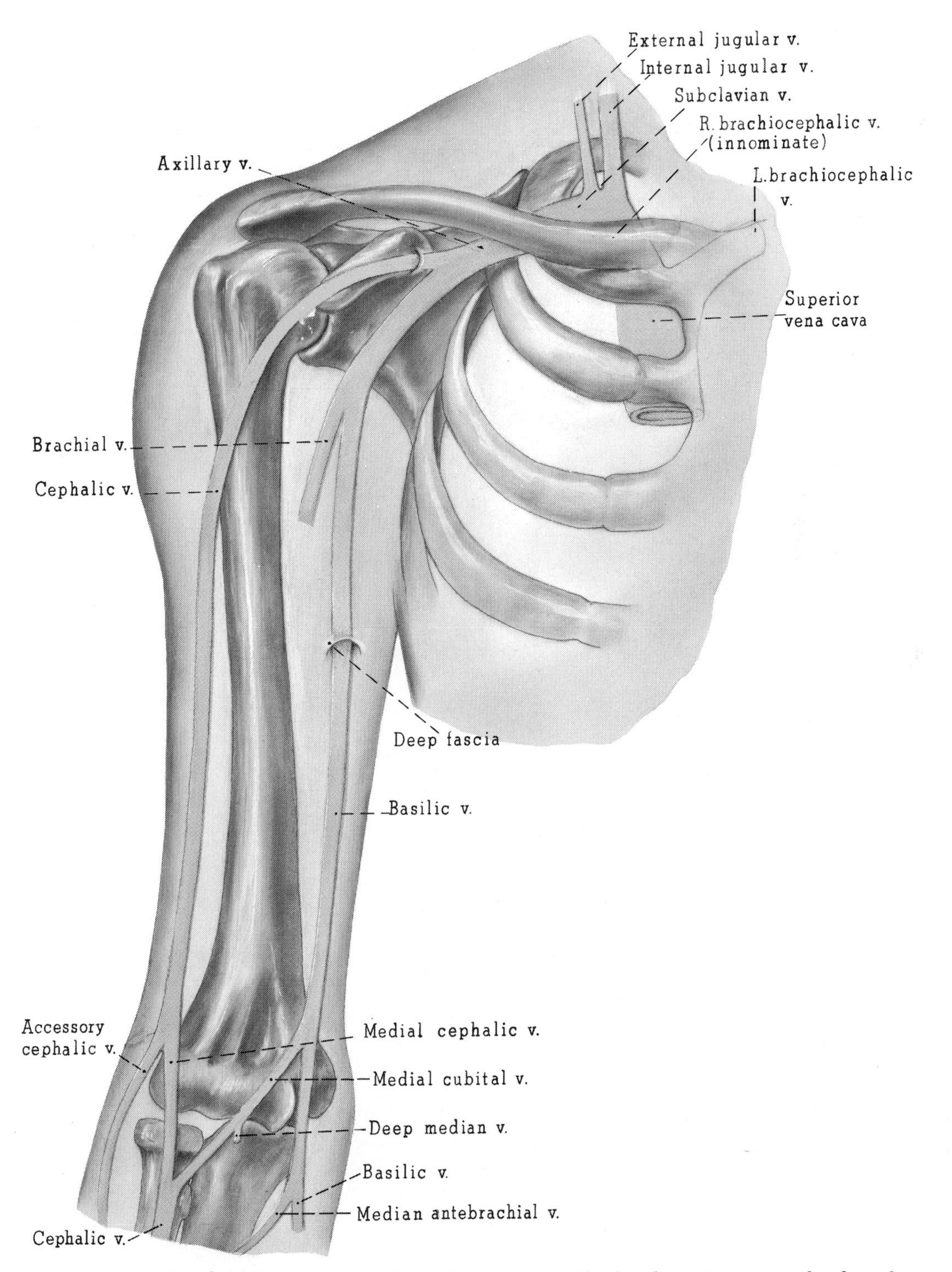

Figure 105. Veins of the right shoulder and upper arm. The basilic vein pierces the deep fascia in the region of the middle of the arm.

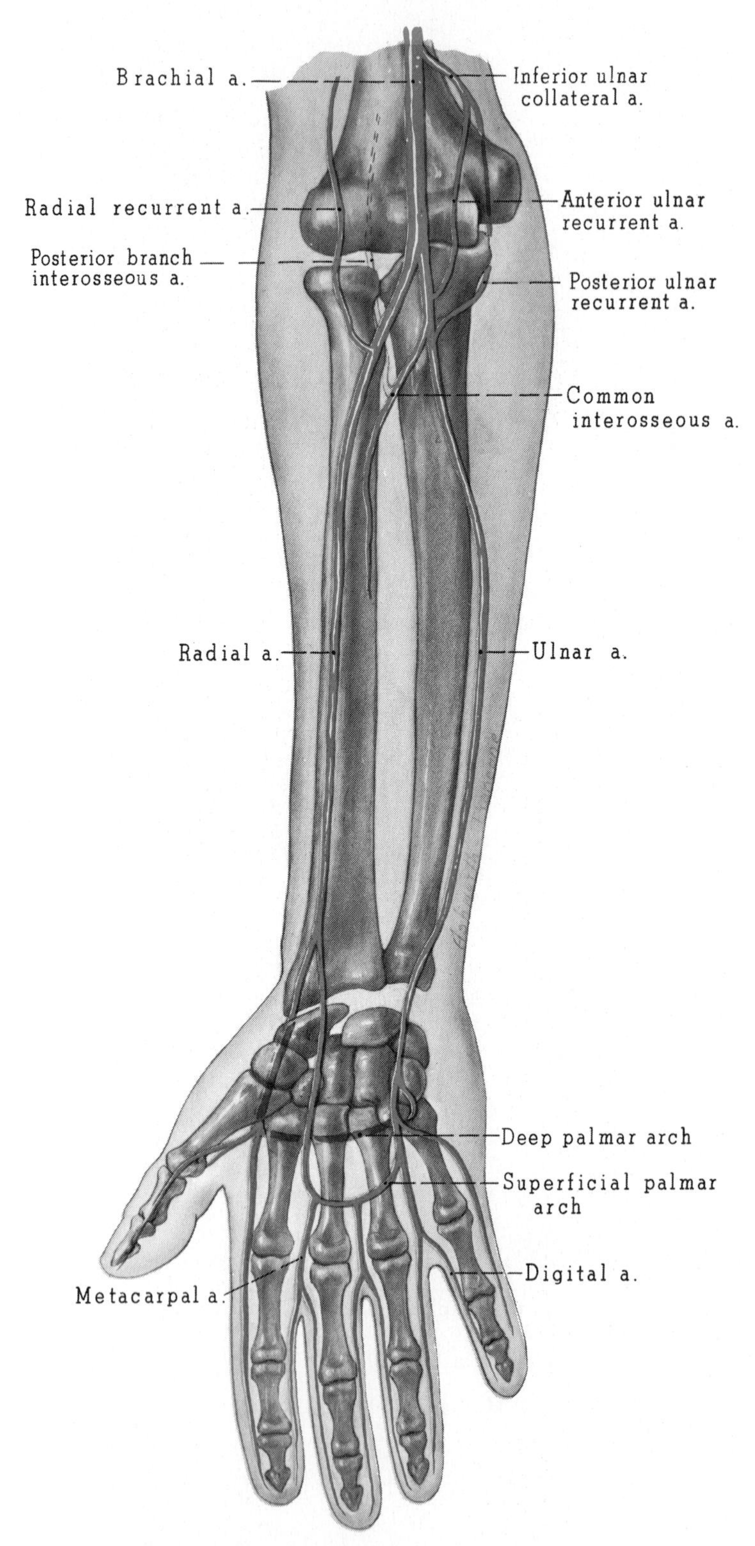

Figure 106. Arteries of the right lower arm.

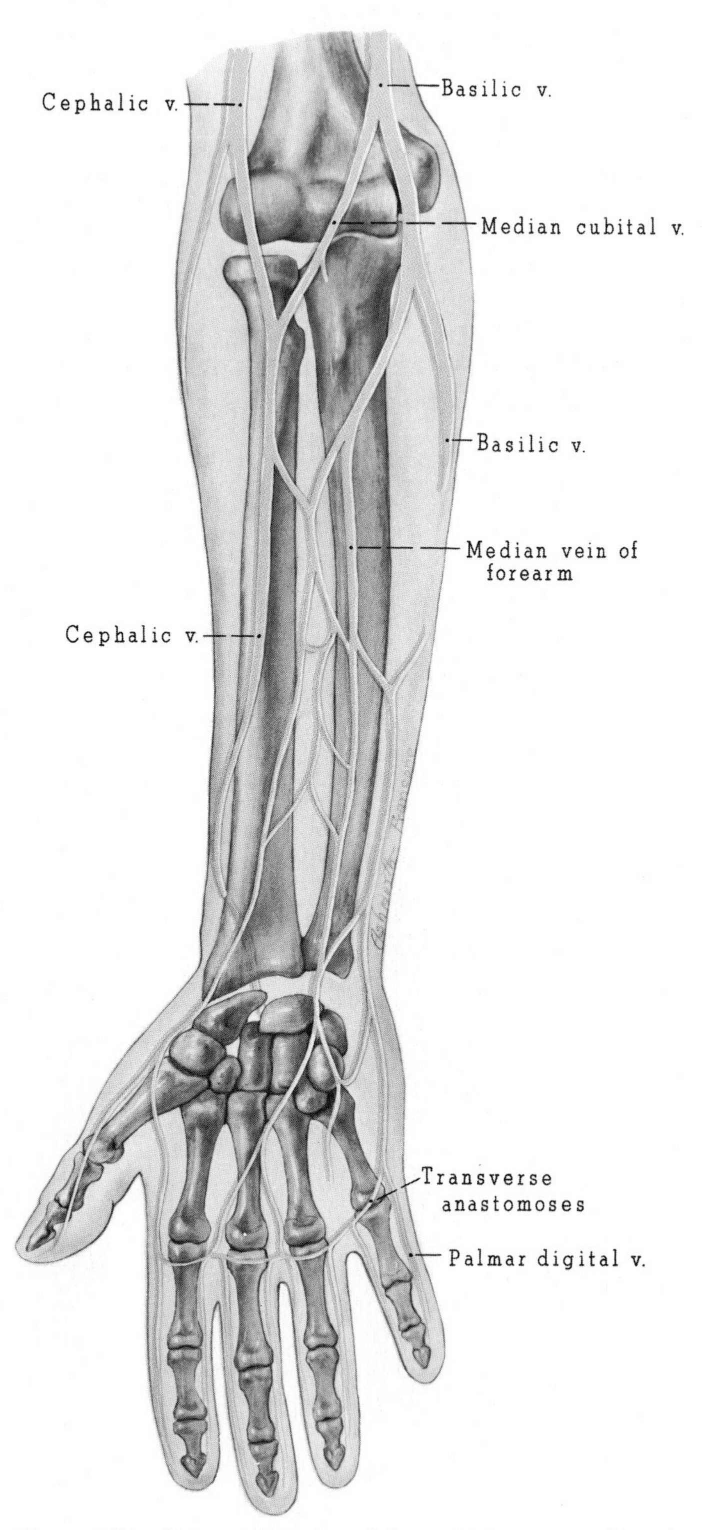

Figure 107. Venous drainage of the right forearm and hand.

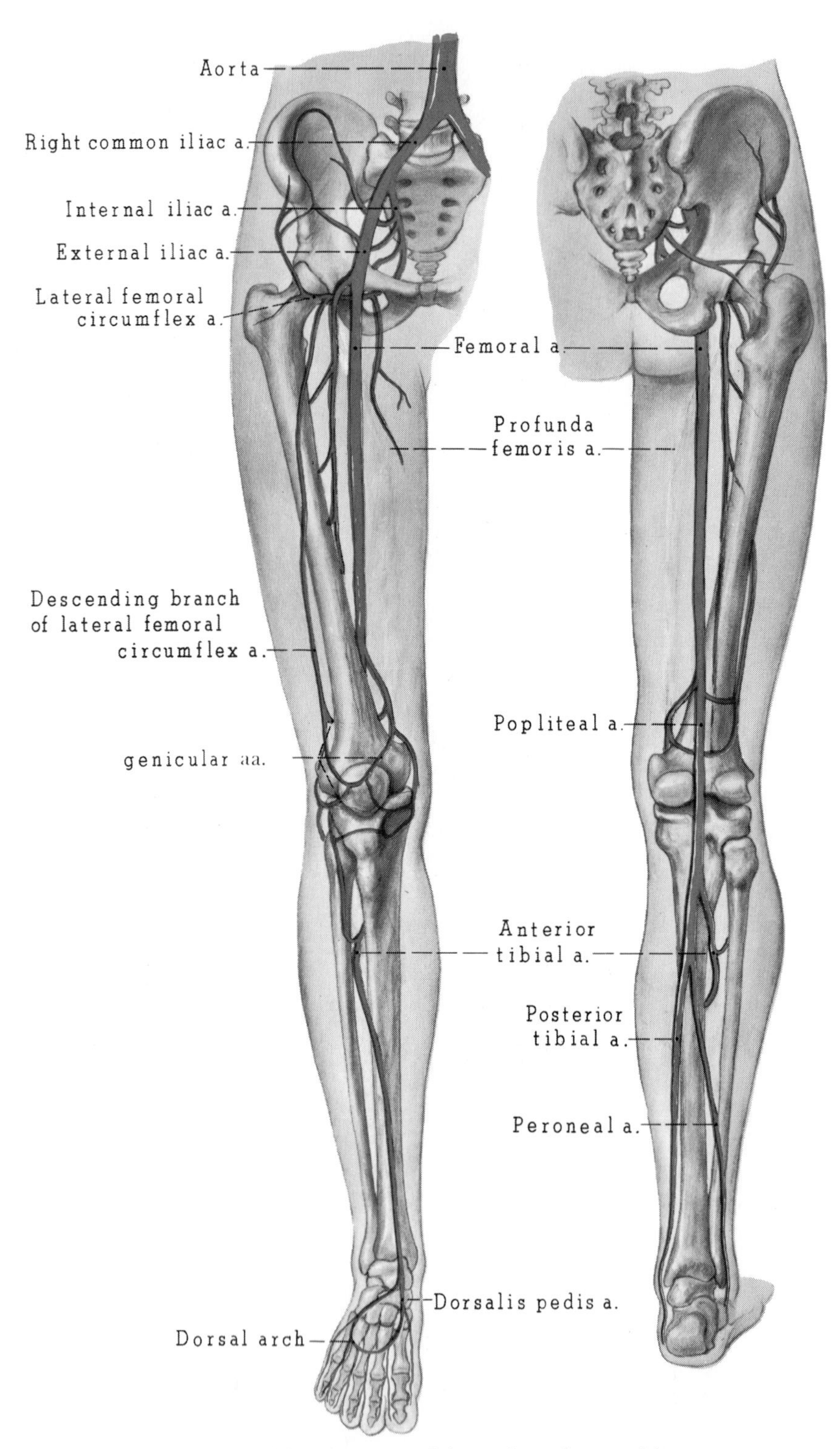

Figure 108. Arteries of the right pelvis and leg.

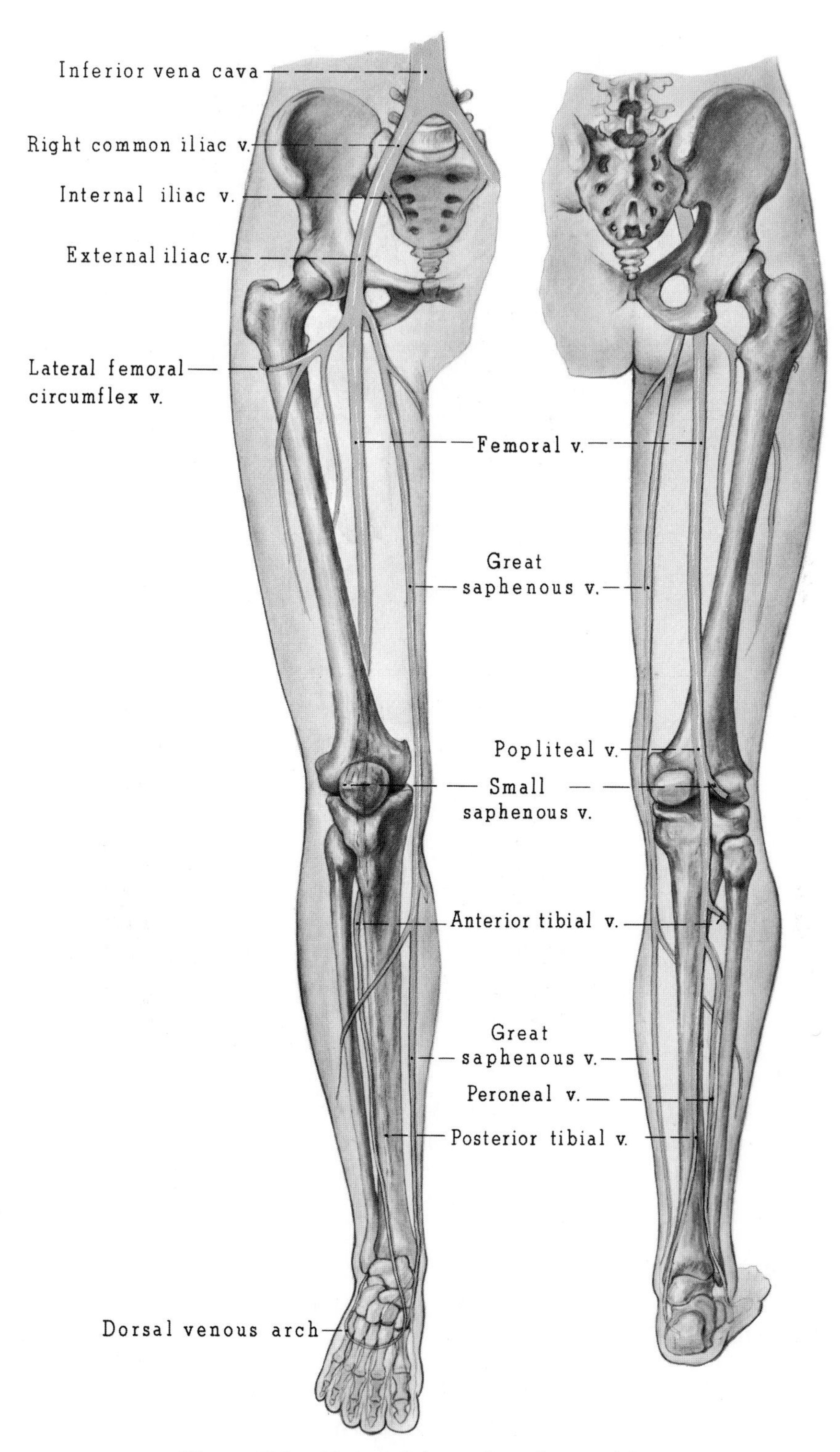

Figure 109. Veins of the right pelvis and leg.

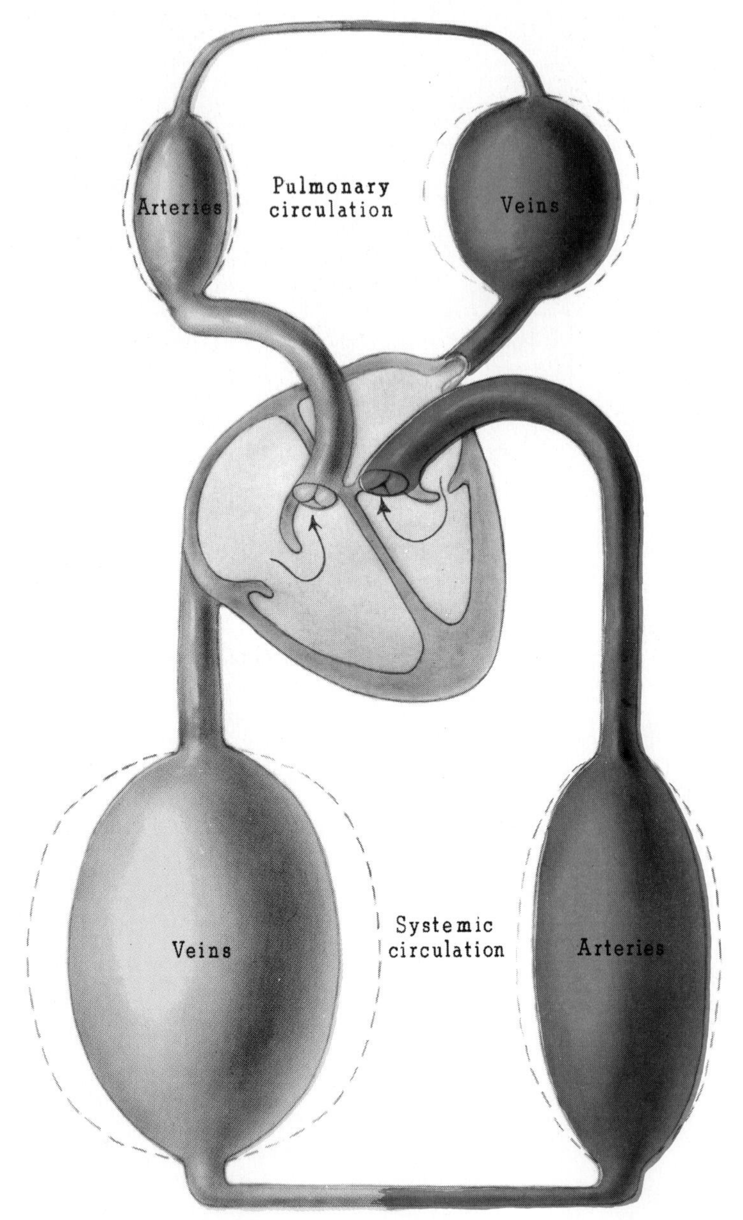

Figure 110. Schematic drawing showing relationship between systemic and pulmonary circulatory circuits.

BLOOD PRESSURE

What Is Blood Pressure?

Blood pressure is the pressure exerted by the blood against the walls of the vessels as it is forced through the circulatory system. The term applies properly to arterial, capillary, and venous pressure.

Most commonly the expression "blood pressure" refers to the pressure existing in the large arteries—usually the *brachial* (bra'ke-al) *artery,* just above the elbow, where external measurement is relatively easy. The blood pressure is highest in the brachial artery at the time of contraction of the ventricles—ventricular systole (recall systole means contraction). This level is known as *systolic* (sis-tol'ik) *pressure.* Pressure during ventricular *distole* (Greek

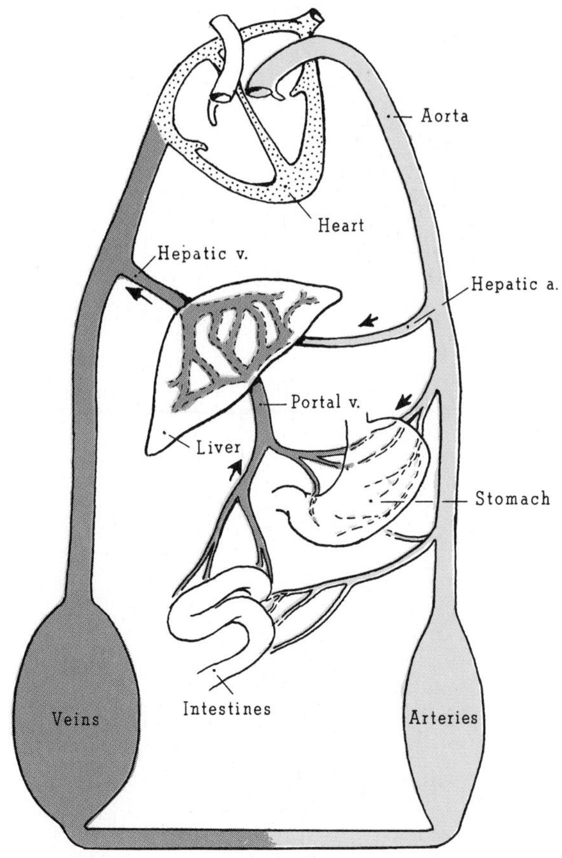

Figure 111. Schematic drawing showing the relationship between systemic and portal circulatory circuits. Notice that the liver receives blood from two sources: (1) from the hepatic artery, which is a direct arterial supply of fully oxygenated blood; and (2) from the portal vein, which drains from the stomach and intestine. This arrangement ensures that the nutrients and electrolytes absorbed into the blood from the digestive system are processed by the liver, before the blood moves on to the heart and lungs.

for dilation or relaxation) is called *diastolic* (di″ah-stol′ik) *pressure.* Blood pressure is usually expressed as a fraction—for example, as 120/80, in which 120 represents systolic pressure and 80, diastolic pressure. The units are in terms of millimeters of mercury, abbreviated mm.Hg.

The blood pressure decreases as the blood passes from the aorta into the arteries to the arterioles and capillaries and venules and veins, until finally, at the left atrium when it enters the heart, the blood pressure is close to zero (Figure 112).

What Is a Normal Blood Pressure?

Blood pressure is subject to fluctuations. In general, the healthy individual has a systolic pressure of 100 to 120 mm. of mercury and a diastolic pressure of 60 to 80 mm. of mercury. Variations in systolic blood pressure are expected in normal persons. Exercise may cause a rise in systolic pressure.

A blood pressure difference of 10 to 15 mm. of mercury often exists between the two arms of an individual. The higher pressure is usually found in the right arm. A pressure difference greater than 10 to 15 mm.Hg should arouse suspicion, since it might be the result of a congenital (from the Latin word meaning born with) narrowing of the aorta between the beginnings of the right and left subclavian arteries. This narrowing condition is known as a *coarctation* (ko″ark-ta′shun), from the Latin word meaning to press together.

The upper limits of normal blood pressure are usually defined as 140 mm.Hg systolic, and 90 mm. Hg diastolic. Pressures above this level (hypertension) shorten life expectancy.

What Are the Factors that Influence Blood Pressure?

The blood pressure is influenced by a number of factors. Normal regulation is the responsibility of the nervous system and, somewhat independently, the heart.

The nervous system can bring about rapid changes in blood pressure and blood flow by stimulating the muscular contraction of the blood vessels, particularly the arterioles. When the vessels constrict they become smaller in diameter, and this causes the blood pressure to increase. The *vasomotor* (*vaso* means vessel, *motor* indicates an activating center) *center* in the medulla of the brain is the area that activates vessel constriction. It is also capable of influencing heart rate and strength of contraction. However, the heart itself can affect blood pressure by increasing its rate

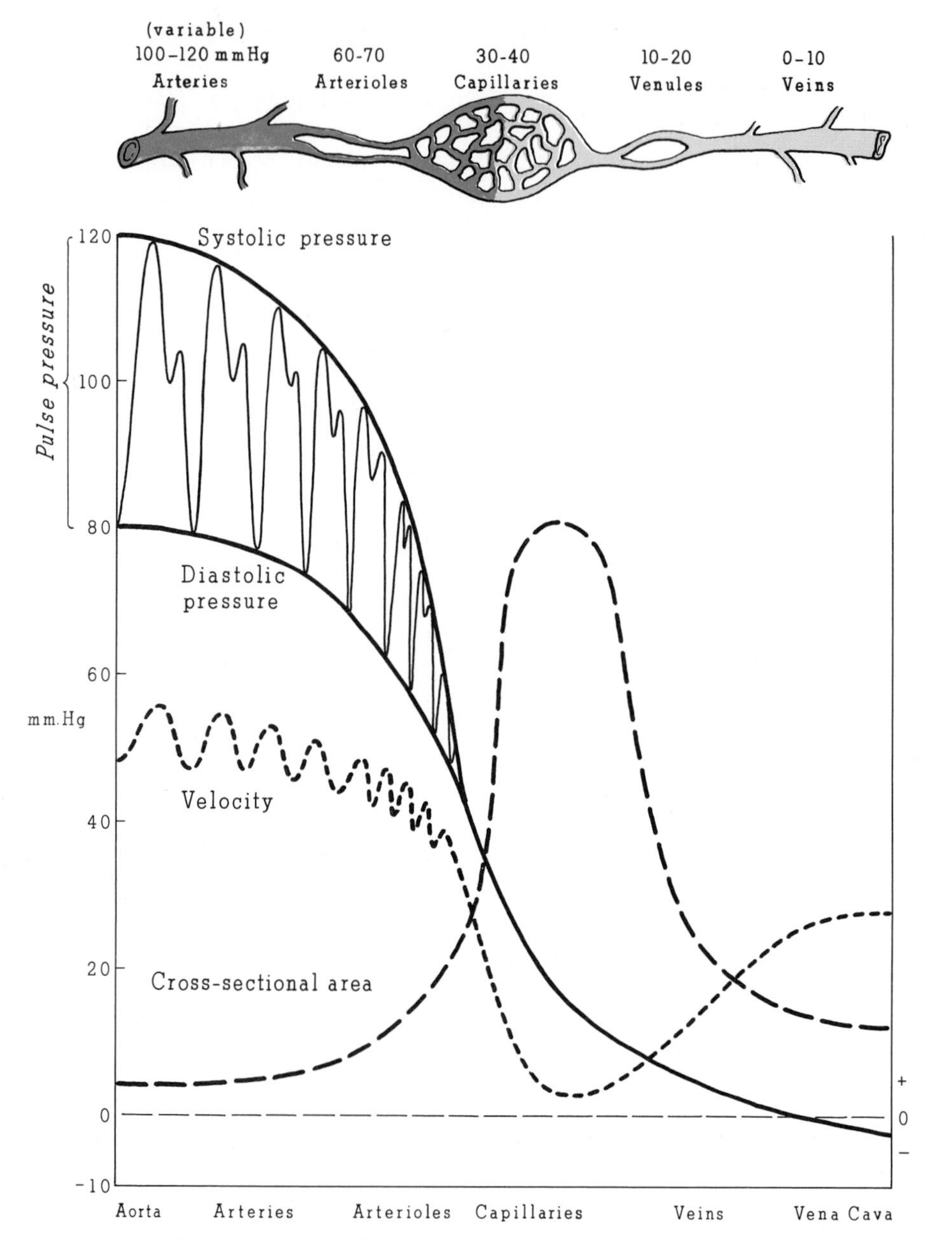

Figure 112. Blood pressure, blood velocity, and cross-sectional area of the vascular tree in various segments of the circulatory system. (From Zoethout and Tuttle: *Textbook of Physiology,* twelfth edition. St. Louis, The C. V. Mosby Co., 1955.)

of strength of contraction, and this seems to occur independently of the vasomotor center activity.

The vasomotor center receives sensory signals from a number of sources that influence its regulatory action on the blood pressure. The *baroreceptors* (pressure sensitive receptors) located in the large arteries monitor blood pressure for the vasomotor center. The baroreceptors are sensitive to the stretching of the artery walls caused by changes in blood pressure inside the arteries. Other receptors, *chemoreceptors,* are sensitive to the concentrations of various substances in the blood, such as oxygen, carbon dioxide, and pH (acid content). If carbon dioxide concentration becomes too high, this will cause the vasomotor center to increase cardiac output and blood pressure, which serves, along with an increased respiration, to accelerate carbon dioxide release in the lungs.

CLINICAL CONSIDERATIONS OF THE BLOOD CIRCULATORY SYSTEM

What Is Shock?

Shock is an impairment (disorder) of the circulation resulting from stress or injury; the damage reduces the output of blood from the heart to a level below that needed for normal cellular function. Another way to characterize the condition is as a loss of effective circulating volume in the heart and vessels, leading to a sort of circulatory panic. The symptoms of shock are apprehension, cold skin, reduced blood pressure, shallow respiratory activity, sweating, and rapid pulse.

What Conditions May Lead to Shock?

A reduction of blood volume frequently produces shock. A loss of over 40 per cent of the blood volume causes collapse of the arteries. This condition frequently does not respond to blood transfusions and is sometimes referred to as "irreversible shock." Sudden losses of small quantities of blood can produce consequences more serious than slow losses of larger volumes. Massive injury to the heart (as in heart attack), leading to an inadequate cardiac output is an important cause of shock.

In the absence of hemorrhage or actual damage to the heart, the reason for loss of effective circulating volume—typical of shock—is not clearly understood. Three widely held views that attempt to explain this blood loss are increased permeability of the capillary walls, actual plasma loss, and, popularly, the pooling of blood in the peripheral capillary beds.

What Is Arteriosclerosis?

Arteriosclerosis (ar-te″re-o-skle-ro′sis) is a condition characterized by a depositing of material, called plaque (plak), on the walls of the arteries, especially at the junctions. The disease frequently leads to a progressive blockage of blood vessels. Occasionally, part of the deposit will break off and rapidly block off a vessel (Fig. 113). This can produce an acute coronary or a cerebral seizure, if the blocked vessel is in the heart or brain, respectively.

Excessive fat in the diet can be an important factor leading to arteriosclerosis, although the disease also seems to run in families. This disease is the number one medical problem in the United States.

Why Take the Pulse?

The pulsation felt when the fingertips are placed over an artery close to the body surface is generally referred to as simply the *pulse*. Pulse rate corresponds to heart rate, one pulse for every ventricular systole (Fig. 114).

The pulse is described according to several characteristics: *rate*, fast or slow; *size*, large or small; *type of wave*, abrupt or prolonged; and *rhythm*, regular or irregular. An increase in pulse rate is normal during and after exercise and after eating. It is decreased during sleep. In most diseases associated with fever the pulse rate is increased, usually at an average of five beats for every degree Fahrenheit. An increased pulse rate is usually present in severe anemias, and becomes markedly increased after severe hemorrhage.

What Are Two Forms of Heart Disease?

If the myocardium receives an inadequate oxygen supply, it cannot function properly. A primary sign of this inadequate

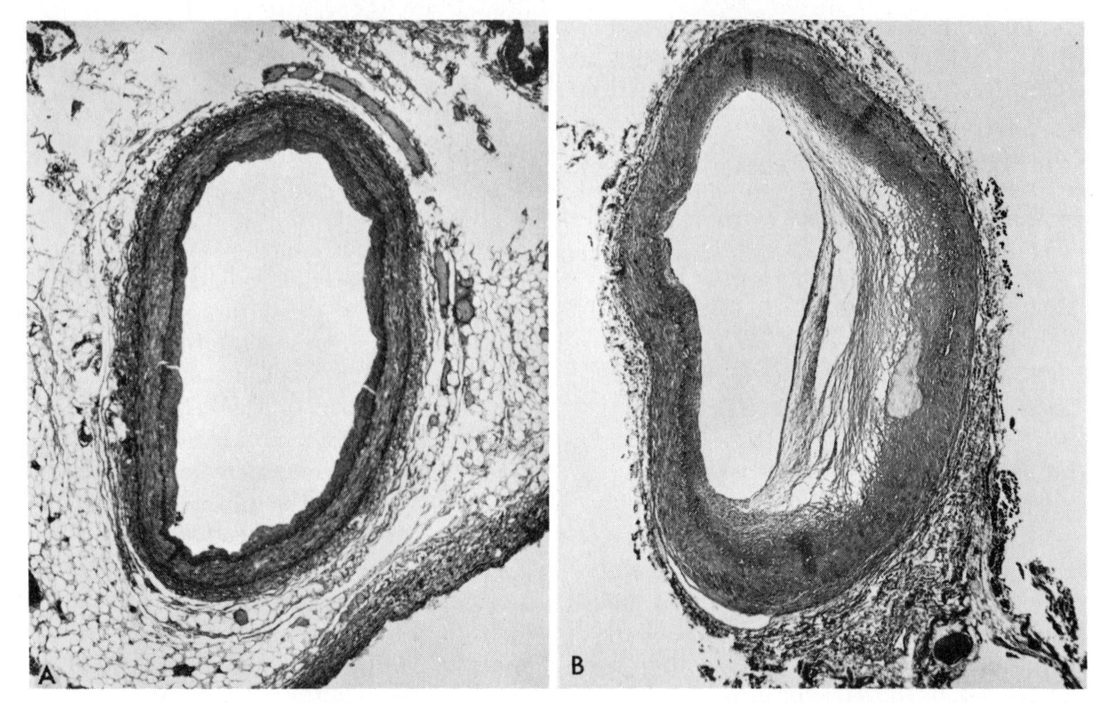

Figure 113. These photomicrographs show (*A*) a normal artery seen in cross section and (*B*) a diseased artery in which the channel is partially occluded by atherosclerosis. Fibrous scar tissue with fatty deposits (clear areas) and other materials have thickened the arterial wall, reducing the blood carrying capacity. (By permission: David M. Spain, M.D., previously published: Scientific American, August, 1966).

supply is a peculiar, severe type of chest pain called *angina pectoris.* Angina pectoris is relieved by rest or the administration of nitrites. Attacks of angina pectoris are usually precipitated by exertion or emotional tension. The pain, frequently described as a feeling of tightness, pressure, aching, or heaviness below the sternum, may radiate to the left shoulder and arm and neck. The pain is of short duration, usually lasting 3 to 5 minutes.

The pain of *myocardial infarction* (from the Latin for disorder) causes more prolonged chest pain than angina pectoris. In this case, there is actual blockage of some part of the arterial supply to the myocardium, resulting in death of a portion of the heart muscle.

THE LYMPHATIC SYSTEM

What Is the Lymphatic System?

The lymphatic system is made up of a network of vessels, beginning with *lymphatic capillaries* which lead into larger and larger lymphatic vessels. This network of vessels drains and filters tissue fluid, which is then returned to the blood stream. Tissue fluid in the lymphatic vessels is called *lymph.* The term "lymph" is from the Latin word *lympha,* meaning clear spring water. The larger lymphatic vessels have valves that permit lymph to flow in only one direction. All lymph vessels are directed toward the thoracic cavity. They converge into either the *right lymphatic duct* or the *thoracic (left lymphatic) duct.* Both ducts empty into large veins in the upper chest area (Fig. 115).

What Are Lymph Nodes?

Lymph nodes are small, oval bodies found at intervals in the course of the lymphatic vessels (Fig. 116).

Lymph passes through several groups of nodes before entering the blood. Within

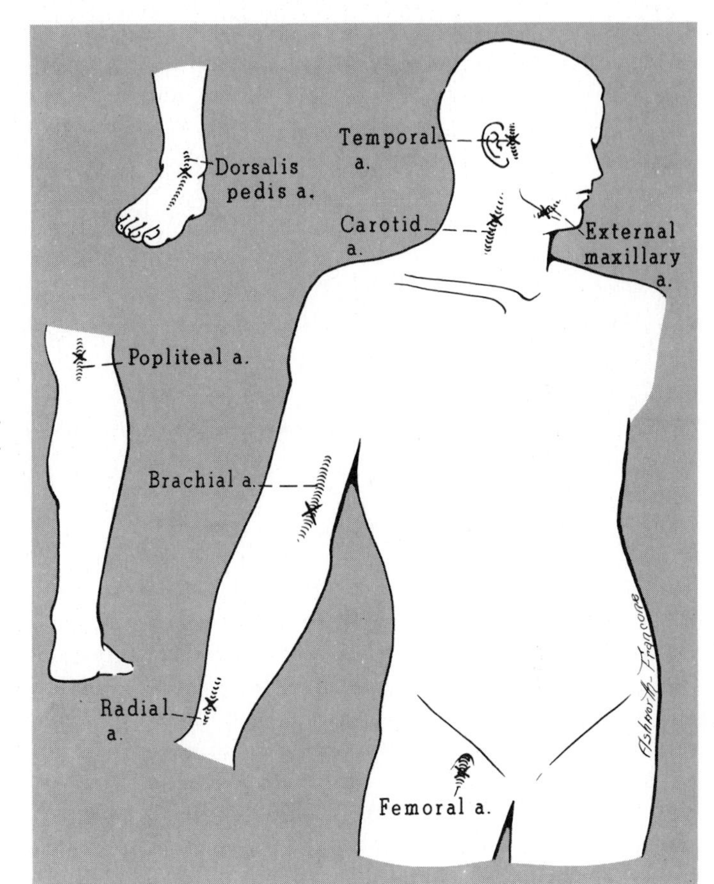

Figure 114. The pulse is readily distinguished at any of the pressure points shown.

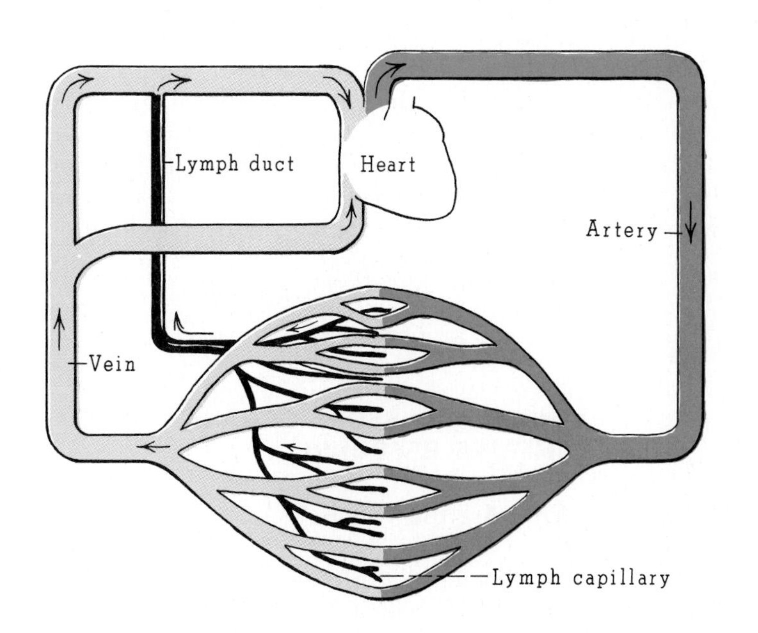

Figure 115. Diagrammatic representation of lymphatic system, showing its relationship to the circulatory system.

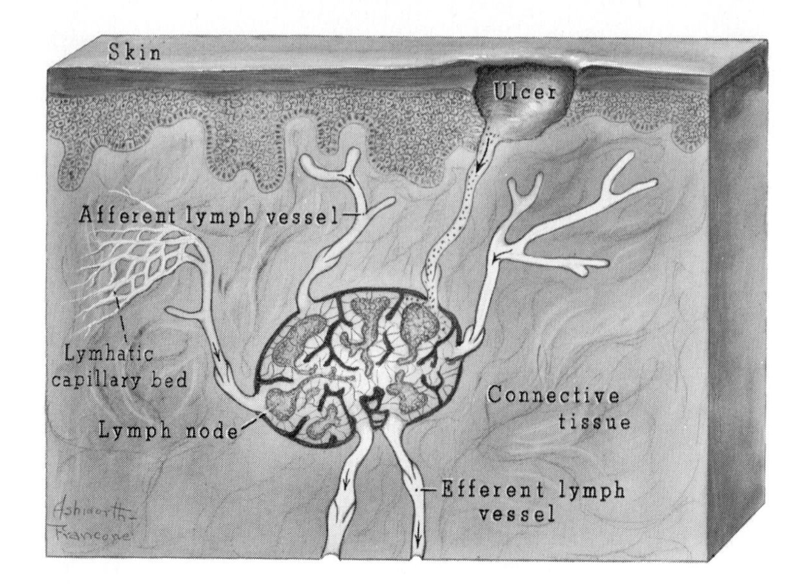

Figure 116. Diagrammatic drawing of a lymph node in the area of an infected ulcer.

the nodes, the lymph is filtered and receives lymphocytes, globulin, and antibodies, which are manufactured by the specialized tissue of the lymph node. Lymph nodes serve as efficient filters for red blood cells and bacteria, but are ineffective barriers against viruses. Lymph enters the nodes through several channels and leaves through one or two channels.

Where Are the Lymph Nodes Located?

Lymph nodes usually appear in groups. The *superficial nodes* are located near the body surface in the neck, armpit, and groin. The *deep nodes* are located in the internal groin area and adjacent to the lumbar vertebrae, at the root of the lungs, attached to the tissue surrounding the small intestines, and in association with the liver (Fig. 117).

What Are the Specific Functions of the Lymphatic System?

The most important function of the lymphatic vessel system is that it returns to the blood stream vital substances—chiefly proteins—that have leaked out of the blood capillaries.

Lymph vessels also provide drainage channels into lymph nodes. The lymph nodes themselves have several specific functions. They filter and isolate products resulting from bacterial and non-bacterial inflammation, and prevent these products from entering the general circulation. This process often produces tenderness and swelling in nodes in an infected area. The special tissue of the lymph nodes also produces lymphocytes, globulin, and antibodies and releases them into the blood, where they function in producing immunity to disease.

The intestinal lymph vessels have an additional special function. Lymph is generally a clear (plasma-like) liquid, but lymph from the intestines becomes milky after a meal. The milky appearance results from the presence of minute fat globules collected from the digestive tract. The lymphatic system distributes these fat globules to the tissues where they are stored, occasionally in excessive amounts.

THREE ACCESSORY ORGANS (IMMUNE RESPONSE)

What Are the Tonsils?

Several groups of *tonsils*, forming a ring of lymphatic tissue, guard the entrance

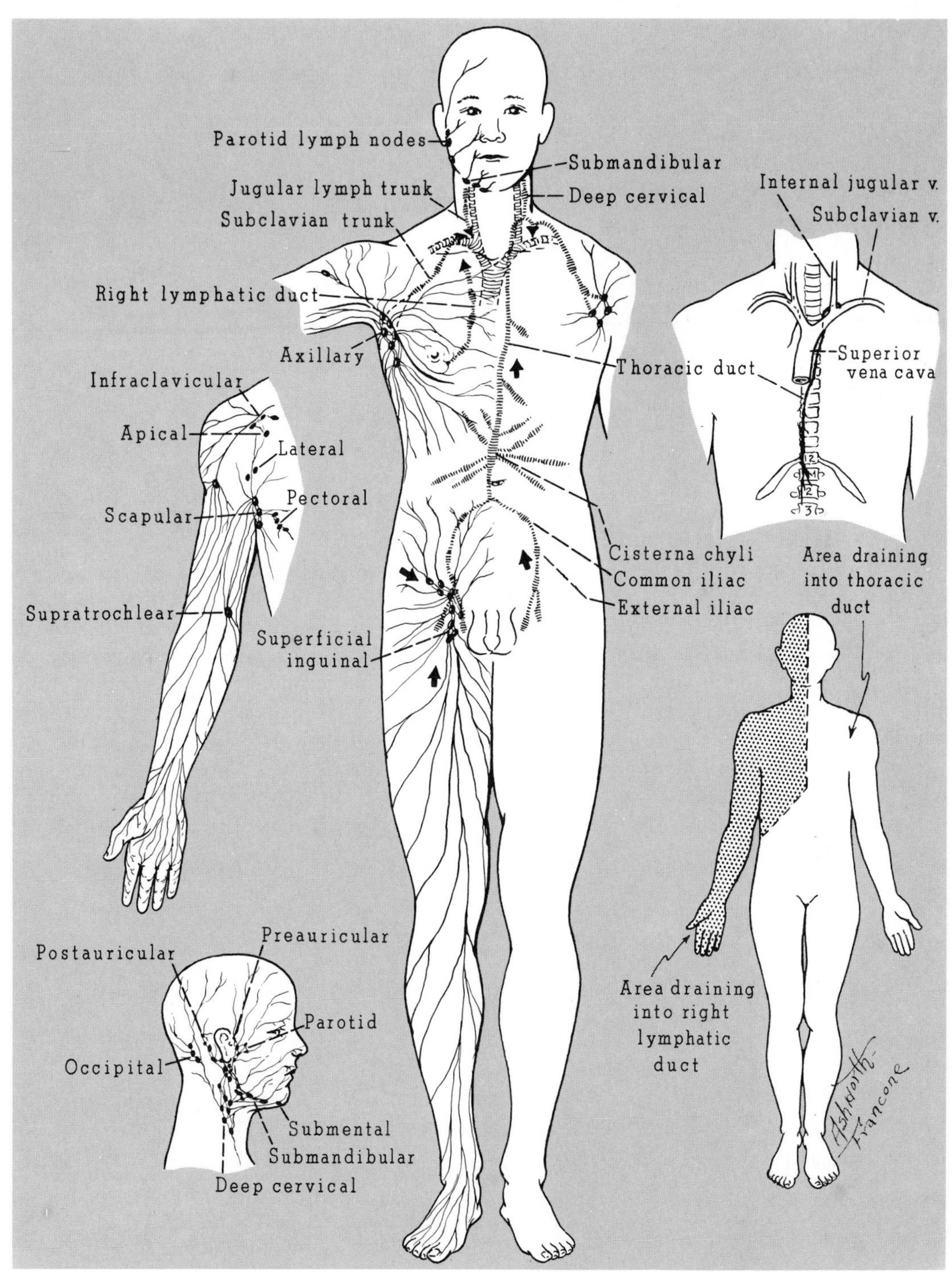

Deep collecting channels and their lymph nodes

Superficial collecting channels and their lymph nodes

Figure 117. The lymphatic system and drainage.

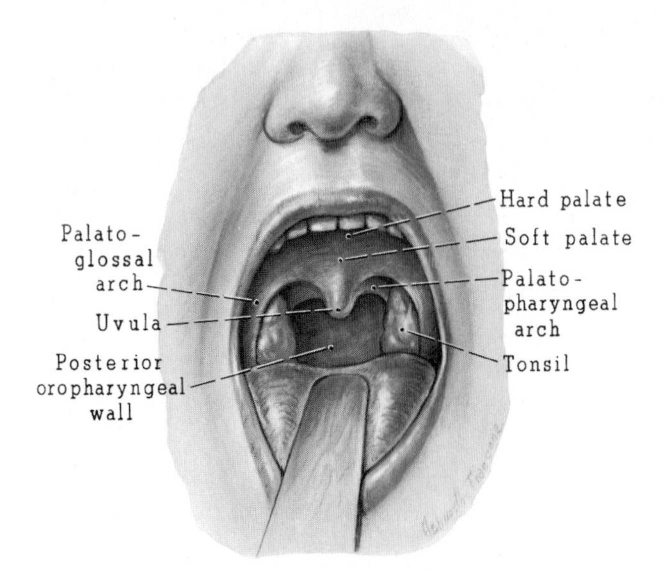

Figure 118. Relationship of tongue, uvula, and tonsils.

to the digestive and respiratory tracts from bacterial invasion (Figs. 118 and 119).

Chronic infection of the tonsils is not so common as was once suspected. The term "chronic tonsillitis" is frequently misused to indicate any type of sore throat occurring when the tonsils are still present. With tonsillitis, enlargement and tenderness of the anterior lymph nodes in the neck are common. The tonsils may be enlarged and red or covered with pus. If both tonsils and adenoids are infected, the lymph nodes in the posterior triangle of the neck enlarge.

Fewer tonsillectomies (*-ectomy* means removal) and adenoidectomies are being performed today than were done 30 years ago. This is because recent knowledge indicates that removal of tonsils and adenoids may not significantly lower the incidence of upper respiratory infection unless the tonsils themselves have been infected. Tonsils form a protective barrier for the mouth, throat, larynx, trachea, and lungs.

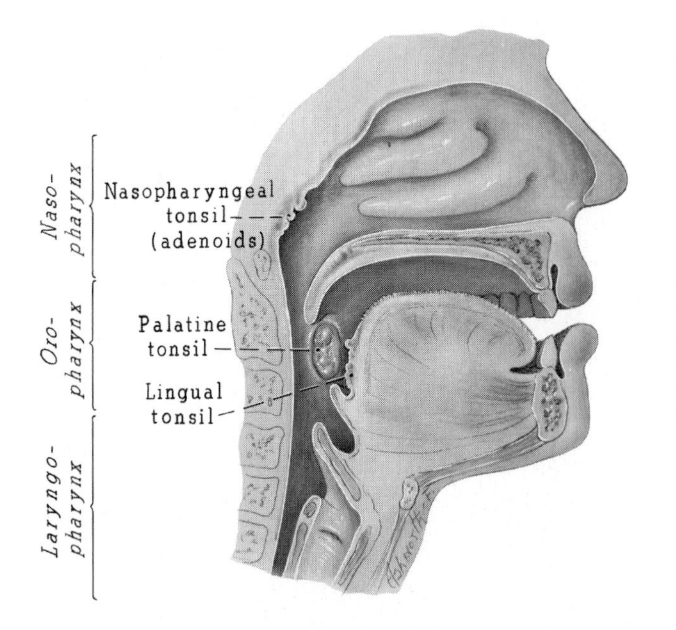

Figure 119. The nasopharyngeal tonsils extend from the roof of the nasal pharynx to the free edge of the soft palate; the palatine tonsils are attached to the side walls of the back of the mouth between the anterior and posterior pillars; the lingual tonsils are located on the dorsum of the tongue from the vallate papillae of the tongue to the epiglottis.

They may also be important in the development of immune bodies; however, true recurrent infection of the tonsils is still an indication for their removal by operation.

What Is the Thymus?

The *thymus* is a flat, pinkish-gray, two-lobed organ lying high in the chest under the sternum and in front of the aorta (Fig. 120).

This gland is one of the central controls of the body's immunity system. The term "immunity" derives from the Latin word meaning exempt or free from. If someone is immune to a disease, it means that they have the necessary protective devices to fight the disease off easily. Recent evidence indicates that the thymus performs two functions. First, early in the life it releases a substance that prepares the lymphatic tissue throughout the body for immune response. Second, the thymus manufactures lymphocytes, which help to populate the other lymphatic tissues of the body.

It was reported in 1961 that removal of the thymus from an animal just after birth resulted in great impairment of its immunological responses to antigens in later life, and also impaired its ability to reject skin grafts.

What Is the Spleen?

The spleen is a soft, vascular (having many blood vessels), oval body, 5 inches long and 3 inches wide, weighing approximately 7 ounces. It lies in the left upper abdomen beneath the diaphragm and behind the lower ribs (Fig. 121).

The spleen has five major functions:

1. Blood Destruction. Old red blood cells, having reached their normal life span of approximately 120 days, are destroyed and digested by special cells located in the spleen.

2. Blood Production. The spleen exerts an effect on production and release of blood cells from bone marrow.

3. Immunologic Function. The spleen is a source of production of antibodies

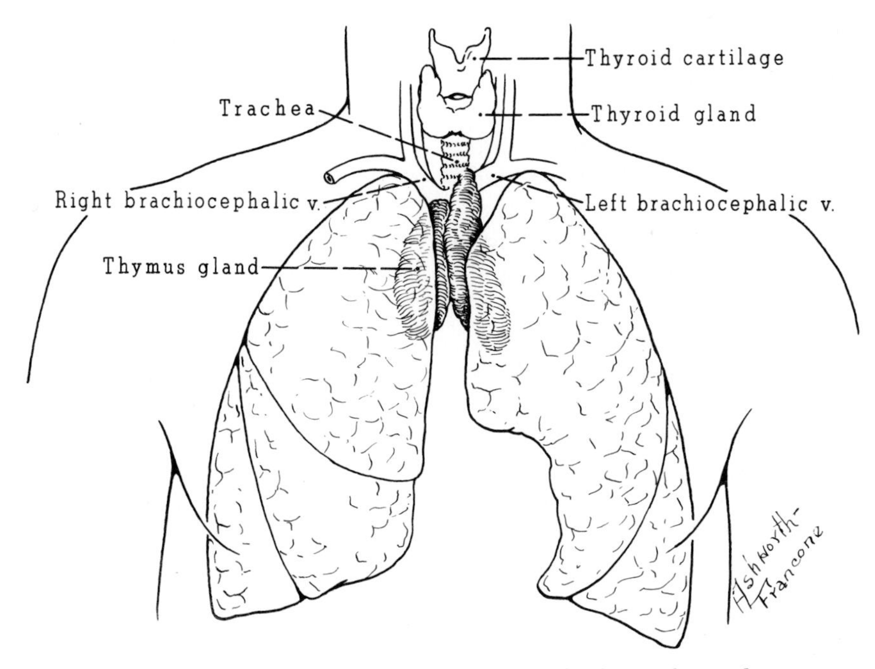

Figure 120. Location of thymus gland and relationship to lungs.

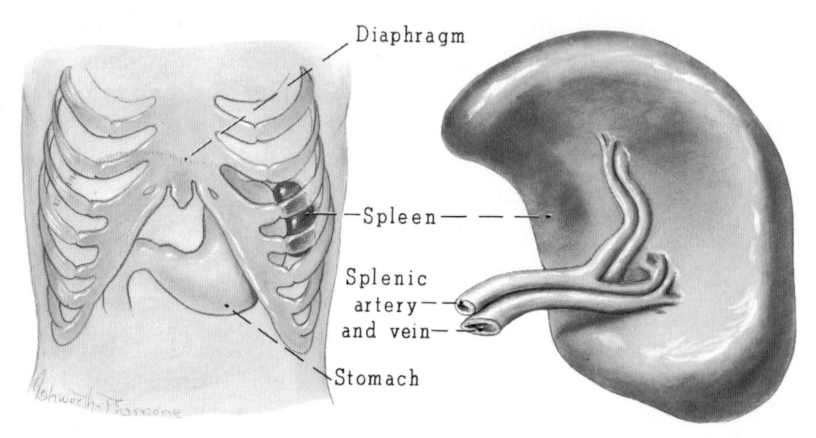

Figure 121. The spleen and its relation to the stomach and rib cage.

and contains a large mass of lymphatic tissue.

4. Blood Storage. The spleen serves as a reservoir for blood. It undergoes rhythmic variations in size in response to physiologic demands such as exercise and hemorrhage, and thus influences the volume of circulating blood.

5. Blood Filtration. The spleen serves as one of the body's defense mechanisms by filtering microorganisms from the blood.

LEARNING EXERCISES

1. Draw a diagram of the heart, lungs, and major vessels; label the chambers and indicate the locations of the valves. Use arrows to trace the flow of blood through one cardiac cycle. Also, identify and label the SA and AV nodes.
2. Following Figures 110 and 111, draw your own schematic diagrams, showing the relations between the systemic and pulmonary circuits, and the systemic and portal circuits.
3. Construct an outline of the body and indicate the major locations of the lymph nodes, the thymus gland, and the spleen. Construct an outline of the head and throat, and indicate the location of the tonsils (and adenoids).
4. Locate the following arteries and veins in the diagrams: trace the origin of the arteries back to the aorta, and trace the path of the veins to the right atrium.
   a. Internal carotid artery
   b. Vertebral artery
   c. Maxillary artery
   d. Superior sagittal sinus
   e. Internal jugular vein
   f. Deep cervical vein
   g. Right brachiocephalic artery and vein
   h. Descending aorta
   i. Common illiac artery
   j. Femoral artery
   k. Subclavian artery and vein
   l. Renal artery and vein
   m. Axillary artery and vein
   n. Cephalic and basilic veins
   o. Radial and ulnar arteries
   p. Femoral artery and vein
   q. Lateral femoral circumflex artery and vein
   r. Anterior tibial artery
   s. Great saphenous vein
   t. Dorsal venous arch

CHAPTER 8

THE RESPIRATORY SYSTEM

FUNCTION OF THE RESPIRATORY SYSTEM

What Does the Respiratory System Do?

The maintenance of life depends on a continuous supply of oxygen to, and removal of carbon dioxide from, the cells of the body. The transportation of these gases between the cells and the respiratory system (specifically the lungs) is accomplished by the circulatory system. The respiratory system functions in the exchange of gases between the blood circulating through the lungs and the external environment. It serves to bring in oxygen and eliminate excess carbon dioxide.

A second related function of the respiratory system is to regulate the acid-base balance in the blood. This happens because carbon dioxide, when dissolved in blood, is an acid. Thus, when there is too much carbon dioxide in the blood, the blood becomes too acidic. And when there is too little carbon dioxide in the blood, the blood becomes too basic (or alkaline). Variations on the rate and depth of respiration serve to control the amount of carbon dioxide in the blood within a rather narrow range of concentrations.

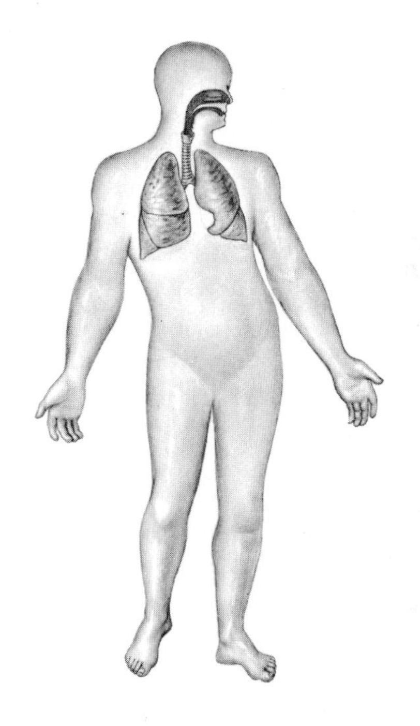

STRUCTURES OF THE RESPIRATORY SYSTEM

The specific function of the respiratory tract is to bring air close enough to the blood to allow oxygen to get into the blood and carbon dioxide to get out. The *upper respiratory tract,* made up of the nose, pharynx, larynx, trachea, and bronchi, forms an open passage between the lungs and the exterior. The *lungs* are the essential organs of respiration. With their extensive network of capillaries, the lungs provide an adequate surface for a high volume of exchange of gases between the body and the external environment. The *muscular diaphragm* and *intercostal muscles* of the chest produce the mechanical force needed to fill and empty the lungs in the process called *breathing.*

What Is the Thoracic Cavity? The Pleural Cavity? The Mediastinum?

The *thoracic cavity*, or chest, is separated from the abdominal cavity by the diaphragm, a large sheet of muscle. The *mediastinum* (mee-dee-as-tye'num) is the middle compartment of the chest and is located between the two pleural cavities, which contain the lungs. The heart and its associated structures are located in the mediastinum.

Each *pleural cavity* contains one of the lungs. The wall of the pleural cavities is lined by membrane called the *pleura*. Each lung is covered by a second layer of *pleura*. These two pleural membranes (pleura) are moist and slippery, and between the two layers is found a special lubricating *serous fluid* (Fig. 122).

This double pleural membrane arrangement allows respiration with minimal friction. In the condition known as *pleurisy*, the pleura become irritated and inflamed, and breathing becomes painful.

What Is the Nose?

The term *nose* includes both the external nose—that part of the upper respiratory tract that protrudes from the face—and the *nasal cavity*. Only a small part of the nasal cavity is in the external nose, most of it lies over the roof of the mouth. The nasal cavity is composed of two wedge-shaped cavities separated by a *septum* (from the Latin word meaning partition or wall) (Figs. 123 and 124).

The nasal cavity is lined with a thick mucous membrane containing a rich supply of tiny capillaries. This mucous membrane serves to warm and moisten the air on its way to the lungs. If the air is not warmed the tissue lining the lower respiratory tract will function poorly. Absence of moisture for even a few minutes can destroy the delicate *cilia* (hairlike) in the lining of the respiratory tract.

The mucous membrane also captures bacteria and dust particles. Air currents passing over the moist mucosa in curved pathways deposit fine particles, powder, and smoke against the walls. These fine particles are subsequently moved to the pharynx by the wavelike action of the cilia. These cilia wave back and forth about 12 times per second, helping to move mucus toward the thorat to be swallowed.

The uppermost portion of the nasal cavity is lined with special nerve tissue containing *olfactory* (ol-fak'to-re) *cells*, which function in the sensation of smell.

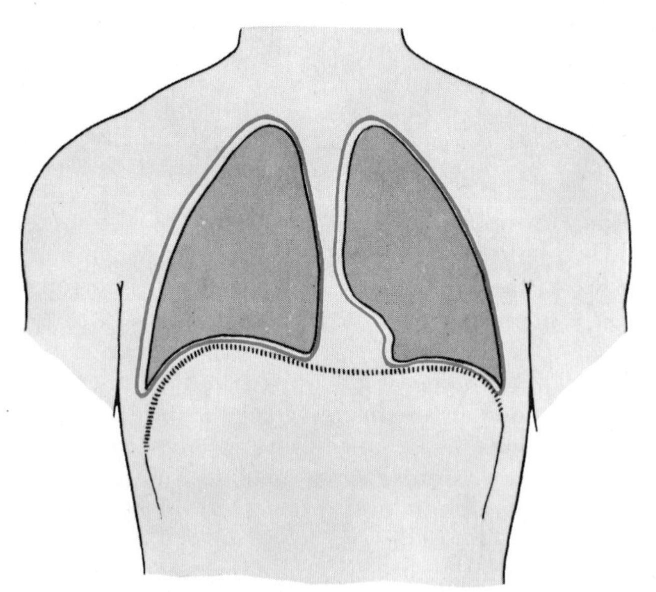

Figure 122. Lungs and associated visceral and parietal pleura.

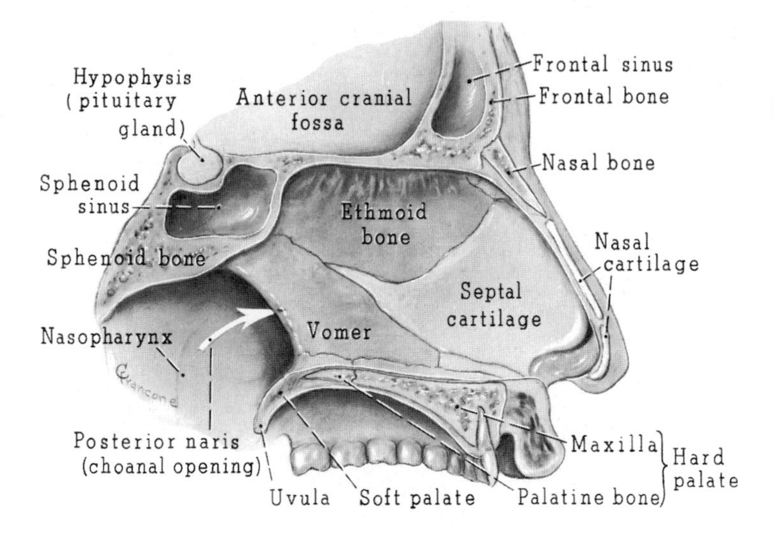

Figure 123. Sagittal section through nose showing components of nasal septum.

The location of the olfactory cells at the beginning of the upper respiratory tract is important in that it allows the brain to be alerted to the presence of poisonous substances in the air *before* they are inhaled.

What Are the Paranasal Sinuses? What Is Their Function?

The *paranasal sinuses* are air-containing spaces lined with mucous membrane that are connected at various points to the nasal cavity. These *paired* sinuses include the *maxillary, frontal, ethmoid,* and *sphenoid* sinuses (Fig. 125).

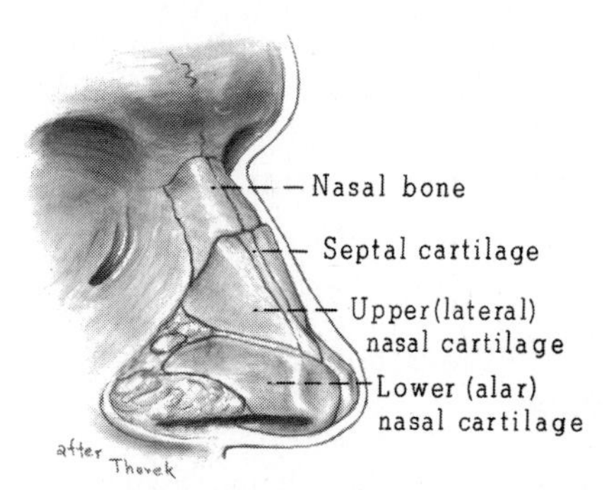

Figure 124. The lower portion of the external nose has a cartilaginous rather than a skeletal framework consisting of a septal cartilage, two lateral cartilages, and a series of smaller cartilages.

The primary function of the paranasal sinuses is to manufacture mucus for the air-cleansing activity of the nasal cavity. They serve secondarily to lighten the bones of the skull and act as sound chambers for the production of sound.

What Is the Pharynx? What Is Its Function?

The *pharynx* (far′inks), or throat, serves as a passage for two systems—the digestive system and the respiratory system. Air can enter the pharynx either from the two nasal cavities or from the mouth. At the lower end, air proceeds to the larynx, while food is swallowed into the esophagus.

The right and left *eustachian* (auditory) *tubes* open into the pharynx, connecting the middle ear with the upper respiratory tract. The tonsils and adenoids are located along the walls of the pharynx.

What Is the Larynx? What Is Its Function?

The *larynx* or "voice box" connects the pharynx with the trachea. The larynx is broad at the top and shaped like a triangular box (Fig. 126). It is made up of nine pieces of cartilage united by muscles and ligaments. The *thyroid cartilage,* or Adam's

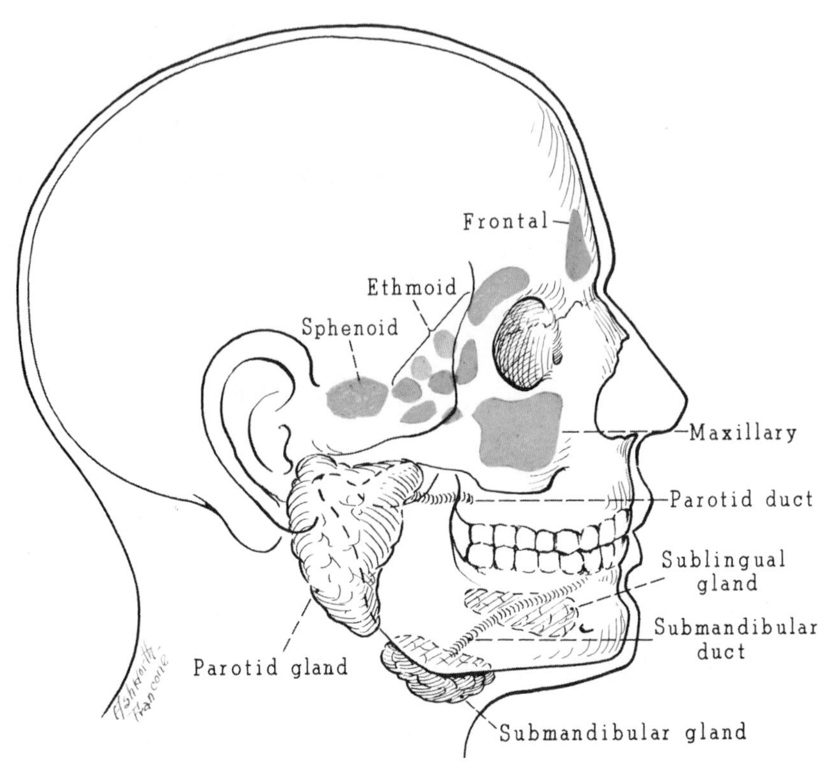

Figure 125. Lateral view of head showing sinuses and salivary glands.

apple, is the largest cartilage in the larynx. In the male, the thyroid cartilage increases in size at puberty. The leaf-shaped *epiglottis* (from *epi,* meaning on, and *glottis,* meaning mouth of the windpipe) is attached to the top border of the thyroid cartilage. It has a hinged, doorlike action at the entrance to the larynx. During swallowing it acts as a lid to prevent food from entering the larynx.

The chief function of the larynx is the production of sounds. Two short, fibrous bands called *vocal cords* stretch across the interior of the larynx chamber (Fig. 127). The pitch (high or low) of the sound produced is determined by the shape and tightness of these cords. Long, loose cords produce low-pitched tones, and short, tense cords give higher tones. The voice is modified by the nose, mouth, and throat (pharynx), as well as by the sinuses, which act as sounding boards and vibrational (echo) chambers.

What Is the Trachea? What Is Its Function?

The *trachea,* or "windpipe," is a cylindrical tube about 4 to 5 inches in length, consisting of circular rings of cartilage separated by fibrous and muscular tissue. The trachea functions as a simple passageway for air to reach the lungs. When it becomes blocked from swelling or from *aspiration* (breathing in) of some material, blocking the passage of air, a *tracheotomy* is necessary (Fig. 128).

What Are the Bronchi? What Is Their Function?

The two primary bronchi or *bronchial tubes* split from the trachea, each leading to one of the lungs. The right bronchus differs from the left in that it is shorter and wider and takes a more vertical course.

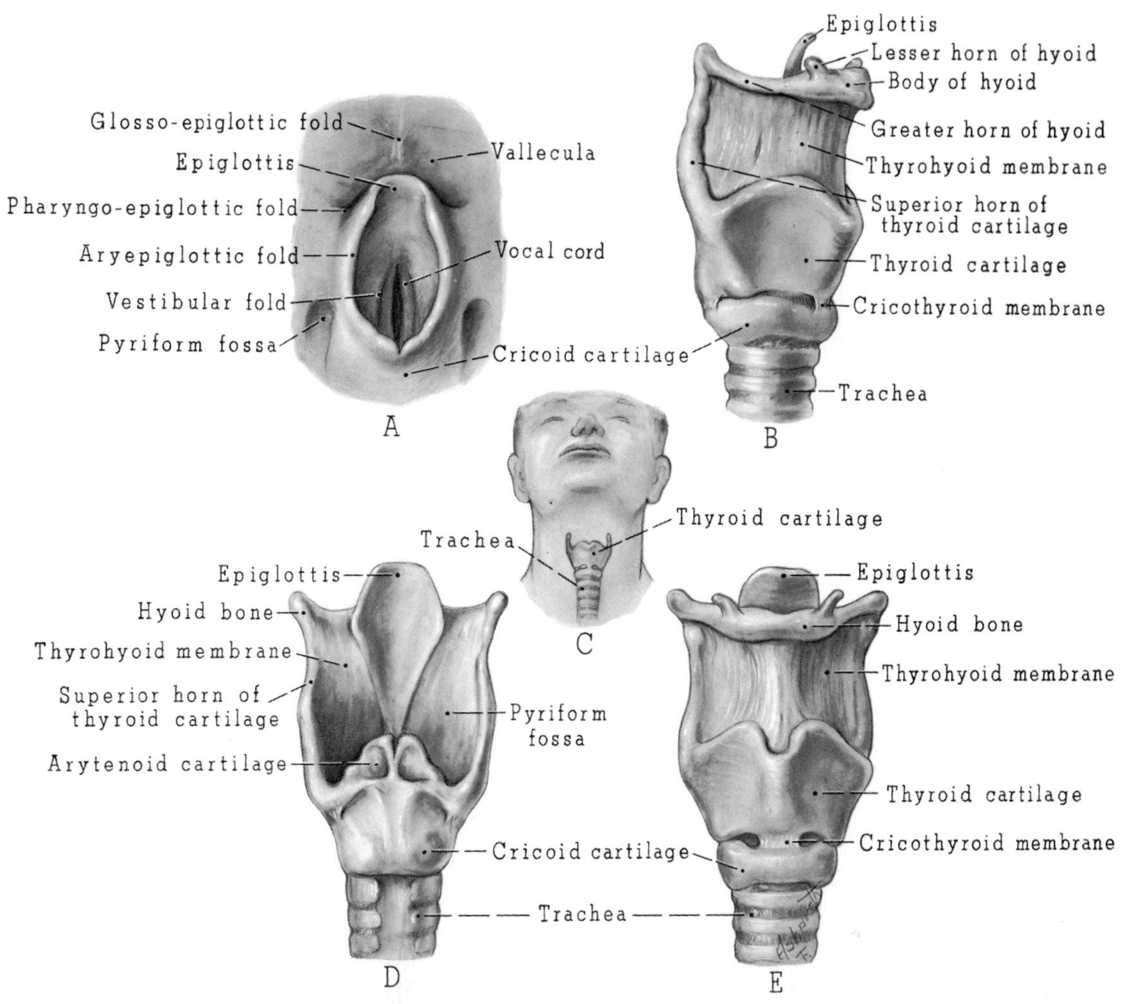

Figure 126. The larynx as viewed from above (*A*), from the side (*B*), in relation to the head and neck (*C*), from behind (*D*), and from the front (*E*).

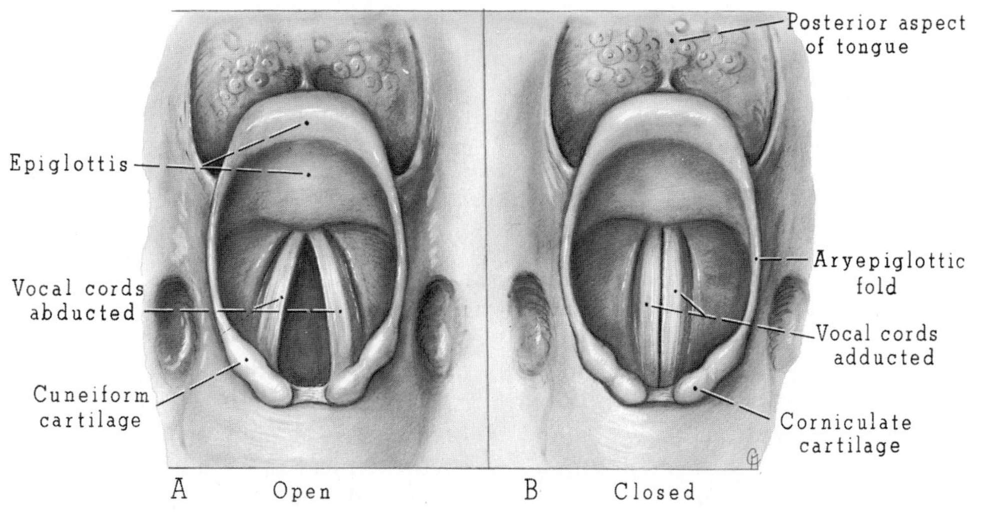

Figure 127. Superior view of vocal cords.

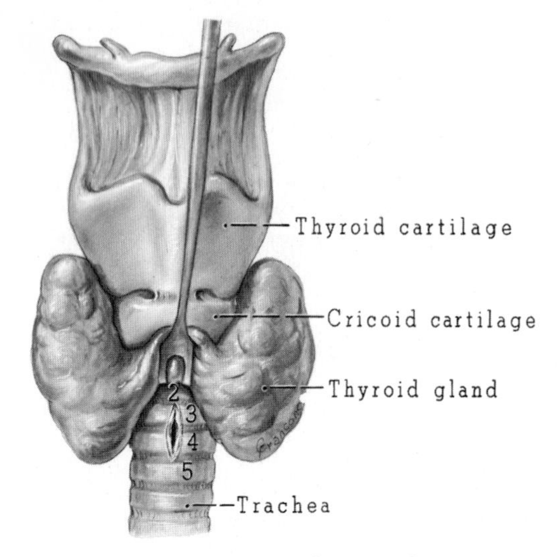

Figure 128. Incision for a tracheotomy.

Because of this vertical characteristic, foreign bodies entering the trachea usually enter the right bronchus (Fig. 129).

Moving down from the trachea through the primary bronchial tubes and into the secondary bronchial tubes, the walls of the passage are composed of less cartilage and more smooth muscle.

What Are the Lungs? How Do They Work?

The lungs are two cone-shaped organs extending from the diaphragm to about 1½ inches above the clavicle (Fig. 129). The large primary bronchi and the pulmonary arteries from the heart enter a slit in each lung called the *hilum.* This entry point at the root of each lung is the only real connection the lungs have to the body itself.

The right lung has three lobes and the left lung has two lobes. The adult lung is a spongy mass and is frequently blue-gray in color because of inhaled dust and soot lodged in the respiratory lymphatics. In contrast, the lung of a baby is pink, since no foreign substances have yet entered. At birth the lungs are filled with fluid. When the first breath is taken the lungs begin to become spongy and eventually fill with air to a degree similar to that of an adult.

What Is An Alveolus?

The thousands of air passages resulting from the repeated branching of the primary bronchi into each lung form a structure resembling an upside-down tree. The smallest bronchial tubes (bronchioles) subdivide into very tiny twiglike tubes called *alveolar* (al-ve′o-lar) *ducts.* Each alveolar duct blossoms out into several *alveolar sacs,* resembling clusters of grapes. Each cluster is made up of numerous *alveoli,* each resembling a single grape in a cluster. The term "alveolus," which is the singular of alveoli, is from the Latin word meaning a small chamber or cavity.

A rich network of capillaries surrounds each alveolus, and it is at this point in the respiratory tract that the exchange of gases between the blood and inhaled air takes place (Figs. 130 to 132).

The interior surface of the lung is by far the most extensive body surface in contact with the environment; its area is many times greater than the skin. In the normal adult this internal lung surface area, when laid out, would cover a regular-sized tennis court.

BREATHING

The process of breathing involves two phases: inspiration and expiration. *Inspiration* requires coordinated muscular contractions that increase chest volume, drawing air into the lungs. *Expiration* is ordinarily entirely passive; the chest volume decreases and air is pushed out of the lungs.

What Are the Mechanisms of Breathing?

There are actually two different mechanisms for breathing; one is called *costal breathing* and the other *diaphragmatic* (di″ah-frag-mat′ik) *breathing.* Costal breathing is shallow and can be identified by the typical upward and outward movement of the chest. It is seen in runners at

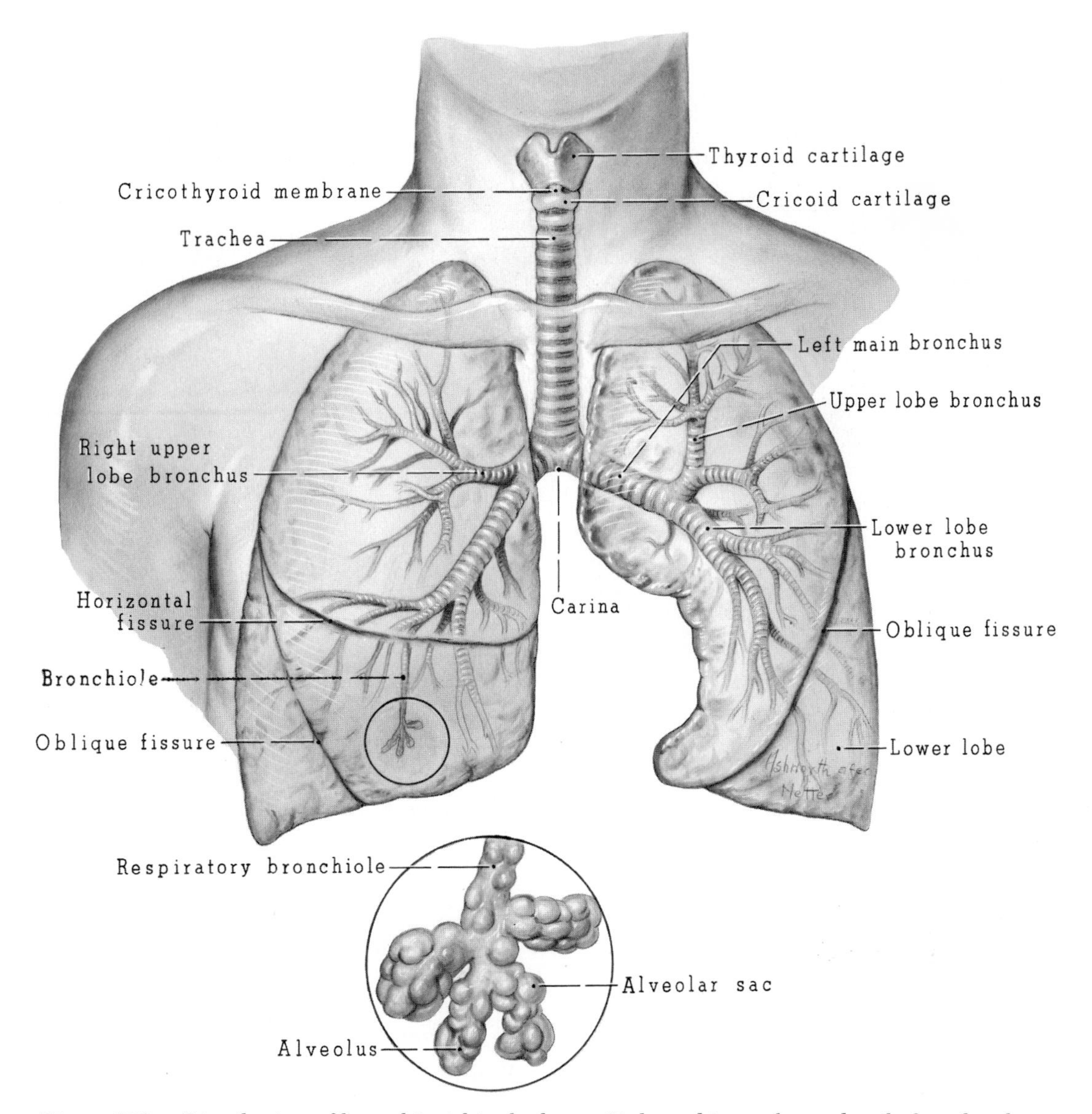

Figure 129. Distribution of bronchi within the lungs. Enlarged inset shows detail of an alveolus.

the conclusion of a race. Costal breathing involves primarily the use of the intercostal (between the ribs) muscles of the chest.

Diaphragmatic breathing, on the other hand, involves the use of the diaphragm rather than the intercostal muscles. Diaphragmatic breathing is deep, and is identified by movement of the abdominal wall, caused by the contraction and descent of the diaphragm. This type of breathing is usually seen during sleep.

What Is Tidal Volume? Vital Capacity?

The normal resting chest volume in a man of average size is about 3 liters (there are 947 milliliters in a quart). Normal inspiration increases this volume by approximately 500 ml. Because this normal breathing volume of 500 ml. comes and goes regularly like the tides of the sea, it is referred to as the *tidal volume* (TV). Forced maximum inspiration raises the

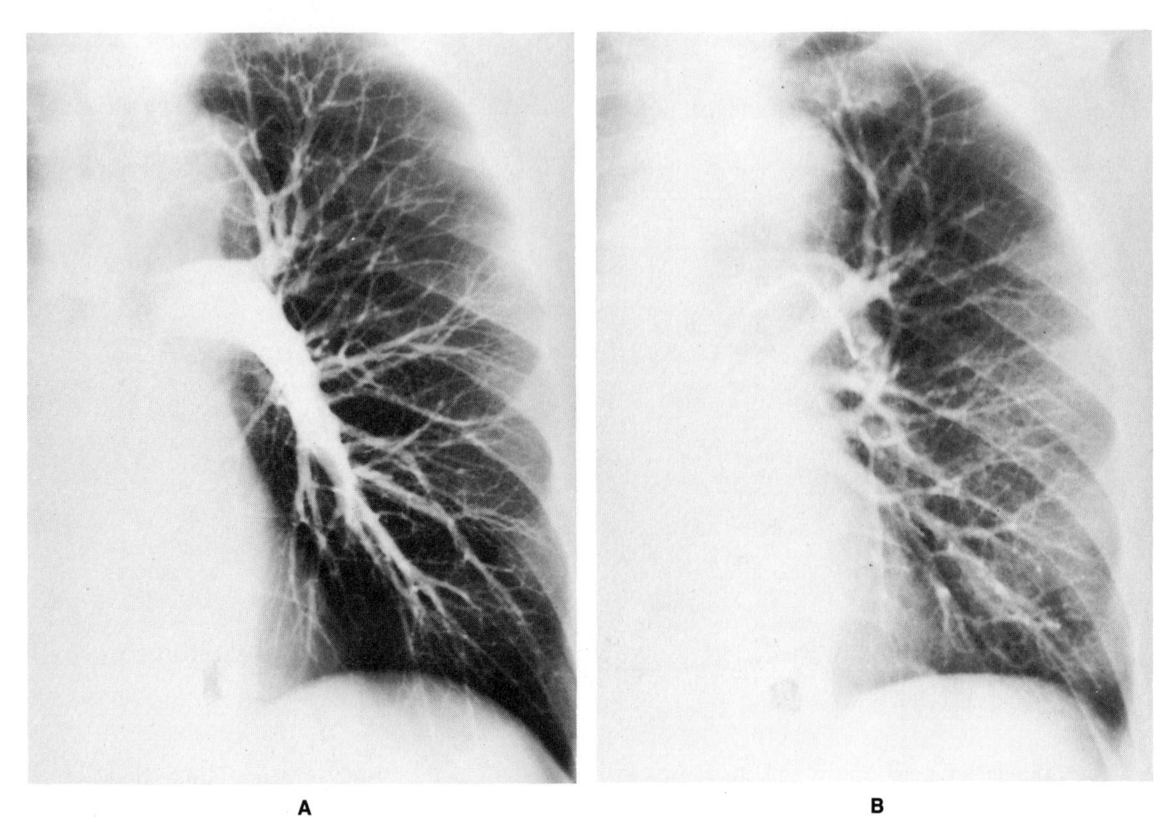

Figure 130. *A*, Angiograph of pulmonary arterial system. *B*, Angiograph of pulmonary venous system.

overall chest capacity to 6 liters. Forced maximum expiration lowers the chest volume to approximately 1 liter. The largest amount of air that we can breathe in and out in one inspiration and expiration, about 5 liters, is known as the *vital capacity* (vc). This is about ten times the tidal volume (Fig. 133).

The total volume of air in the lungs upon maximal inhalation is called *total lung capacity*. The air remaining in the lungs even after maximal forced expiration is known as *residual volume*. The volume of air capable of being inspired at the end of a quiet respiration plus the tidal volume is called the *inspiratory capacity*. The *inspiratory reserve volume* is the volume capable of being inspired after quiet inspiration. The *expiratory reserve volume* is the volume of air capable of being expired at the end of a quiet expiration. *Functional residual capacity* is the expiratory reserve volume plus residual volume.

Clinically, respiratory volumes and capacities are measured by using a *spirometer* (from the Latin words *spirane*, meaning to breathe, and *metrum*, meaning to measure). Spirometric studies provide a graphic record of the volume that can be expelled from the lung after maximum inspiration (vital capacity), and how fast the air can be expired. The spirometer can also be used to record the rate and depth of normal respiration and of activity-related respiration.

How Are Gases Exchanged in the Lungs?

Upon entering the lungs the venous blood moves through the thousands of tiny lung capillaries that surround the thousands of alveoli. During the passage through the capillaries, where the contact between the blood and the air in the lungs is closest, carbon dioxide is released from the blood

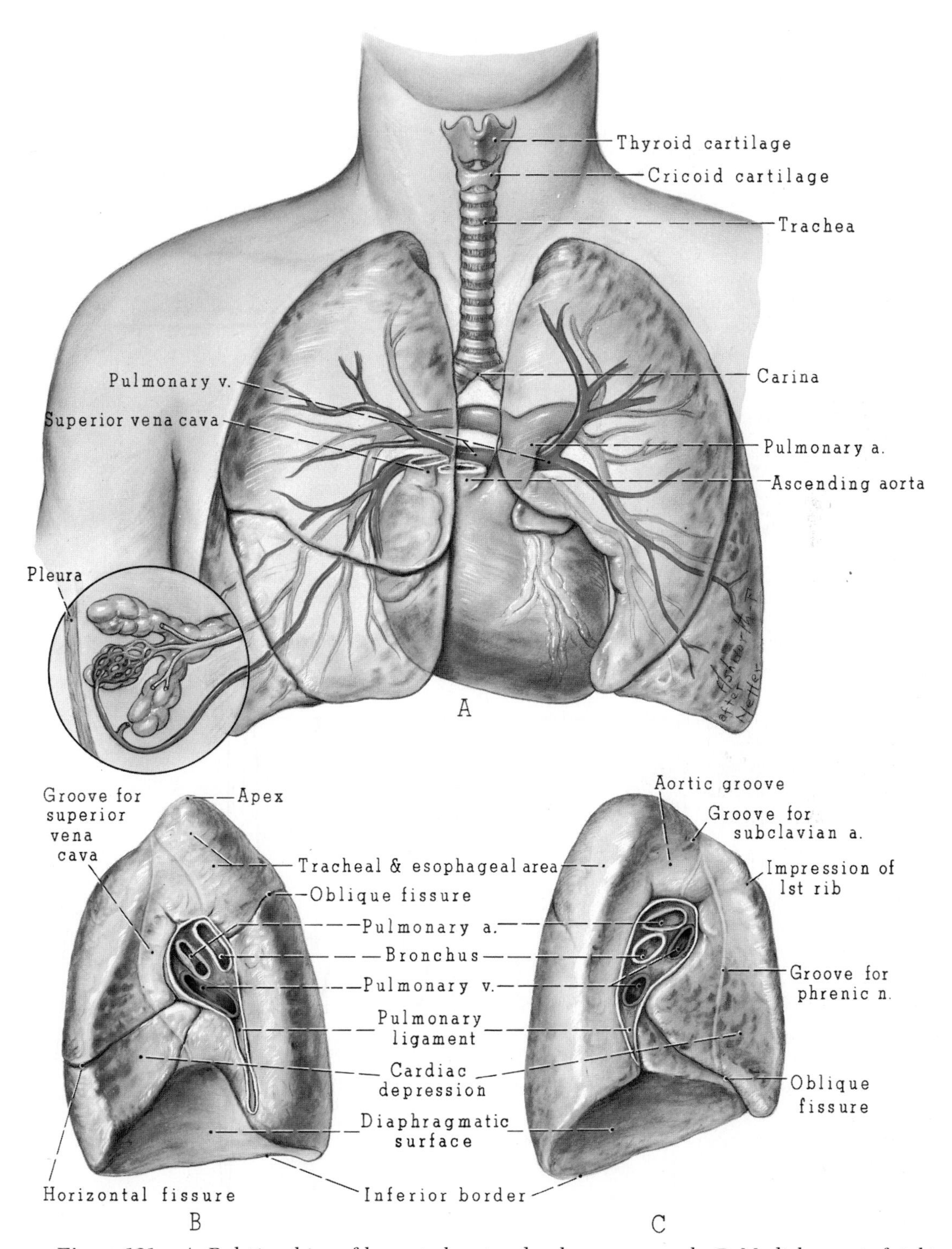

Figure 131. *A*, Relationships of lungs to heart and pulmonary vessels. *B*, Medial aspect of right lung. *C*, Medial aspect of left lung.

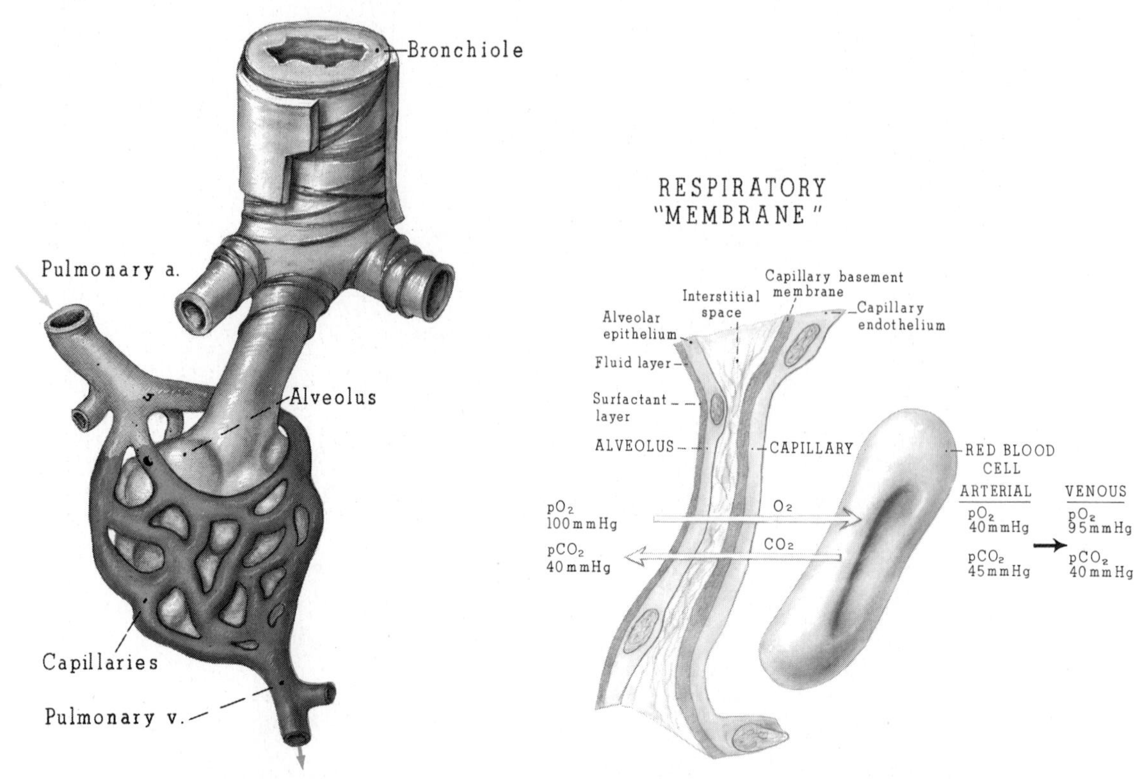

Figure 132. Basic microscopic functional unit of the lung. (Courtesy of Roche Laboratories.)

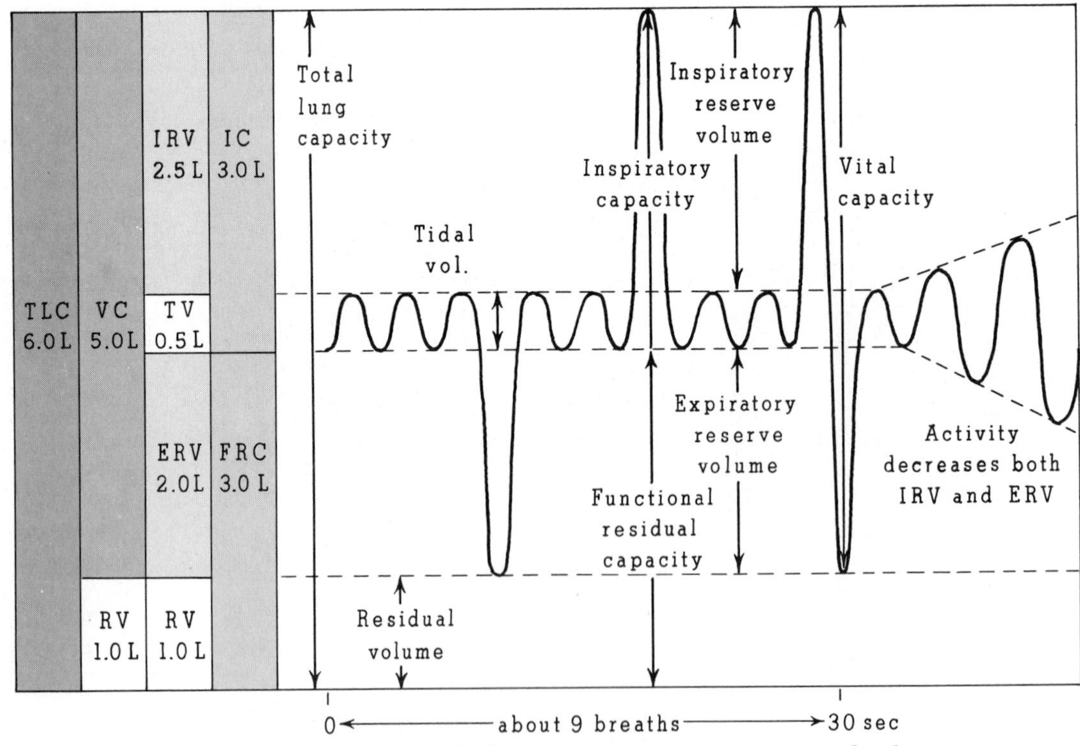

Figure 133. Spirometric graph showing respiratory capacities and volumes.

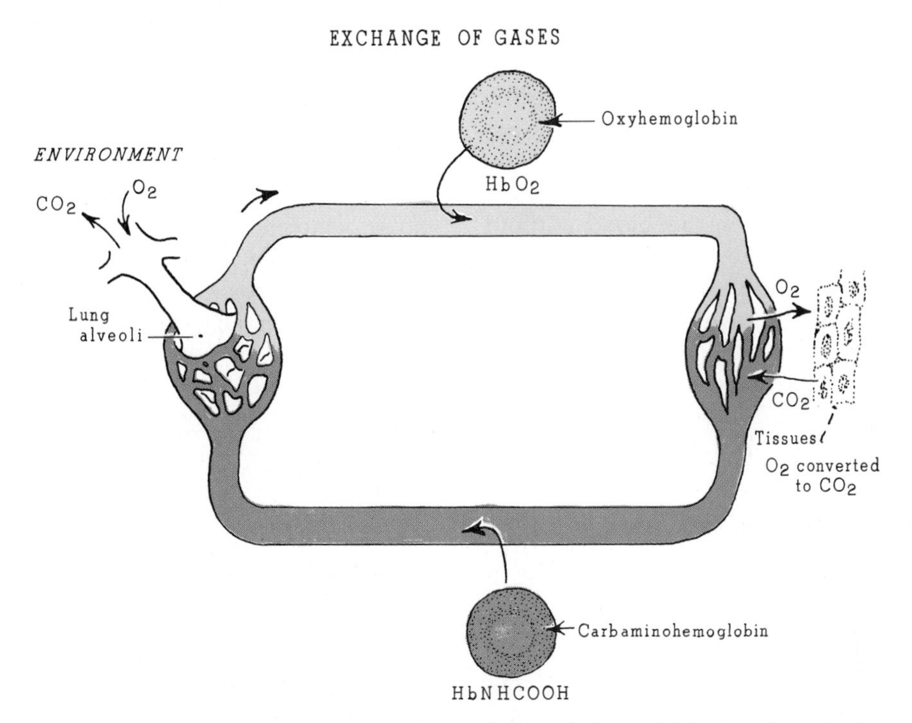

Figure 134. Oxygen combines with the hemoglobin of the red blood cell in the lungs to form oxyhemoglobin. Oxygen is carried to the tissues by the red blood cells in this form. When the red blood cell reaches the tissues, it releases the oxygen and picks up carbon dioxide. The carbon dioxide combines with the hemoglobin to form carbaminohemoglobin. The red blood cell carries the carbon dioxide to the lungs, where it releases it into the air in the lung. The lung operates to exchange gases with the environment.

into the alveolar space in exchange for oxygen, which is picked up (Fig. 134).

The most important mechanism in this exchange is simple diffusion. The concentration of carbon dioxide in the blood is much greater than in the alveolar spaces. As a result, there is a natural diffusion of carbon dioxide out of the blood into the air spaces. Similarly, the concentration of oxygen in the blood entering the lung capillaries is very low compared to the concentration of oxygen in the air spaces. So, again, oxygen diffuses from the lungs into the blood. Of course, if the air drawn into the lungs is high in carbon dioxide and low in oxygen, there will be little or no difference between the blood entering and the blood leaving the lungs. Nature (primarily the green plants), however, provides us with an environment which has very little carbon dioxide and an adequate amount of oxygen to sustain life.

The natural process of diffusion of gases between the alveolar spaces and the blood in the capillaries of the lungs is greatly aided by the action of *hemoglobin.* Hemoglobin has the effect of "fooling" the diffusion process, because when oxygen enters the blood from the alveolar space, the hemoglobin combines with it and "hides" it. As a result, the diffusion process just keeps working, trying to even up the concentration of oxygen in the air and in the blood. This keeps happening until all the hemoglobin is used up in hiding oxygen molecules. Then the concentration of free oxygen in the blood rises until it equals the concentration in the alveolar air. By this rather ingenious process (of hiding oxygen molecules) the blood is able to absorb—by diffusion—about 60 times more oxygen than it would without hemoglobin.

The exchange mechanism for carbon dioxide is quite similar. Some of the carbon dioxide from the cells combines with a special amine molecule and then with hemoglobin in the blood. The large molecule, *carbaminohemoglobin* (kar-bam″i-no-he″-mo-glo′bin), quickly releases the carbon dioxide when it reaches the lungs, the hemoglobin then combining with an oxygen molecule. Some of the carbon dioxide, of course, just dissolves in the blood without being carried from the tissues by a hemoglobin combination. This dissolved carbon dioxide turns out to have much greater significance than the similarly dissolved oxygen, because dissolved carbon dioxide is *acidic* (oxygen is neutral); the importance of this will be discussed below.

How Are the Gases Exchanged in the Tissues?

The exchange of gases in the tissues of the body is a very simple reverse of the exchange of gases in the lungs. Oxygen is released from the blood to the tissue fluid, and carbon dioxide is picked up. The exchange occurs in the capillaries and again the mechanism is simple diffusion, with the amplifying action of hemoglobin.

The concentration of carbon dioxide is normally high in the tissue fluid, and the concentration of oxygen is low; this is because the cells rapidly turn oxygen into carbon dioxide in their normal metabolic activities. Notice that these concentrations are the exact reverse of what is found in the air of the alveolar space (also environment). This simply serves to reverse the direction of diffusion.

What Are Acidosis and Alkalosis?

Acidosis (as-i-do′sis) is the condition in which there is too much acid in the blood. *Alkalosis* (al-kah-lo′sis) is the condition in which there is too little acid in the blood. The normal pH of the blood is 7.4.

Since carbon dioxide is acidic, a condition of acidosis will occur if there is too much carbon dioxide in the blood. This can occur when respiration is restricted and the carbon dioxide given off by the cells begins to build up.

Alkalosis occurs less frequently but can occur when respiration is overactive. For instance, in *hyperventilation* (too much breathing or ventilation) the exchange of gases occurs too rapidly, there is too little carbon dioxide in the blood, and the pH moves (increases to 7.5 or 7.6) toward alkaline. The chemistry of the blood is in a delicate balance in many complex ways, so that the acid-alkaline balance is very important. Most immediately, the acid level of the blood affects the concentration of sodium and potassium, so that an abnormal acid level can produce very serious side effects in a short period of time.

The respiratory system has the responsibility for helping to maintain (homeostasis) the acid level of the blood at the normal pH 7.4. If for any reason the blood becomes too acidic, the respiratory system is stimulated to increase activity to give off more carbon dioxide and thereby to lower the acid level back to normal. If there is too little acid in the blood, respiration slows down, allowing carbon dioxide to build up until a normal acid level is reached again.

By holding your breath for a minute or so you can produce a mild acidosis, and by breathing rapidly (hyperventilation) you can produce a mild alkalosis. However, dangerous abnormalities in acid level of the blood cannot be brought about by these voluntary means, because the automatic parts of the nervous system override the voluntary, and "force" normalization.

What Factors Are Important in the Control of Respiration?

The parts of the nervous system that control respiration are located in the *medulla* (the bulb of the spinal cord) and *pons* in the brain. This general area is known as the *respiratory center*.

Several sets of sensory nerves bring different types of information into the respiratory center. For example, stretch receptors located in the pleura monitor the inspiration and expiration process. The underlying reflex arc, involving the stretch receptors, prevents overexpansion or collapse of the lungs as the extremes of inspiration and expiration are approached.

Several sensory receptors keep track of the CO_2, O_2 and pH levels in the blood. These receptors are known as *chemoreceptors*. If CO_2 concentration gets too high, it stimulates an increased rate of respiration. A similar action occurs if pH gets too low (acidity increase). When O_2 concentration gets too high, the rate and depth of respiration are decreased.

Age is another factor that influences respiratory rate. At birth the rate is rapid—from 40 to 70 times per minute. This decrease with age, so that at about one year, 35 to 40 times per minute is normal; at five years, about 25 times per minute is normal; at ten years, a rate of 20; at 25 years, 16 to 18. With old age, however, the rate can increase again to more than 20 times per minute.

CLINICAL CONSIDERATIONS

The most important respiratory disorders are those in which the blood fails to become oxygenated. Nearly all respiratory problems tend toward *hypoxia* (decreased amount of oxygen in the tissues). Hypoxia (hi-pok′se-ah) sometimes results in *cyanosis* (si″ah-no′sis). Cyanosis refers to the fact that the skin, mucous membranes, and surfaces under the nails turn blue because of an increased presence of deoxygenated hemoglobin in the capillaries. The term "cyanosis" is from the Greek word meaning blue.

Orthopnea (or-thop-ne′ah; from the Greek, meaning straight and breathing) is the inability to breathe in a horizontal position, a condition that often arises in patients with pneumonia. *Hyperpnea* (hi″-perp-ne′ah) is an increased depth of breathing. *Tachypnea* (tak″ip-ne′ah) is excessively rapid and shallow breathing. Figure 135 explains the method of mouth-to-mouth respiration. Figure 136 shows a modern hospital artificial respirator.

What Is Emphysema?

Emphysema (em-fi-se′mah) is a condition in which the alveoli of the lungs are dilated (expanded) and the walls of the alveoli are thin and deteriorated. The condition may result from any factor producing repeated distention (stretching) of the lungs, particularly during expiration. This commonly occurs in patients with asthma or chronic bronchitis, which is an inflammation of the bronchial tubes. Heavy smokers have a high incidence of emphysema. A patient with emphysema has a barrel-shaped chest, and in severe cases the patient exhibits *clubbing* (a broadening and thickening) of the ends of the fingers.

What Is Asthma?

Asthma is a condition of recurrent labored respiration caused by intermittent obstruction of the bronchi; it is characterized by wheezing and prolonged expiration. The outflow of air through the bronchioles is obstructed more than the inflow, permitting the asthmatic to inspire more easily than to expire. With longstanding asthma, the chest becomes barrel-shaped. In children asthma is commonly caused by sensitivities to food; in adults it is frequently caused by sensitivities to pollen. Asthmatic attacks can result from emotional crises as well as from an allergen (substance causing an allergic reaction).

What Is Pneumonia?

Pneumonia is an inflammation of the lungs in which the alveoli are partially filled by fluid and white blood cells. The alveolar walls become inflamed, and there is generalized *edema* (collection of fluid in tissue spaces) of the lung tissue. As in many other pulmonary diseases, carbon

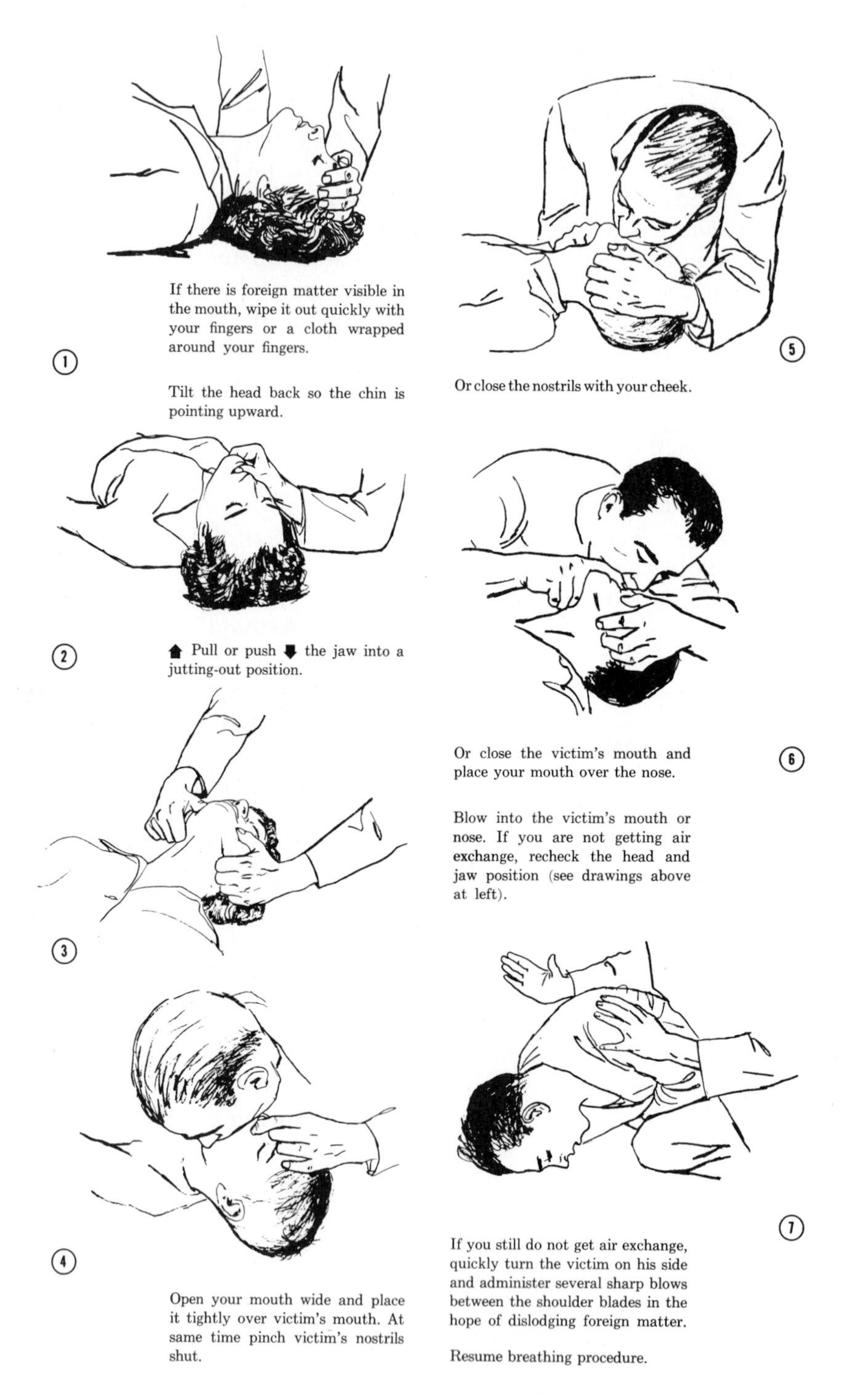

Figure 135. Mouth-to-mouth respiration. (Courtesy of the American National Red Cross.)

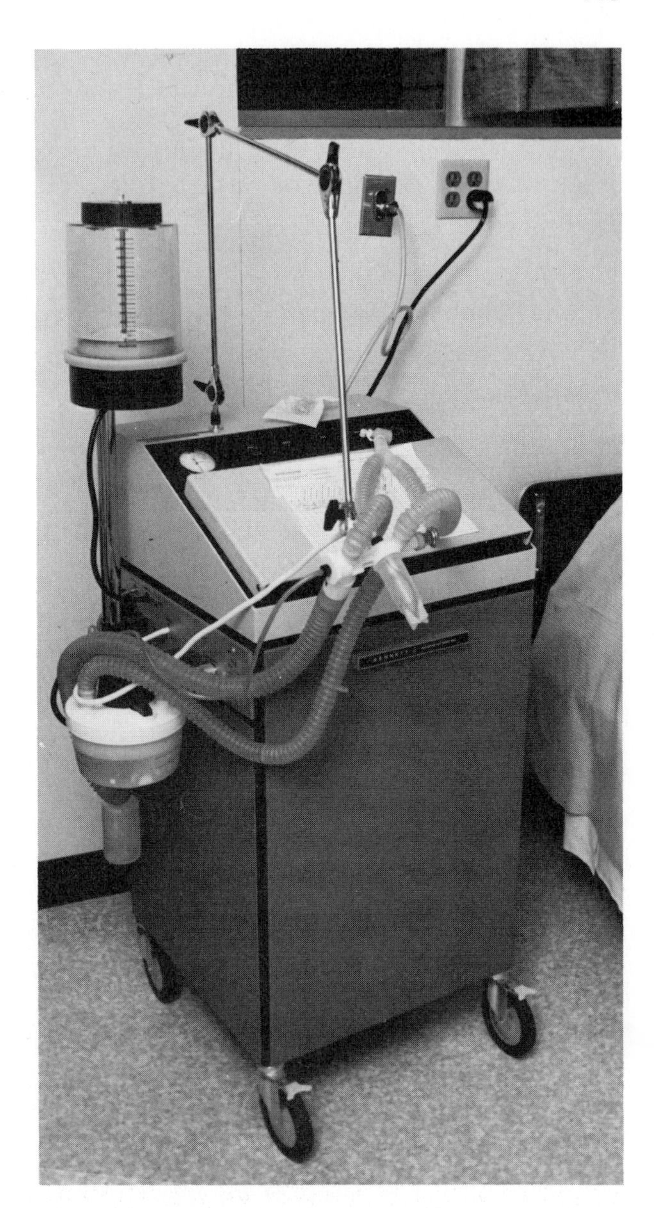

Figure 136. Artificial respirator.

dioxide is adequately excreted, but oxygenation of the blood is diminished. This is caused by the fact that carbon dioxide passes through the alveolar walls about 20 times as readily as does oxygen. Pneumonia occurs most frequently in young children and in the aged. In many cases, the disease is due to an organism, the *pneumonococcus* (nu″mo-no-kok′us); the remainder are due to a variety of common viruses. The word pneumonococcus derives from the Greek words *pneumon,* meaning lung, and *kokkus,* meaning berry.

What Is Tuberculosis?

In tuberculosis (tu-ber″ku-lo′sis), the tubercle bacilli invade the lungs, producing a local tissue reaction. Initially, the area is invaded by macrophages and then becomes walled off by fibrous connective tissue. Thus, a characteristic *"tubercle"* (from the Latin word *tuber,* meaning a swelling) is produced. In the late stages, secondary infection by other bacilli is present, and more *fibrosis* (abnormal building-up of fibrous connective tissue) results. The fibrosis reduces the ability of the lung to expand and decreases the vital capacity. Fibrosis also decreases the working surface area of the lungs, thus decreasing the capacity for exchange of gases by diffusion.

What Is Pulmonary Edema?

Pulmonary edema, the collection of fluids in the tissue spaces of the lung, influences respiration in much the same way as pneumonia does. It is caused by an insufficiency of the left heart in pumping blood from the lungs to the rest of the body. Blood tends to collect in the lungs, and fluid leaks out of the capillaries into the tissue spaces, owing to the increased blood pressure in the lungs. The condition is generally the result of heart failure, in which the blood supply to the heart muscles is impaired, as in atherosclerosis.

LEARNING EXERCISES

1. Construct a diagram of the chest,showing the relation between the pleural cavities and the mediastinum.
2. Draw an outline of the head, and show the locations of the sinuses.
3. Construct a diagram of the upper respiratory tract, and label as many parts as you can.

CHAPTER 9

THE URINARY SYSTEM

FUNCTION OF THE URINARY SYSTEM

What Does the Urinary System Do?

The *urinary system* functions to *eliminate* the wastes of protein metabolism (mostly urea) and to *regulate* the amount of water in the body and the concentrations of a variety of salts in the blood, including sodium, potassium, calcium, phosphate, and chloride. These tasks are accomplished through the formation and elimination of urine. Urine is mostly water—about ninety per cent—however, the precise makeup depends on what substances are in excess in the blood at the time.

The organs responsible for the formation of urine are the two *kidneys*. The *urinary bladder* receives and holds the urine from the kidneys, and periodically eliminates it from the body. The term *renal* is from the Latin word *renalis*, meaning kidney; it is a very common term used normally to refer to anything directly associated with the kidneys.

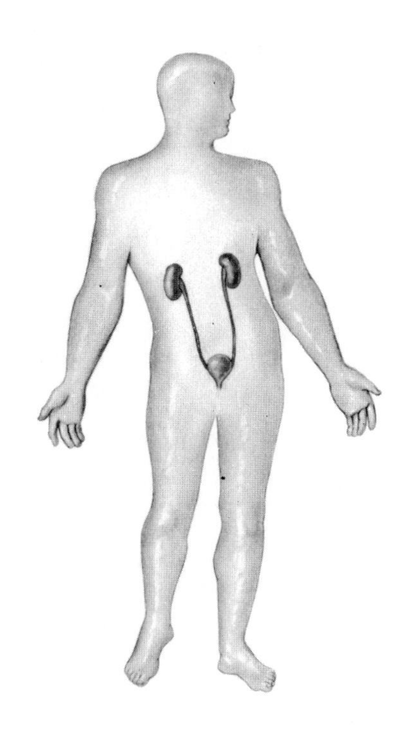

STRUCTURES OF THE URINARY SYSTEM

What Are the Kidneys?

The *kidneys* are two large bean-shaped organs lying behind the abdominal organs against the muscles of the back (Fig. 137). A renal artery, vein, and nerves, as well as lymphatic vessels, enter and leave the concave (inward curving) surface of each kidney through a notch called the *hilus* (hi′lus). The cavity located at the hilus is a urine-collecting portion called the *renal pelvis*. The renal pelvis also forms the expanded upper portion of the *ureter*. The ureter is a long tube that carries the urine from the kidney to the urinary bladder.

What is the Internal Structure of the Kidney?

In cross section, the kidney is seen to have an inner darkened area called the *medulla* and an outer pale area called the *cortex* (Fig. 138). (The word "cortex" comes from the Latin word for bark [of a tree] or

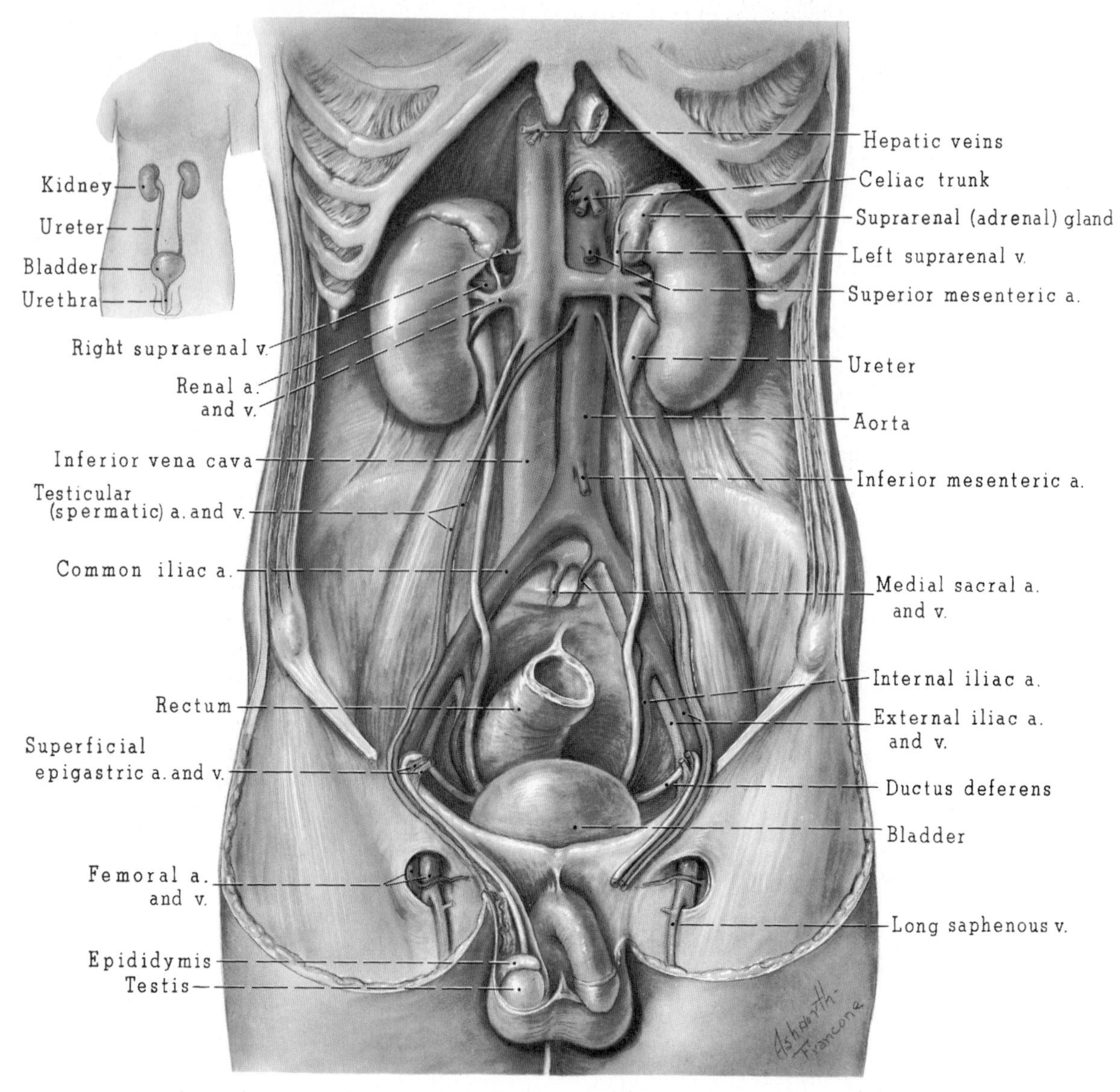

Figure 137. Posterior abdominal wall, showing relationship of urinary system, genital system, and great vessels.

rind [of an orange] so the cortex of an organ—whether it be the kidney, the brain or the adrenal glands—is its outer layer. Similarly, the medulla of an organ is always its inner portion.)

The *cortex* contains the working parts of the kidney, the *nephrons* each of which is a tiny urine producing factory. Nephrons have three main parts: a glomerulus, a Bowman's capsule, and a tubule, or renal tubule (Fig. 139).

The *glomerulus* (the Latin word *glomero* means to wind into a ball) is a tightly interwoven network (or tuft) of blood capillaries fitted inside a cuplike structure called *Bowman's capsule.* Extending from Bowman's capsule is a long *renal tubule.* The term "tubule," pronounced too′byool, means little tube.

The first segment of each renal tubule is called the *proximal convoluted tubule*—proximal (next to) because it lies nearest the

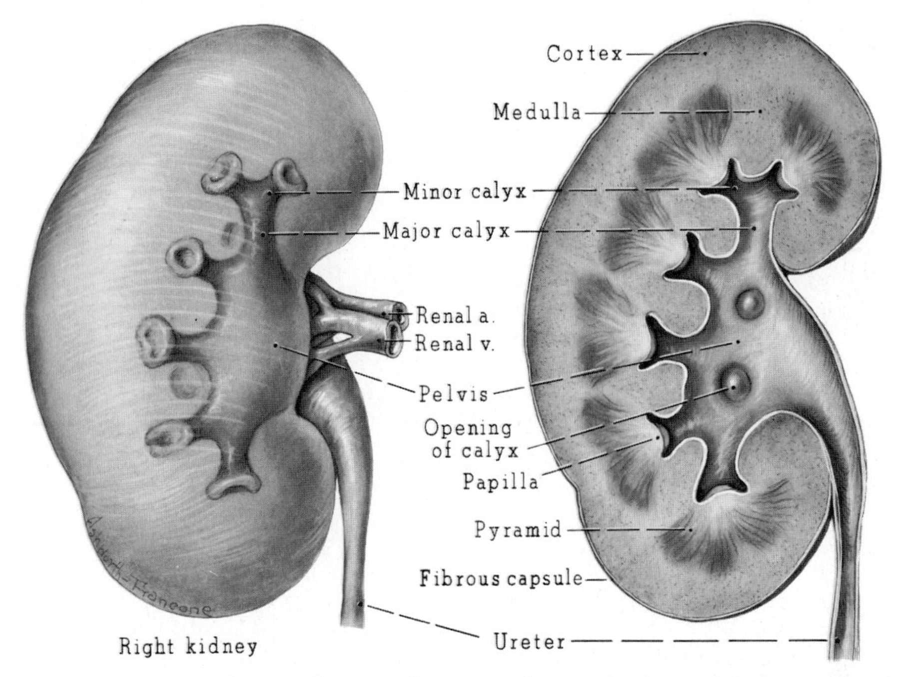

Figure 138. Entire and sagittal views showing relation of calyces to kidney as a whole.

tubule's origin from Bowman's capsule, and *convoluted* (Latin meaning, rolled together) because it twists around to form several coils. Next, the tubule forms a loop called the *loop of Henle* (hen′le), after its discoverer. After that comes a larger, collecting tube serving many nephrons (Fig. 140).

In the *renal medulla* one finds a branching out of the renal artery which feeds into the (outer) cortex. Separately, the urine collecting tubes from the cortex come together in the medulla, finally forming the renal pelvis, which extends to form the ureter.

How Does the Nephron Work?

Three processes are involved in the formation of urine by the nephron. The simplest to understand is the first step, glomerular filtration. But the real magic of the nephron occurs in the next two processes, which take place in the renal tubules, tubular reabsorption and tubular secretion (Fig. 141).

Glomerular Filtration. The thin walls of the capillaries of the glomerulus act like a semipermeable membrane. They permit a *protein-free plasma filtrate* to pass out of the blood in the capillaries into Bowman's capsule. The mechanisms underlying the process of glomerular filtration are essentially passive, and can be explained mostly in terms of simple diffusion under pressure.

The specific rate of glomerular filtration varies directly with filtration pressure, which is dependent on blood pressure. If blood pressure falls too low, glomerular filtration will stop; the kidneys will shut down. Normally, approximately 1200 ml. of blood or about one fourth of the total cardiac output is filtered through the kidneys each minute. Of this, the outflow of protein-free plasma from all the glomerate of both kidneys is about 125 ml., or about 1 ml. of filtrate for every 10 ml. of blood filtered per minute. If all this became urine, your bladder would be completely filled every three to five minutes. Surely most of this must be reabsorbed back into the blood stream. But how?

Tubular Reabsorption. Of the 125 ml.

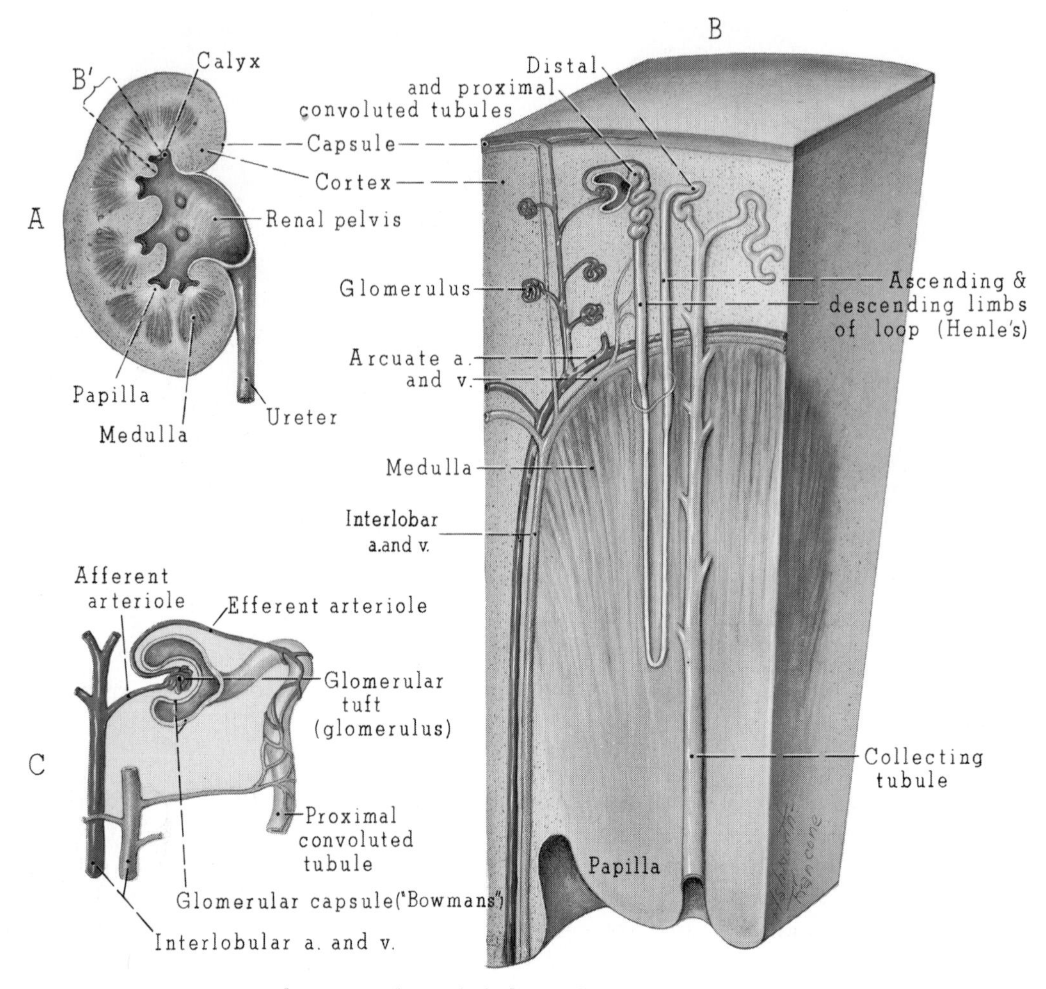

Figure 139. *A*, Sagittal section through kidney showing gross structure (note pelvis, calyces, medulla, cortex). *B*, Nephron and its relationship to medulla and cortex. The dotted lines in *A* show the area of the kidney from which this section was taken. *C*, Magnified view of nephron.

of glomerular filtrate normally formed each minute, approximately 124 ml. is reabsorbed by the cells of the renal tubules and transported back into the blood capillaries surrounding the tubules (see Fig. 140). Only 1 ml. eventually passes on to the bladder as urine. Different parts of the renal tubule specialize in reabsorbing different substances from the filtrate. Also, the rate at which reabsorption of specific substances takes place can be increased or decreased, depending on various factors. For example, when the amount of water in the body is low (dehydration), antidiuretic hormone (ADH) is secreted by a gland in the midbrain (the pituitary), which stimulates the cells of the distal tubule to reabsorb even more water than normal from the filtrate. More water then flows back from the nephron (more specifically, from the tubular filtrate) into the blood, and less urine, or more concentrated urine, is formed. In this way the body prevents further loss of water (and further dehydration) through the urine.

Tubular Secretion. Tubular secretion is the reverse of tubular reabsorption. However, the substances secreted are most often different from those that are reabsorbed. In the process of tubular secretion

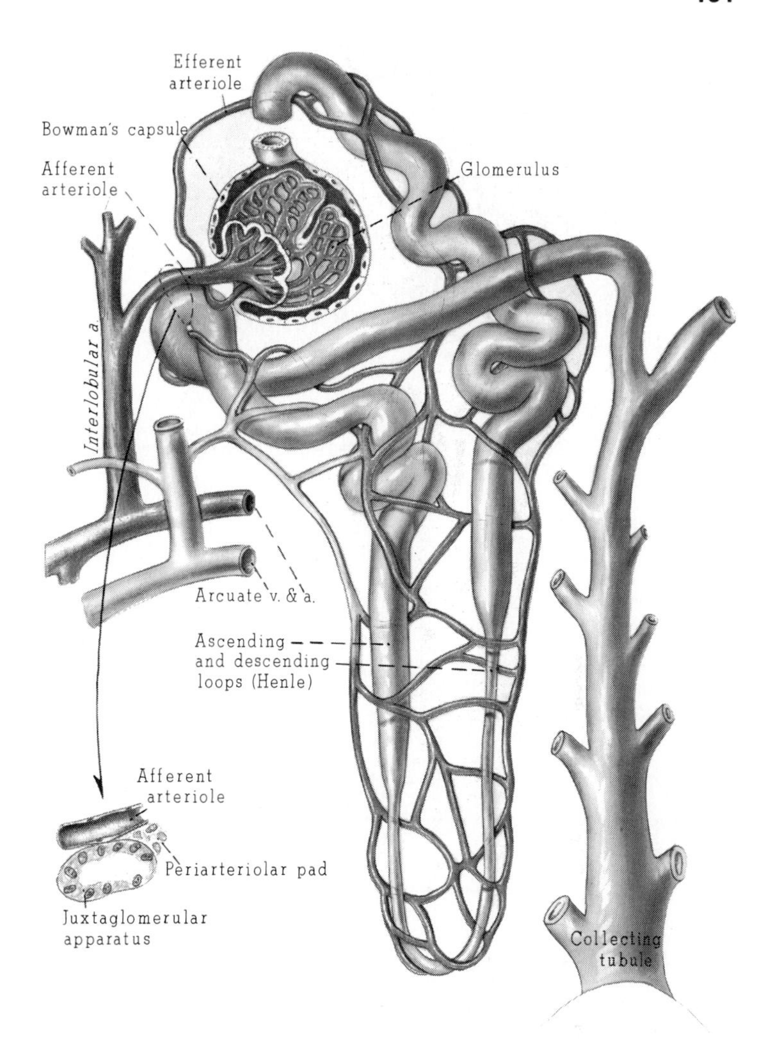

Figure 140. Detail of nephron showing vascular supply, juxtaglomerular apparatus, and tubule.

the renal tubular cells—that is, the cells forming the tubules—take substances from the surrounding capillaries and add them to the filtrate inside the tubules. This increases the concentration of these particular substances in the filtrate (soon to be urine). One of the most important substances secreted into the filtrate by the tubular cells is *hydrogen ion.* This is the active part of all acid substances. When the blood becomes too acidic the proximal tubules are very active in increasing the elimination of hydrogen ion (the ion of acid) from the blood. The filtration of the blood at the glomerulus, of course, captures many hydrogen ions, but the cells of the renal tubules go even farther by actually drawing these ions out of the blood capillaries in the area and then secreting them into the filtrate fluid. This process can make the urine very acid, but this is much better than having an acidic blood.

The renal elimination of acid from the blood is slower to respond to quick increases in blood activity than is the respiratory system. Recall that the respiratory system can increase its activity to eliminate carbon dioxide through the lungs, and since carbon dioxide has an acidic effect while in the blood ($CO_2 + H_2O \rightarrow H^+ + HCO_3$) this helps to reduce the acidity. Over a period of time the kidneys can eliminate a larger amount of acid than can the respiratory system, since the respiratory system is

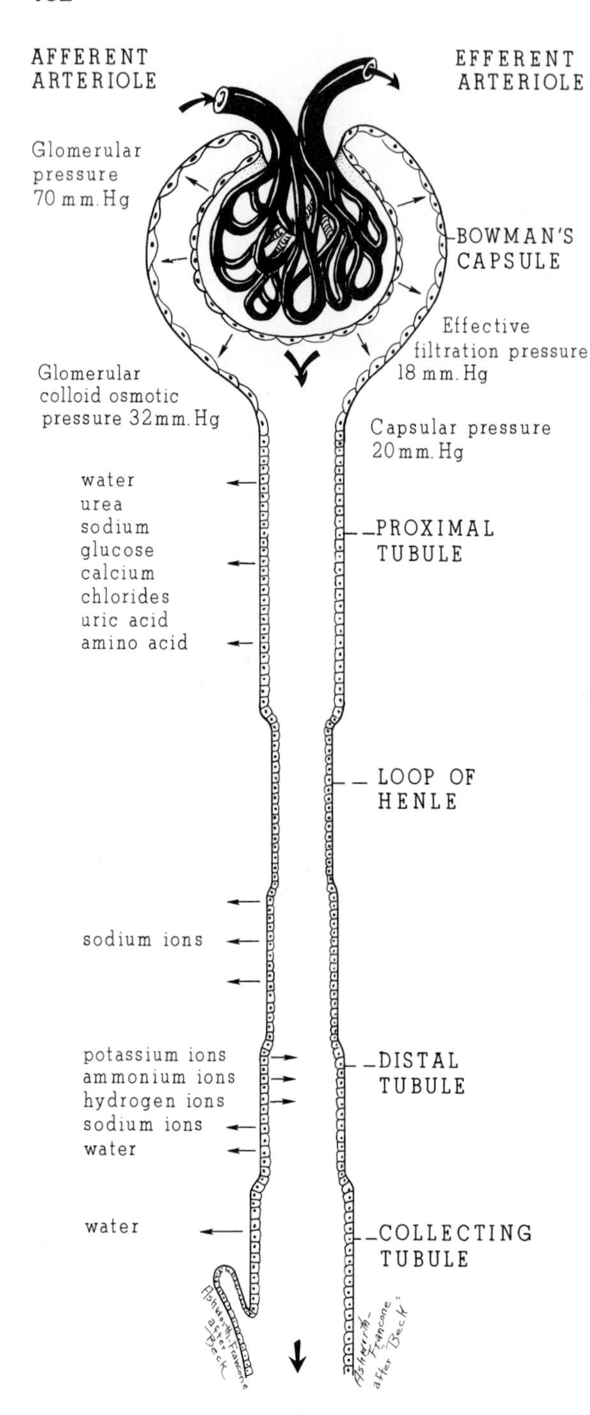

Figure 141. The normal filtration pressure is about 18 mm. of mercury. Glomerular hydrostatic pressure (70 mm. Hg) minus glomerular colloid osmotic pressure (32 mm. Hg) minus capsular pressure (20 mm. Hg) equals filtration pressure, 18 mm. Hg. The passage of substances in and out of the tubule varies in different portions of the tubule and collecting duct.

limited by the amount of carbon dioxide that can be safely eliminated. (Table 7 summarizes the function of different parts of the nephron. Table 8 summarizes the composition of the urine.)

What Controls Kidney Function?

Since it is the function of the kidneys to regulate the concentration of many substances in the blood, it is necessary for the

Table 7. FUNCTIONS OF DIFFERENT PARTS OF NEPHRON

PART OF NEPHRON	FUNCTION
Glomerulus	Produces filtrate of protein-free plasma
Proximal convoluted tubule and loop of Henle	Absorption of Na^+, K^+, and glucose by active transport; Absorption of Cl^- by diffusion; Obligatory water absorption by osmosis Secretion of certain drugs such as penicillin and Diodrast, by active transport
Distal convoluted tubule	Absorption of Na by active transport; Secretion of H^+, K^+
Collecting tubule	Water absorption by osmosis; ADH-controlled absorption of water

tubules to be highly selective in what they reabsorb and what they secrete. Some of the selective elimination by the nephrons is simply passive. For instance, when there is an excess of phosphate in the blood, more phosphate is naturally eliminated, and when there is too little, a correspondingly smaller amount is found in the urine. This occurs naturally through the general filtering action of the glomerulus.

The body also has ways of actively influencing both the amount and composition of the urine that is produced by the kidneys. This active influencing occurs primarily with the control of the amount of water and sodium that are reabsorbed (from filtrate back to the blood) by the renal tubules. For instance, as mentioned before, when the body needs to retain water more than it needs to form urine (as in dehydration), the pituitary gland (see Chapter 12, The Endocrine System) secretes a chemical stimulant or *hormone* (specifically, antidiuretic hormone or ADH). This hormone stimulates the cells of the renal tubules to reabsorb even more water than normal. ADH is sometimes referred to as the "water-retaining hormone".

Another hormone, *aldosterone,* is secreted by the adrenal gland when sodium concentration in the blood is low; for example, when a person drinks a lot of water without taking salt after perspiring all day. Aldosterone stimulates the cells of the renal tubules to reabsorb almost all sodium from the glomerular filtrate. This prevents any further loss of sodium by passage into the urine. Aldosterone is sometimes referred to as the "sodium-retaining hormone".

A *diuretic* (from *dia,* meaning intensive, and *uresis,* meaning urination) is any substance that causes a significant increase in the amount of urine formed. Diuretics also have the opposite effect of ADH, since in order to form more urine you need

Table 8. COMPOSITION OF URINE

Solutes 60 gm. daily	Organic wastes 35 gm.	Urea	30 gm.
		Creatinine	1–2 gm.
		Ammonia	1–2 gm.
		Uric acid	1 gm.
		Others	1 gm.
	Inorganic salts[*] 25 gm.	Chloride	Sodium
		Sulfate	Potassium
		Phosphorus	Magnesium

[*] Sodium chloride is the chief inorganic salt in urine.

more water. Diuretics are often used to treat patients who have abnormal water retention. For example, *edema* (from the Greek word meaning swelling) is a condition in which abnormal blood pressure and/or osmotic forces lead to a buildup of water in the tissue spaces between cells. Administration of a diuretic can help to restore a more proper water distribution. In such a case, however, it is important to understand that the diuretic does not cure whatever caused the edema in the first place but rather only helps to alleviate the serious resulting symptom.

What Are the Ureters?

The *ureters* are two tubes that function to convey urine from the kidneys to the urinary bladder. Urine first drains out of the renal tubules into the larger, collecting tubes and finally into the *renal pelvis*. The renal pelvis forms the funnel-shaped upper end of the ureter. The ureter proper passes from the renal pelvis to the backside of the urinary bladder. Each ureter is about 1/4 inch in diameter and is 10 to 12 inches long—the distance between the kidneys and the bladder—and consists of outer fibrous, middle muscular, and inner mucous layers. Contractions of the muscular layer produce *peristaltic* (from the Latin words meaning around and constriction) waves carrying urine from the renal pelvis to the urinary bladder.

What Is the Urinary Bladder?

The *urinary bladder* serves as a holding reservoir and organ of elimination for the urine produced by the kidneys (Fig. 142). As the bladder gradually fills, its elastic, muscular walls become *distended* (stretched). In the distended state, the muscular wall partially contracts, and the pressure within the bladder increases. The normal capacity of the bladder is slightly more than 1/2 pint (300 to 350 ml).

As the volume gradually increases, the tension continues to rise. Finally, stretch and tension receptors are stimulated, producing a desire to urinate. Voluntary control can be exerted until the bladder pressure increases to the point at which involuntary voiding occurs. Normal *micturition* (urination) is under voluntary control. Pressure sufficient to accomplish voiding is created by contraction of the *detrusor mus-*

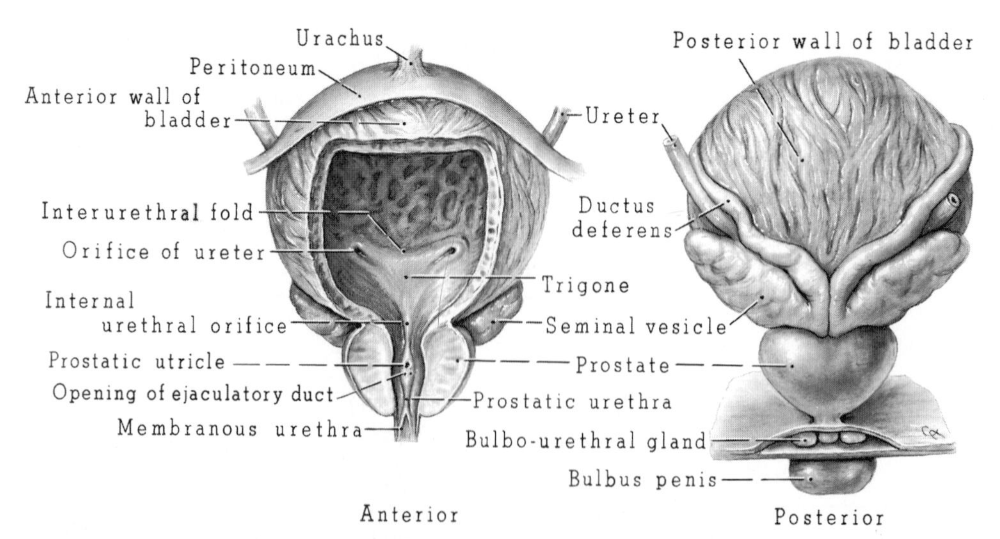

Figure 142. Internal and external aspects of the urinary bladder and related structures.

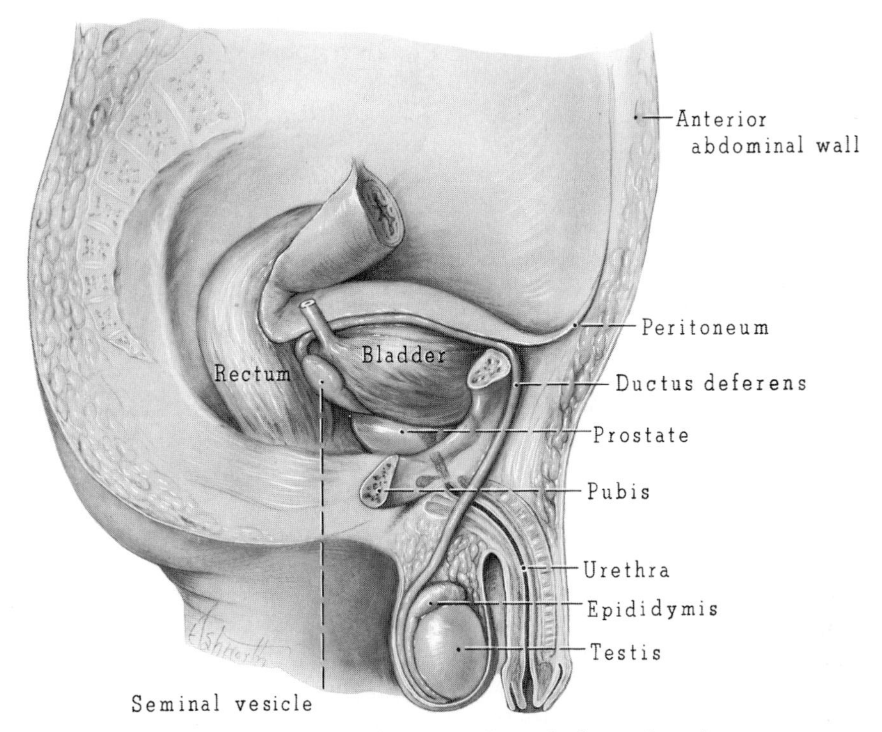

Figure 143. Sagittal section through the male pelvis.

cles (bladder muscles), the abdominal wall, fixation of the chest wall and diaphragm, and relaxation of the urethral musculature.

What Is the Urethra?

The *urethra* is the tube leading from the bladder to the exterior of the body. It is currently believed that the entire urethra in the female, and the prostate and urethra in the male, function as the *sphincter* (muscle that closes off) of the bladder (Fig. 143). When urination begins the musculature of the urethra relaxes and urine is forced down the narrow urethral tube.

In the female, the urethra is 1 1/2 inches in length and serves only a urinary function. The male urethra is about 8 inches long and also serves in the reproductive system as a passageway for semen.

The mucous membrane that lines the renal pelvis, ureters, and bladder also lines the urethra—it is one continuous cellular sheet. This helps to explain the fact that an infection in the urethra may spread upward throughout the urinary tract to the kidneys.

CLINICAL CONSIDERATIONS

The clinical terms used to describe the production, or lack of production, of urine by a patient during a specified period are *anuria,* meaning literally, the absence of urine, *oliguria,* meaning very little urine, and *polyuria,* meaning unusually large amounts of urine.

Uremia, which literally means "urea in the blood," is a symptom of renal insufficiency. One of the characteristic signs of uremia is a "uremic" odor to the breath caused by ammonia (NH_3) produced by the breakdown of urea.

What Are Some Common Disorders of the Kidneys?

Nephritis refers to any inflammation of the kidneys, and is relatively common.

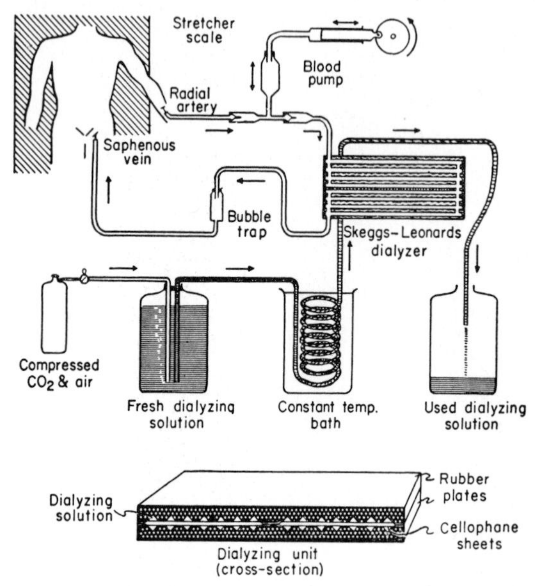

Figure 144. Schematic diagram of the Skeggs-Leonards artificial kidney.

Glomerulonephritis (glo-mer″u-lo-ne-fri′tis), an inflammation that specifically involves the glomeruli, generally develops in persons under 20 years of age, 10 to 20 days after an acute infection. Most patients recover spontaneously (without treatment). Chronic (or recurrent) glomerulonephritis can in later life result in hypertension and uremia. *Pyelonephritis* (pi″e-lo-ne-fri′tis) is an inflammation of the kidney with special involvement of the renal pelvis; *pyelo* is the Greek word for trough or pelvis.

Nephrosis is a degenerative disease of the renal tubules. A *degenerative disease* is one that tends to bring about the gradual destruction of the structures involved.

Cystitis is an inflammation of the mucous lining of the urinary bladder.

What Is Hemodialysis?

Hemodialysis (from the Latin, meaning to separate the blood) or dialysis, for short, is an increasingly common therapy used to maintain patients whose kidneys have either stopped working altogether or are not functioning adequately. It works by the simple principle of diffusion. The dialysis

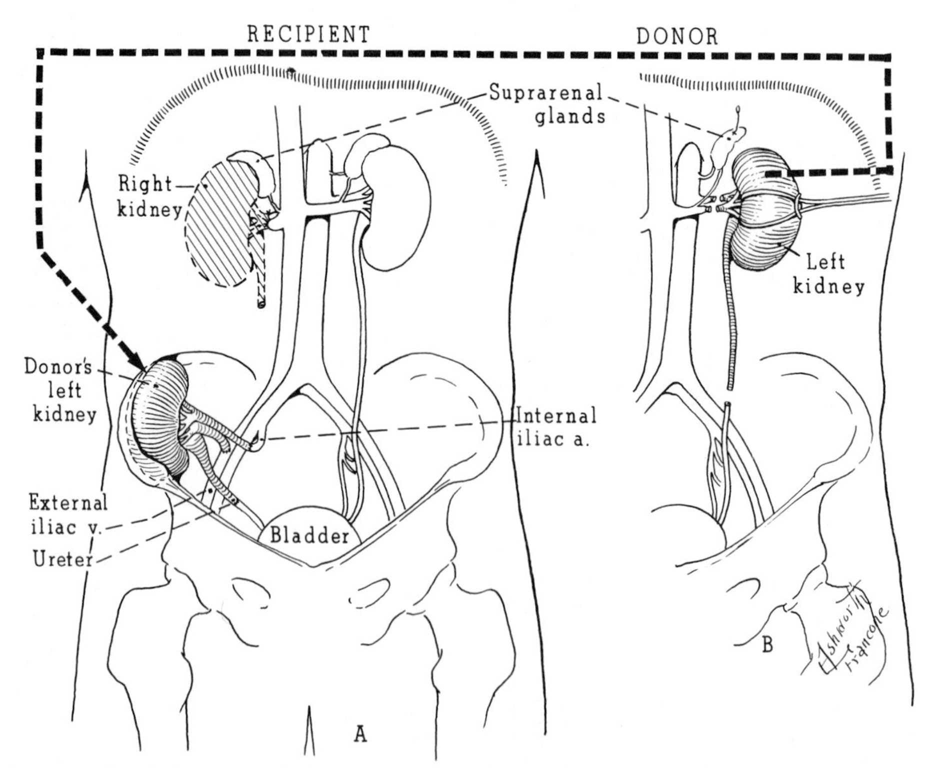

Figure 145. *A* illustrates kidney transplanted to right pelvis. *B* shows kidney of donor.

machine uses a semipermeable membrane between the blood from the patient and a wash solution (Fig. 144).

If there is a relatively high level of any substance in the blood of a patient, and none in the wash solution, that substance will diffuse from the patient's blood into the wash solution. Urea, potassium, phosphate, and other molecules present in toxic (poisonous) quantities in the uremic patient can thus be removed by hemodialysis. It is now possible by hemodialysis to keep patients alive without any functioning kidney tissue for as long as five years.

What About Kidney Transplant?

During the last decade several thousand homologous kidney transplants have been done on patients dying of kidney failure. "Homologous transplants" are those between individuals of the same species who are not identical twins. A system of tissue typing, similar to the ABO blood-type system, has been developed to increase the likelihood of a "take." The use of drugs to suppress the immune response to the inevitable foreign proteins of the transplanted kidney has allowed excellent long term function in a majority of these patients. The best results in homotransplantation occur when the donor is closely related to the recipient; 90 per cent of the kidneys survive for one year or longer. Kidney transplants between identical twins may function almost indefinitely (Fig. 145).

LEARNING EXERCISES

1. Construct a diagram of the urinary system, showing the locations and relations between the kidneys, ureters, bladder, and urethra.
2. Make a schematic diagram of a nephron, and label the major parts.

CHAPTER 10

THE DIGESTIVE SYSTEM

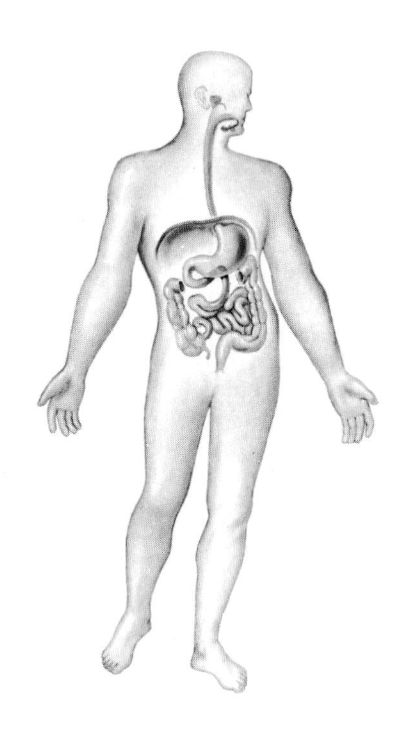

THE DIGESTIVE SYSTEM: AN OVERVIEW

What Does the Digestive System Do?

The digestive system has three primary functions: digestion, absorption, and elimination. *Digestion* (from the Latin word meaning to divide into parts) is the process of breaking down large food molecules into the simple nutrient molecules that can be used by the cells. *Absorption* is the process by which the simple nutrient molecules are transferred from the digestive tract into the blood stream for delivery to the cells. *Elimination* is the process of passing the leftover solid wastes of ingested foods from the body.

What is the Peritoneum?

The peritoneum (per″i-to-ne′um) is a lubricated, double-membrane system. One layer of the peritoneum, known as the *parietal* (pah-ri′e-tal) *peritoneum,* lines the abdominal cavity. The other layer, known as the *visceral* (vis′er-al) *peritoneum,* forms a covering that adheres to the surface of each abdominal organ. Between these two membrane layers is a small amount of lubricating serous fluid. The main function of the peritoneal membrane system is to allow the abdominal organs to slide freely and easily against each other during breathing and digestive movements. Without this system, frictional irritation would develop. Notice the similarity between the peritoneum and the pleural membrane system discussed in Chapter 8, The Respiratory System.

What Structures Make Up the Digestive System?

The digestive system consists of (1) a long, muscular *tube* beginning at the lips and mouth and ending at the anus, and includes the pharynx, esophagus, stomach, and intestines; and (2) certain large accessory glands located outside the digestive tube, including the salivary glands, liver, gallbladder, and pancreas—each of which

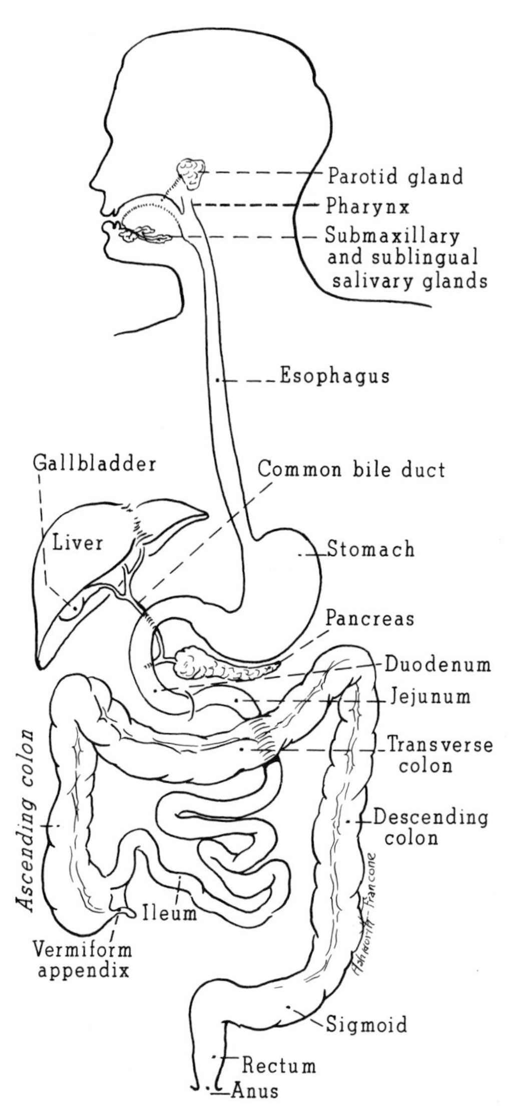

Figure 146. The digestive system and its associated structures.

secretes its special digestive juice into the digestive tube (Fig. 146).

Other terms used frequently to refer to the digestive tube running from the mouth to the anus are the digestive tract and the gastrointestinal tract, or the G.I. tract for short. The term "gastro" or "gastric" is often used to refer to the stomach, its action or products. Alimentary (from the Latin word meaning to nourish) tract is another term referring to the digestive tract that you will come across.

What Are the Digestive Functions of the Upper Portions of the Tract?

In the upper portions of the digestive tract, food is received into the *mouth,* (Fig. 147) where the *tongue* functions to mix it with saliva from the salivary glands and to keep the mass pressed between the *teeth* for chewing (Figs. 148 and 149). In the process of swallowing the tongue pushes the food back into the throat, initiating a wave of muscular contraction that propels the mixture to the stomach. The *pharynx* (far'inks) and *esophagus* are muscular tubes that convey the chewed food from the mouth to the stomach. The passage of food from the upper portions of the digestive tract to the stomach is normally aided by gravitational forces; however, the type and arrangement of muscles in the pharynx and esophagus allow swallowing even in the weightless environment of outer space.

THE STOMACH

What is the Nature of the Stomach?

The *stomach* is the most widened or enlarged portion of the digestive tube. It is located just below the diaphragm on the left side of the body (Fig. 150). Its three major functions are to store food; to mix food with gastric secretions until the semifluid mass of partly digested food called *chyme* (kime), Greek word for juice, is formed; and to permit the chyme to slowly empty into the duodenum at a rate suitable for proper digestion and absorption by the small intestine.

When empty, the stomach is only about the size of a large sausage. After a meal, however, it expands considerably. There are three parts or sections to the stomach: the *fundus,* an upper portion ballooning to the left side of the body; the *body* or central portion; and the *pylorus* (pi-lo'rus), a relatively narrowed portion at the end of the stomach just before the entrance into the duodenum (small intestine). The term "pylorus" is from the Greek

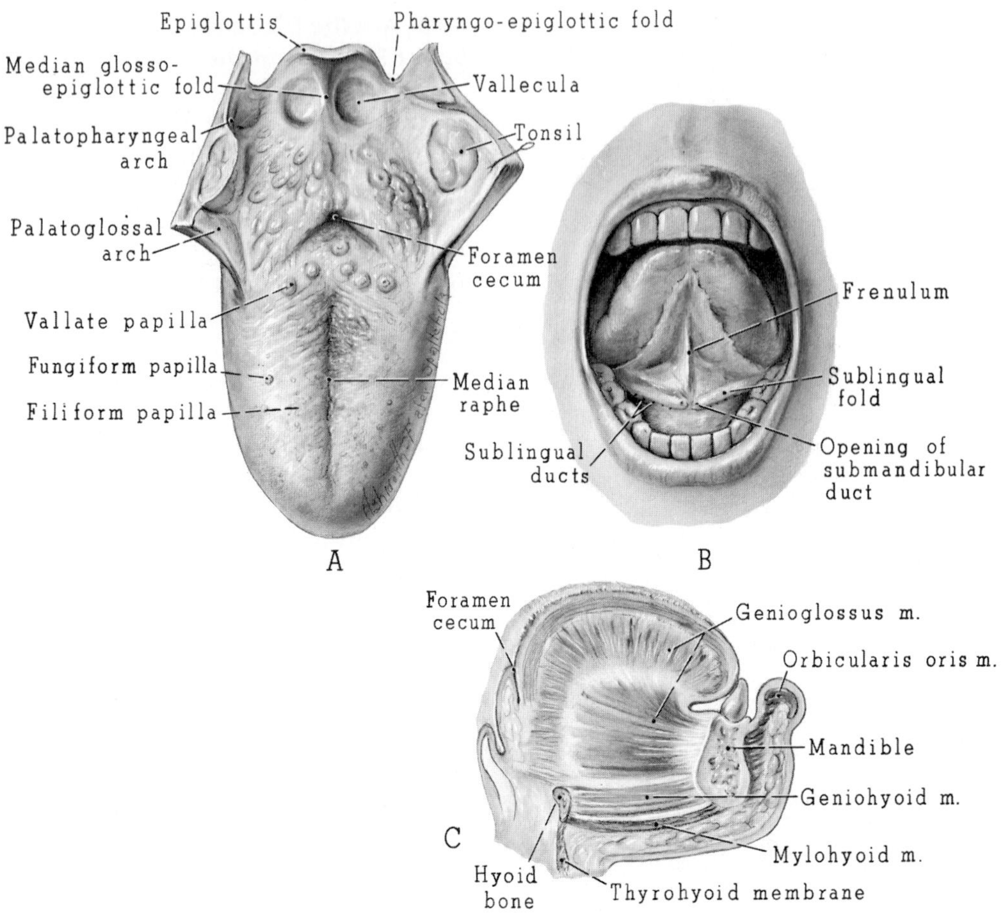

Figure 147. *A*, Dorsal view of the tongue. *B*, Anterior view of the oral cavity with tongue raised. *C*, Mid-sagittal section through the tongue.

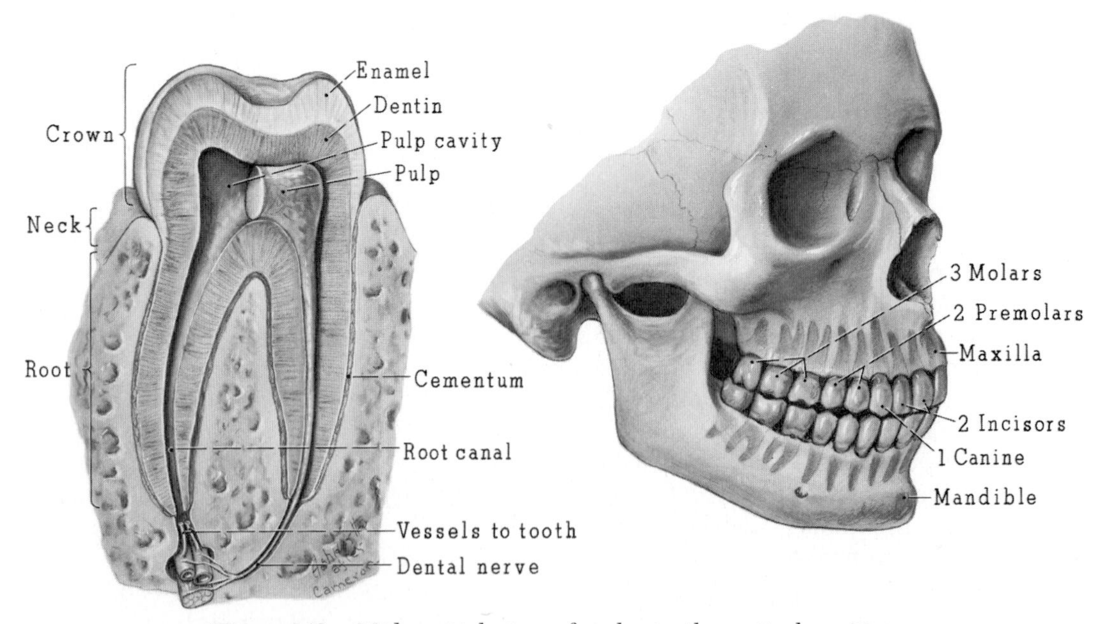

Figure 148. Mid-sagittal view of molar tooth, vertical position.

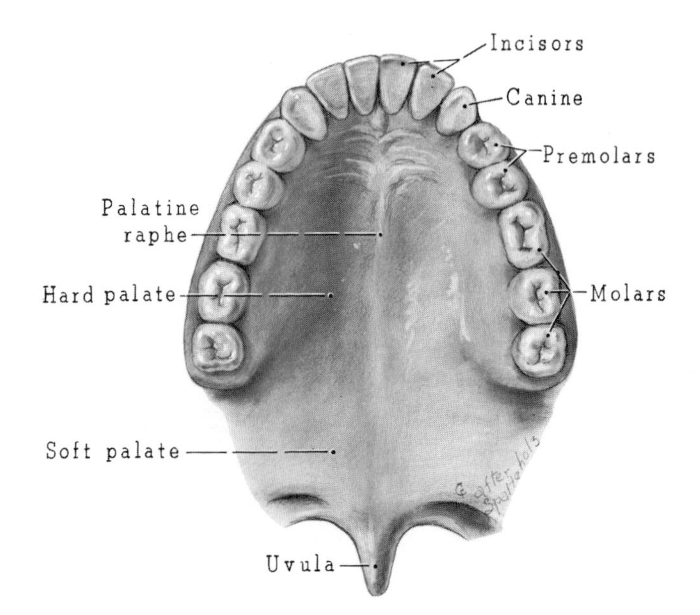

Figure 149. Roof of mouth with adult teeth.

word for gatekeeper. At the very end of the pylorus is the *pyloric sphincter* (sfingk′ter), which opens and closes at appropriate times to allow the flow of chyme into the duodenum. The term "sphincter" always refers to a muscle that closes off some hollow tube or chamber.

The walls of the stomach have an extra

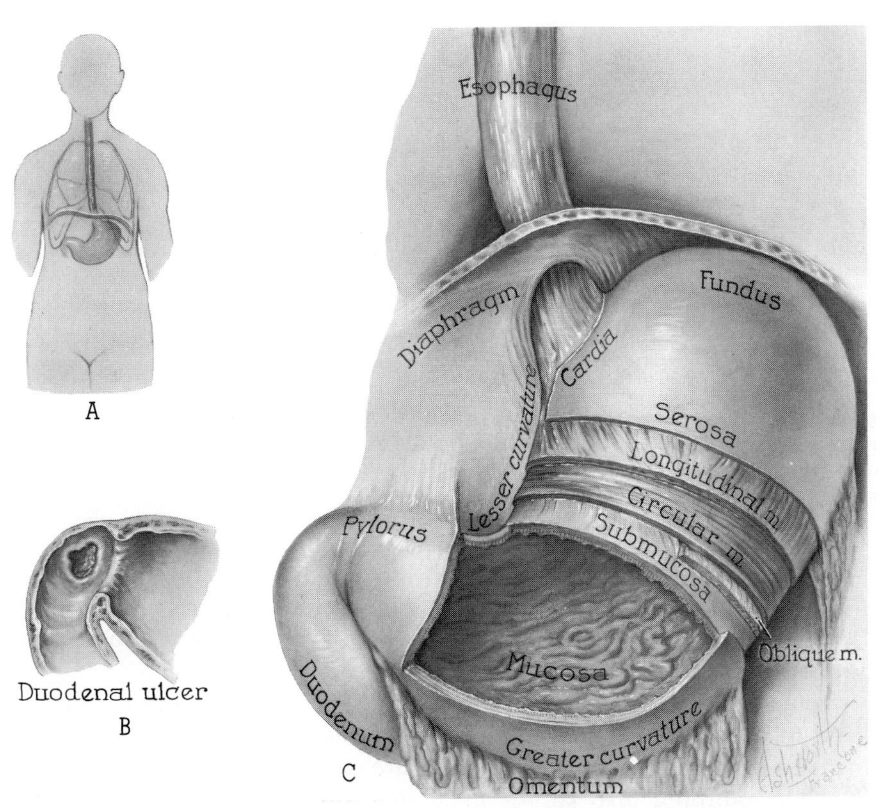

Figure 150. *A,* Anatomic position of esophagus and stomach. *B,* Duodenal ulcer. *C,* Anterior view of the stomach with portion of the anterior wall removed. (Note the various layers which make up the stomach wall.)

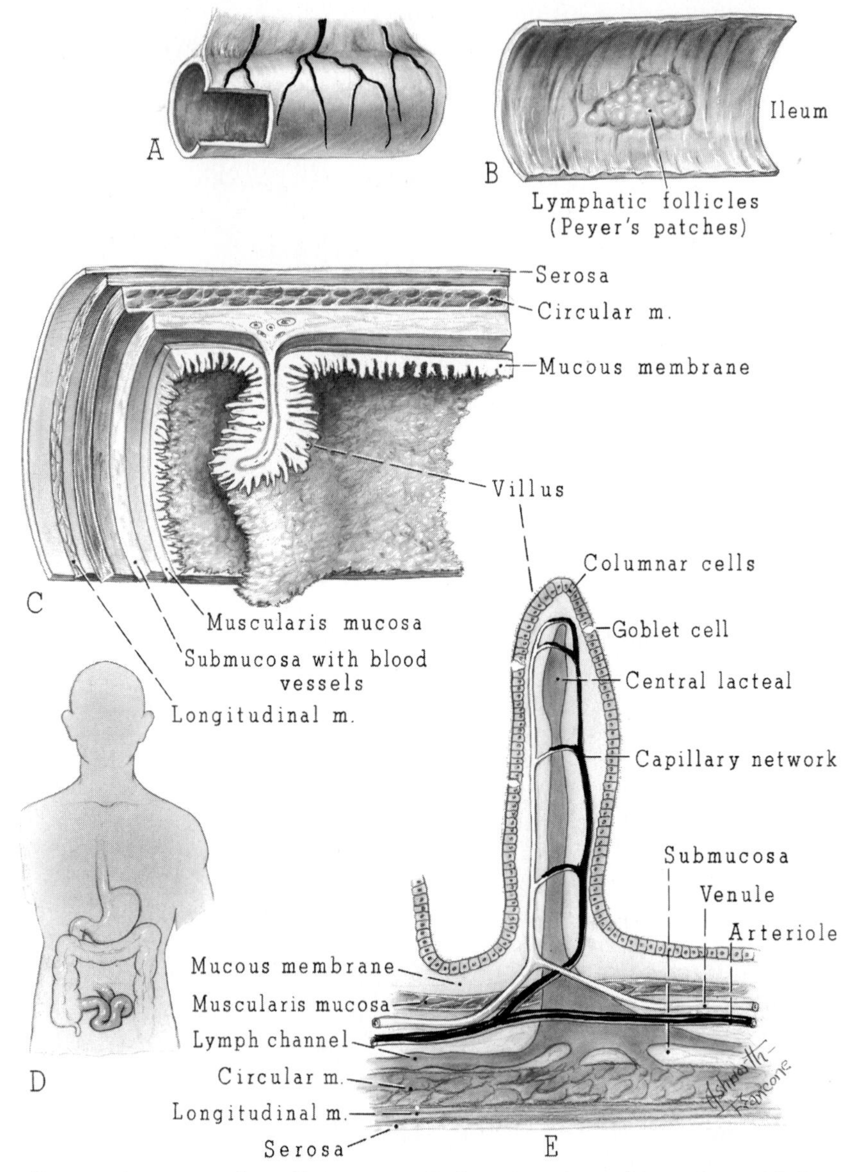

Figure 151. *A*, Segment of small intestine. *B*, Interior view of intestine with Peyer's patch. *C*, Layers composing intestinal wall. *D*, Anatomic position showing stomach and large and small intestines. *E*, Mid-sagittal section through villus.

layer of muscular tissue not found in other areas of the digestive tube. With its *three* layers of smooth muscle, the stomach is one of the strongest organs of the body. It is well suited to the task of mechanically breaking up food through its strong churning actions. This churning also serves to mix the tiny food particles thoroughly with the gastric juice. Recall that the term "gastric" is used to refer to anything directly related to the stomach.

The gastric juices, which begin the chemical breakdown of the food molecules, are secreted by the thousands of microscopic gland cells located in the inner lining of the stomach wall. Hydrochloric acid,

HCl, is one of the more important and effective digestive substances making up the gastric juices.

What is Peristalsis?

The muscular layers of the stomach, and of the intestines, take part in contractions that produce *peristalsis,* the circular, wavelike movement that propels food down the length of the digestive tract. Swallowing is actually an example of a very powerful peristaltic action. The term "peristalsis" is derived from the Greek words *peri,* meaning around, and *stalsis,* meaning constriction. A peristaltic constriction narrows the digestive tube at the point of constriction and then travels down the tube, forcing the contents along.

THE SMALL INTESTINE

What Is the Nature of the Small Intestine?

The *small intestine* is about 18 feet long; however, it is noticeably smaller in diameter than the large intestine. This small diameter gives the small intestine its name. The small intestine has two primary functions: that of *chemical digestion* and that of *absorption* of nutrients into the blood. In other words, two of the three principle functions of the digestive system are carried out primarily by the small intestine. The upper digestive tract and the stomach function to prepare ingested food by mechanical digestion for the chemical digestive and absorptive processes of the small intestine.

The small intestine has three parts or sections; in the order in which food passes through them they are the duodenum, the jejunum, and the ileum.

The *duodenum* (du″o-de′num) (from the Latin *duodeni,* meaning twelve), named because it is about equal in length to the breadth of twelve fingers, is the shortest, widest, and most fixed portion of the small intestine. The secretions of digestive juices from the liver and pancreas enter the digestive tube at the duodenum.

The inner, mucous lining of the small intestine is typical of the digestive tract as a whole, but is distinguished by the thousands of microscopic *villi* or tiny finger-like projections protruding into the *lumen* (the hollow interior) (Fig. 151). The villi absorb the simple, digested nutrients from the small intestine into the blood stream. Inside each villus is a rich network of blood and lymph capillaries. This system of thousands of villi creates a large surface area for contact between the digested nutrients and the blood. This arrangement is similar in function to the alveoli-capillary system in the lungs, which also creates a large surface area of contact for the exchange of gases. The large contact surface in the small intestine, owing to the many tiny villi, increases the rate and efficiency of the (one-way) absorption of nutrients into the blood.

Besides the villi, the inner lining of the small intestine also contains numerous tiny *intestinal glands* that secrete a major portion of the digestive enzymes into the intestinal contents. As will be discussed below under *digestion,* these digestive enzymes work to chemically break down the large food molecules into nutrients that can be easily used by the cells.

THE LARGE INTESTINE

What Is the Nature of the Large Intestine?

The *large intestine,* so named because of its large diameter, differs from the small intestine in several ways. The large intestine neither receives nor secretes digestive juices into its interior. There are no villi on its internal surface.

A considerable amount of fluid moves into the intestinal contents as they pass through the stomach and small intestine. Digestive juices are approximately 95 per cent water. Much of the water is reabsorbed through the walls of the large intestine. The remaining, increasingly solid waste, known as feces (fe'sez), is moved along by peristaltic waves to the *rectum*, where it is eliminated from the body through the *anal canal*.

The major areas of the large intestine in the order in which the contents pass through them are the cecum, the colon, the rectum, and the anal canal (Fig. 152).

The contents of the small intestine enter the large intestine at the *cecum* (se'kum), through the *ileocecal* (from *ileum* + *cecum*) *valve*. Attached to the base of the cecum is the *appendix*. In appendicitis, the appendix becomes damaged and inflamed because a solid obstruction (usually some piece of material) has lodged in its interior; in such cases removal of the appendix is usually indicated.

The *colon* is distinguished into three

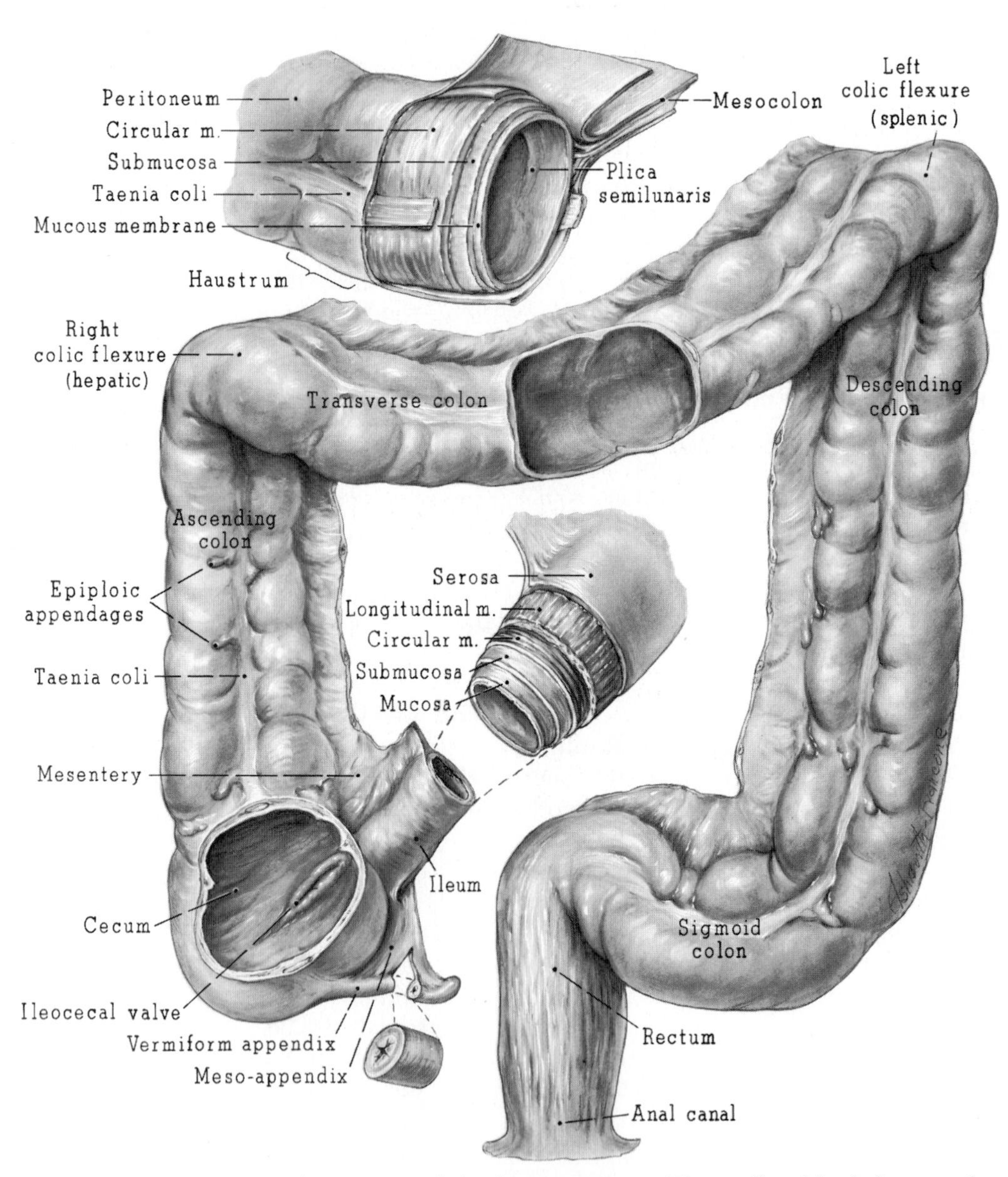

Figure 152. Position and structure of the large intestine. The walls of both large and small intestines have been enlarged and dissected to show their various layers.

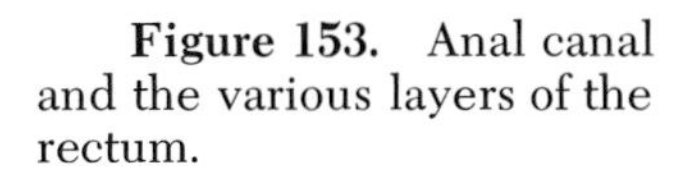

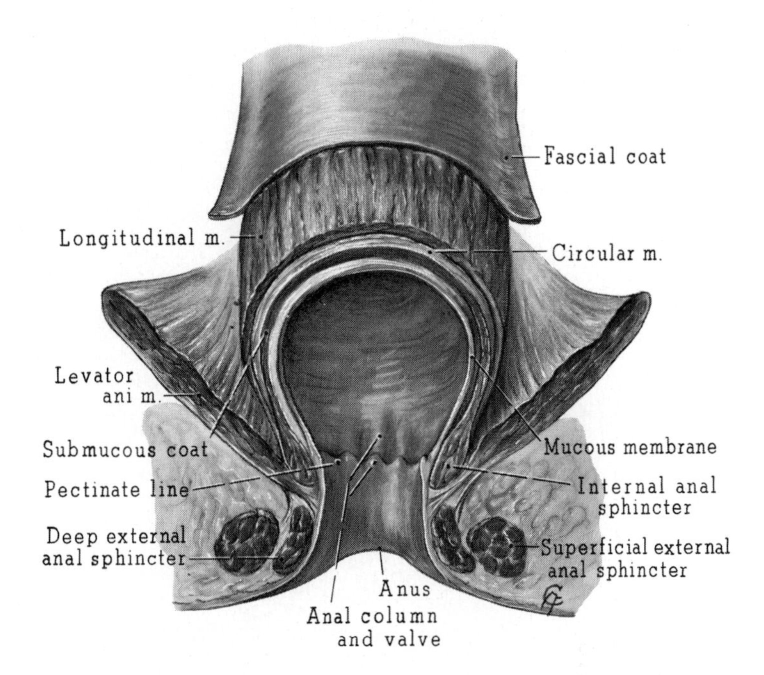

Figure 153. Anal canal and the various layers of the rectum.

areas: the *ascending colon,* the *transverse colon,* and the *descending colon* (Fig. 152). The *rectum* begins at the end of the descending colon and terminates in the narrow *anal canal.* Vertical folds of muscular tissue called rectal columns line the wall of the rectum. These columns are subject to abnormal enlargements known as *hemorrhoids.* Hemorrhoids are either internal (in the anal canal) or external (at the mouth of the anus), and can cause bleeding and pain (Fig. 153).

ACCESSORY ORGANS OF DIGESTION

What Is the Digestive Function of the Pancreas?

The *pancreas* (pan′kre-as) is a large, lobulated (having lobes) gland resembling the salivary glands in structural appearance (Fig. 154). It is a dual purpose gland having both exocrine and endocrine functions. *Exocrine* glands secrete into ducts, and in its exocrine function the pancreas secretes *pancreatic juice* (a digestive juice) by way of the pancreatic duct into the duodenum. This pancreatic juice is produced and secreted into the duct by the *acinar* (as′i-nar) *tissue* of the pancreas. The term "acinar" is derived from the Latin word for grape, and is appropriate because the acinar tissue looks like tiny clumps of grapes. The pancreatic juices are very important in the chemical breakdown of proteins.

Endocrine glands secrete directly into the blood, and the endocrine portion of the pancreas secretes *insulin* and *glucagon* (gloo′kah-gon), two hormones that are important in controlling the metabolism of glucose. This endocrine portion of the pancreas will be discussed in Chapter 12, The Endocrine System.

What Are the Digestive Functions of the Liver and Gallbladder?

The *liver,* which is the largest organ in the body, is located in the upper part of the abdominal cavity under the dome of the diaphragm. The liver is composed of four major *lobes.* Each lobe is divided into numerous lobules (little lobes), which are the functional units of the liver (Fig. 155).

One of the main functions of the liver is to secrete *bile,* a light brown to greenish yellow alkaline fluid that acts to: emulsify fats, neutralize intestinal acid, and remove

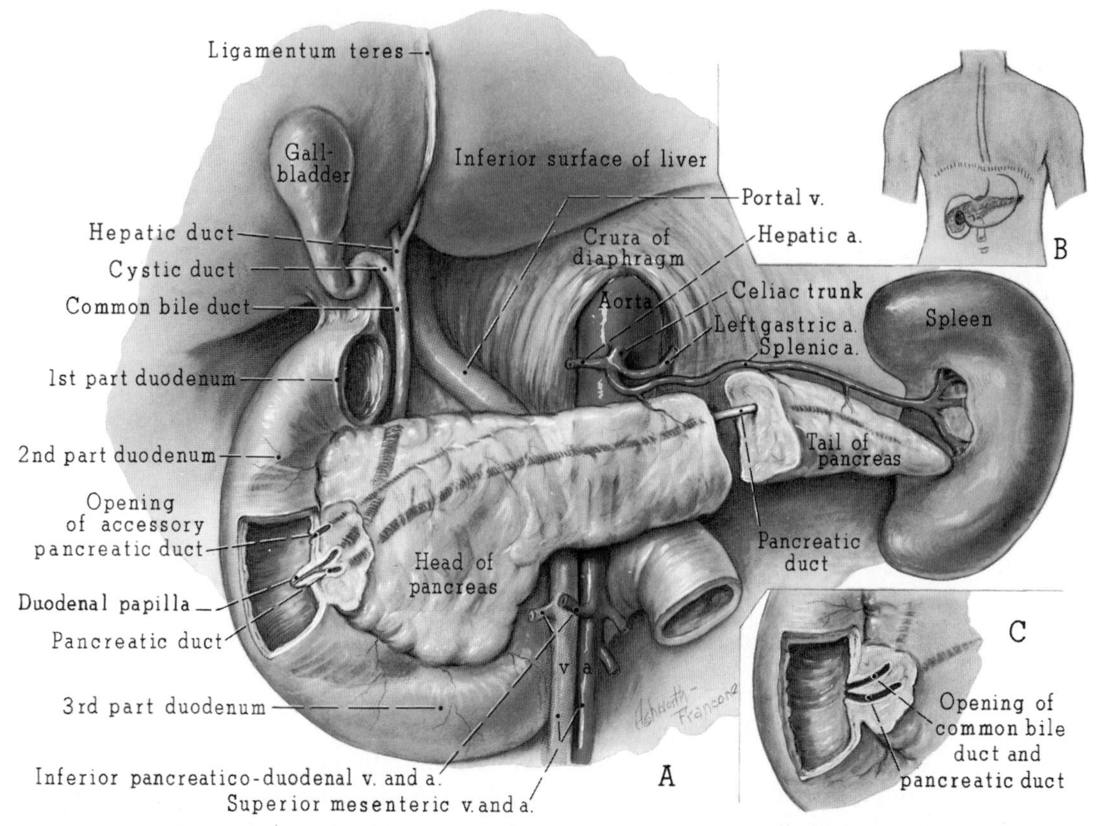

Figure 154. *A*, Relationship of the pancreas to the duodenum, showing the pancreatic and bile ducts joining at the duodenal papilla. A section has been removed from the pancreas to expose the pancreatic duct. *B*, Anatomic position of the pancreas. *C*, Common variation.

toxins from the liver. Some of the bile drains out of the liver by way of the *hepatic ducts.* (The term "hepatic" refers to anything concerning the liver and is used similarly to the term "renal," which refers to anything concerning the kidneys.) Between meals, however, bile goes up the *cystic duct* into the *gallbladder* (Fig. 156) for concentration and storage. "Gall" is the old word for bile. After meals, when fats are prevalent in the duodenum, bile drains from both the gallbladder and the liver proper, joining in the *common bile duct* and entering the duodenum. Bile is very important in the digestive breakdown of fats and is essential for the absorption of fat-soluble vitamins.

Notice all the discussion of ducts; the production and secretion of substances into a duct is the defining character of exocrine glands; and the liver then is an exocrine gland in this functional relation to the digestive system.

What Is the Role of the Liver in Glucose Metabolism?

It has already been mentioned that parts of the pancreas are important in glucose metabolism (production of the hormones insulin and glucagon). The liver cells play a different role in glucose metabolism but one of great, and perhaps equal, importance.

Since glucose is easily absorbed into the blood from the small intestine, its concentration in the blood increases after a meal. As its level in the blood begins to rise, the liver cells take up the glucose and combine it into large molecules of *glycogen* (gli′ko-jen); in this manner glucose is

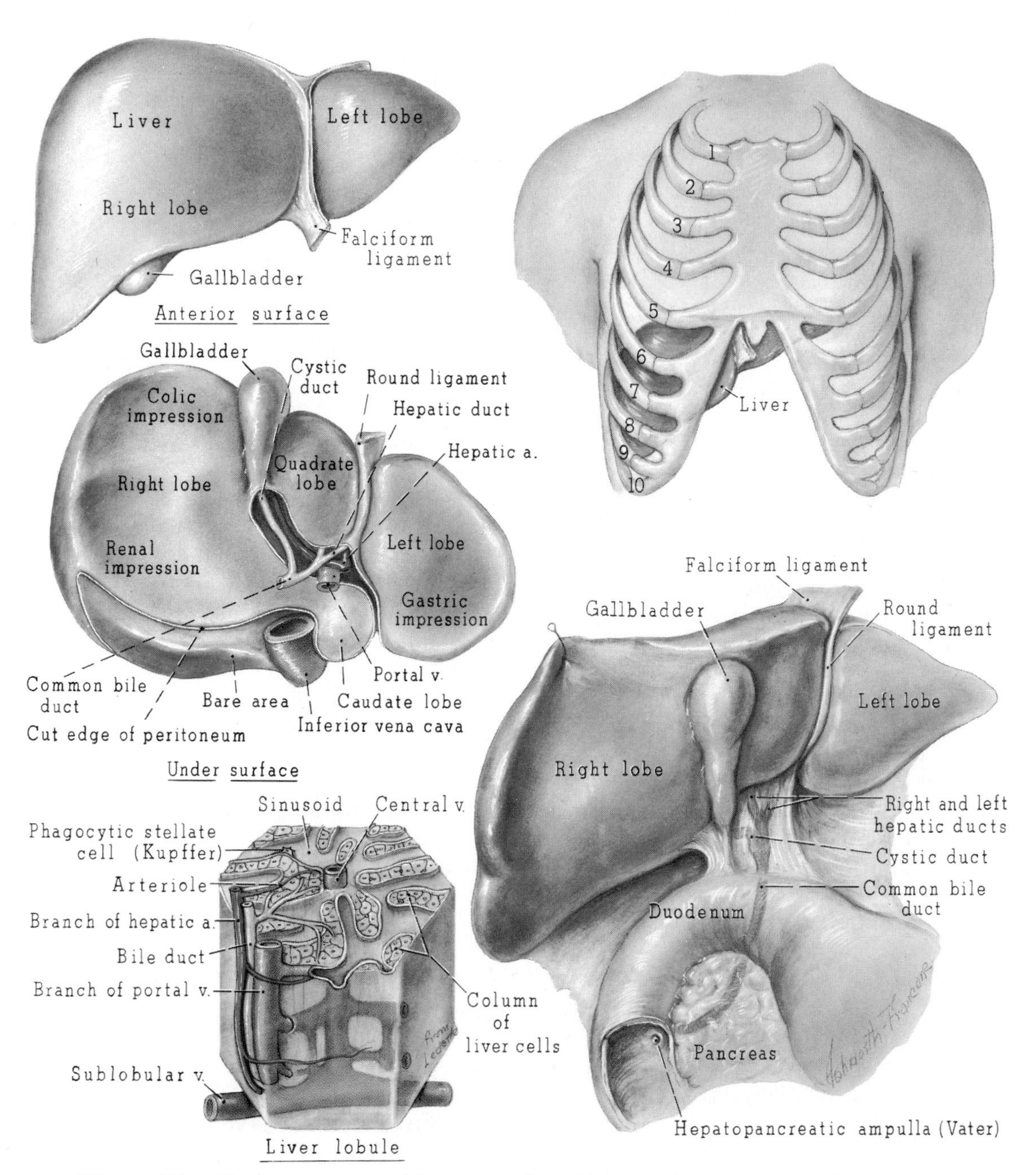

Figure 155. The liver, its normal location, relationships, and unit structure. (*Liver Lobule* section courtesy of Lederle Laboratories.)

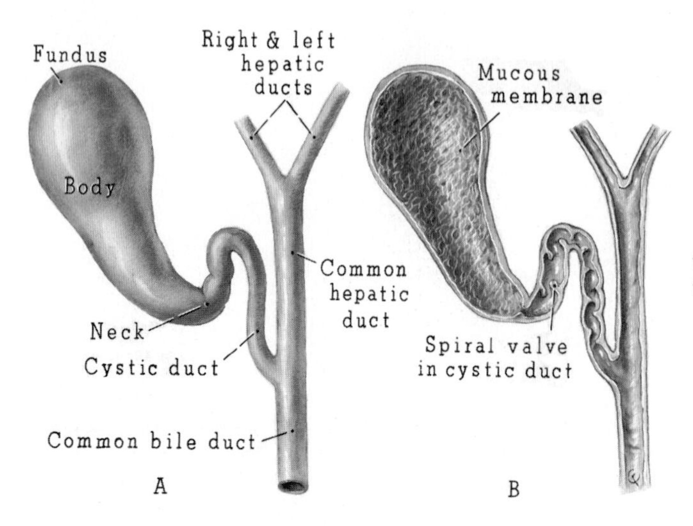

Figure 156. *A*, External view of the gallbladder. *B*, Sagittal section through the gallbladder.

stored for later use. This process is known as *glycogenesis* (*glyco* for glycogen + *genesis*, to make). Later, between meals, the liver cells will reverse the process, in a regulated way, to supply glucose to the blood as the body needs it. This reverse process is known as *glycogenolysis* (*glycogen* + *olysis*, to dissolve into parts).

One other important process occurring in liver cells that is related to glucose metabolism is known as *gluconeogenesis* (*gluco-se* + *neo*, new + *genesis*, to make), which means literally "making new glucose." It consists of a series of chemical reactions performed by the liver cells to produce glucose molecules from proteins and fats. The process involves breaking down the proteins or fats into molecular pieces and then building up glucose molecules from some of the parts.

The liver cells also store vitamins A, D, E, and K, as well as B_{12} and certain other water-soluble vitamins. Table 9 lists the major vitamins, their source, function, and the results of deficiencies.

DIGESTION

What Is Involved in the Process of Digestion?

Digestion is the process by which ingested foods are broken down both mechanically and chemically into simple nutrient molecules, which can later be most easily used by the cells.

The first stage of the process of digestion involves chewing, swallowing, and the mixing of substances in the stomach to produce chyme. This is the stage of mechanical breakdown of the food, although even during this period some chemical breakdown of large food molecules has begun. Special digestive enzymes that begin this chemical breakdown are secreted into the mouth by the salivary glands and into the stomach by the glands in the stomach lining.

The final tasks of chemical digestion take place in the small intestine, where the digestive juices from the pancreas and the intestinal lining, and bile from the liver, are delivered into the duodenum. Figure 157 illustrates the time required for food substances to reach various portions of the digestive tract.

How Are Proteins Digested?

Protein digestion is accomplished by the digestive enzymes from the pancreas and intestinal walls (Fig. 158). Protein molecules are broken down by stages into amino acids. *Amino acids* are sometimes called the "building blocks of proteins." These small, nutritive molecules are then

Table 9. VITAMINS

VITAMIN	SOURCE	FUNCTION	DEFICIENCY
Fat Soluble			
A	Yellow vegetables, fish liver oils, milk, butter, eggs	Essential for maintenance of normal epithelium; synthesis of visual purple for night vision	Faulty keratinization of epithelium; susceptibility to night blindness
D	Egg yolk, fish liver oils, whole milk, butter	Facilitates absorption of calcium and phosphorus from the intestine; utilization of calcium and phosphorus in bone development	Rickets in children; osteomalacia in adults
E	Lettuce, whole wheat, spinach	Essential for reproduction in rats; no definite function has been determined in humans	Sterility in rats; no known effects on humans
K	Liver, cabbage, spinach, tomatoes	Synthesis by the liver of prothrombin; necessary for coagulation	Impaired mechanism of blood coagulation
Water Soluble			
B-Complex:			
B_1 (thiamine)	Whole grain cereals, eggs, bananas, apples, pork	Coenzyme in metabolism of carbohydrate as thiamine pyrophosphate (cocarboxylase); maintains normal appetite and normal absorption	Beriberi, polyneuritis
B_2 (riboflavin)	Liver, meat, milk, eggs, fruit	Coenzyme in metabolism (as flavoprotein)	Glossitis, dermatitis
B_6 (pyridoxine)	Whole grain cereal, yeast, milk, eggs, fish, liver	Coenzyme (as pyridoxal phosphate) in amino acid metabolism	Dermatitis
Niacin	Liver, milk, tomatoes, leafy vegetables, peanut butter	Niacinamide in metabolic processes, especially energy release	Pellagra
B_{12}	Liver, kidney, milk, egg, cheese	Maturation of erythrocytes	Pernicious anemia
Pantothenic acid	Egg yolk, lean meat, skim milk	Necessary for synthesis of acetyl coenzyme A, metabolism of fats, synthesis of cholesterol, and antibody formation	Neurologic defects
Folic acid	Fresh, leafy green vegetables, liver	Production of mature erythrocytes	Macrocytic anemia
Biotin	Liver, egg, milk; synthesized by bacteria in the intestinal tract	Coenzyme in amino acid and lipid metabolism	Not defined in man, since a large excess is produced by intestinal flora
C (ascorbic acid)	Citrus fruits, tomatoes, green vegetables, potatoes	Production of collagen and formation of cartilage	Scurvy; susceptibility to infection; retardation of growth, tender, swollen gums, pyorrhea, poor wound healing

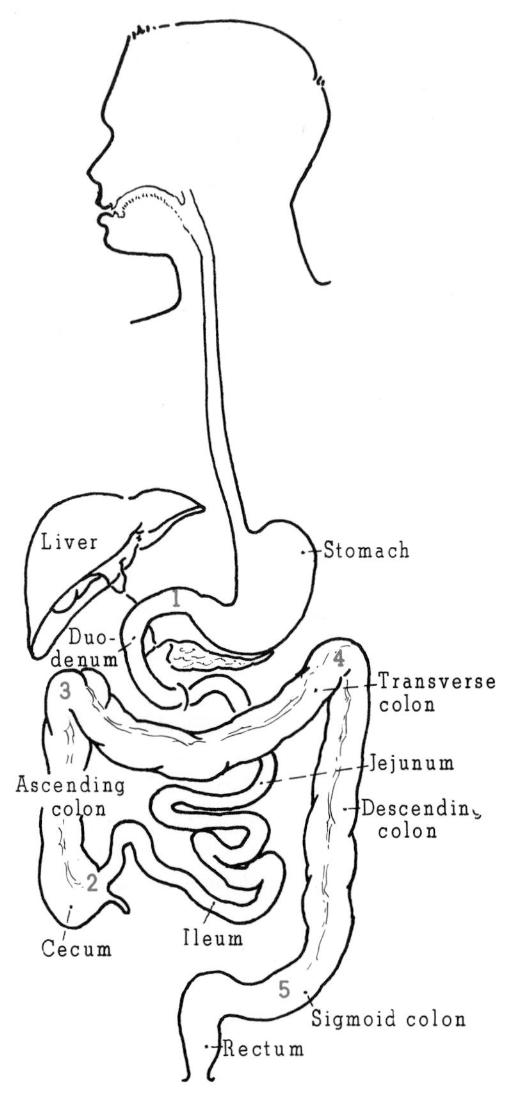

Figure 157. The time required for food substances to reach various portions of the digestive tract.

transported to the cells where they are used to make new proteins that serve specific human body functions. Amino acids are the same in all organisms. Many protein molecules contain 1000 or more amino acids in very specific sequences and arrangements.

In other words, the proteins of one species are almost never identical to those of another species, so that it is necessary for the human body to break down the ingested proteins from meat, fish, eggs, and cheese into constituent amino acids, and then build up new proteins from the amino acids. These new proteins will work in the human body, but would not work in the body of another species.

How Are Carbohydrates and Fats Digested?

Digestion of *carbohydrates,* such as sugars and starches, occurs mainly in the small intestine (Fig. 159). The chemical breakdown is effected by the enzymes of the pancreatic juice and those of the intestine. The main end product of carbohydrate digestion is *glucose* (from the Greek word for sweet). Glucose is the principal source of energy for the cells—it is the "energy molecule." Other end products of carbohydrate digestion are galactose and fructose.

Digestion of *fats* is slow until *bile* from the liver and gallbladder is introduced into the duodenum (Fig. 160). Bile breaks down the large fat globules into tiny particles. Then the pancreatic enzymes split the fat molecules into glycerol and fatty acids. These two substances are the end products of fat digestion. They can be easily absorbed into the blood and picked up and utilized by the cells.

ABSORPTION OF NUTRIENTS

How Are the Nutrient Molecules Absorbed into the Blood?

Absorption is the process by which the end products of digestion are transferred from the interior of the digestive tract into the blood stream. The mechanism of absorption is mostly by *passive diffusion,* but *active transport* of substances into the blood is not uncommon.

Absorption through the wall of the

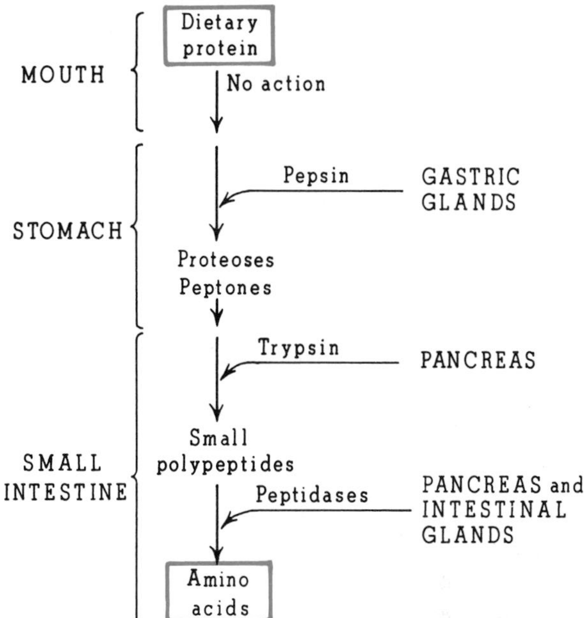

Figure 158. Digestion of protein. Note that protein is broken down to proteoses and peptones in the stomach and to small polypeptides and amino acids in the small intestine.

stomach is limited, but small amounts of water, simple salts, glucose, and alcohol can be absorbed to some extent.

The small intestine, with its large number of *villi,* is the site of most absorption from the digestive tract. The major absorption of carbohydrate, protein, and fat occurs through the capillaries of the villi in the small intestine, (Fig. 161).

If absorption of substances from the digestive tube into the blood did not occur, the cells would never get any of the sub-

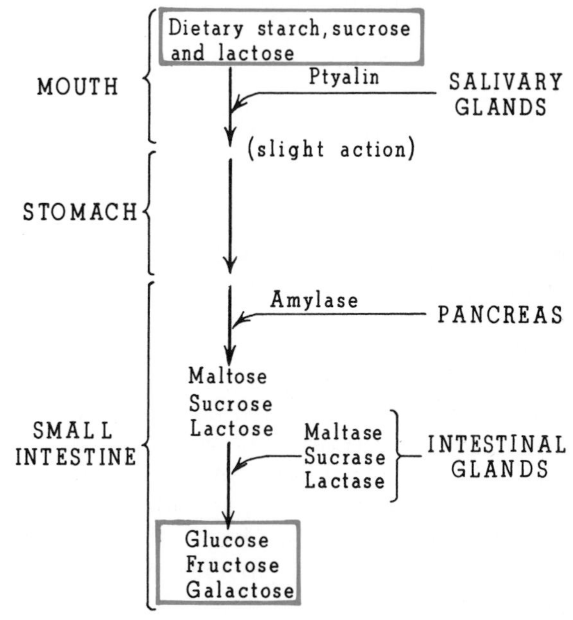

Figure 159. Digestion of carbohydrate. Note that the major digestion of carbohydrate occurs in the small intestine with the end products being glucose, fructose, and galactose.

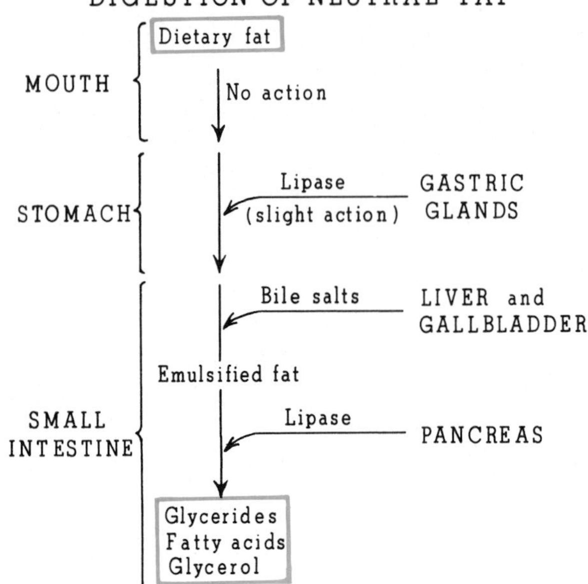

Figure 160. Digestion of neutral fat, the end products being glycerides, fatty acids, and glycerol.

stances that pass down through the digestive tract.

How Are the Specific Nutrients Utilized?

Figures 162, 163, and 164 summarize the specific metabolism of carbohydrates, proteins, and lipids. The illustrations show how, or from what source, the specific substance enters the blood stream, and then (on the right) when it is utilized in the body. All nutrients originally enter the blood through the intestine, but many are stored in the liver or tissues, from where they can be released more slowly into the blood as needed.

What Is Basal Metabolic Rate?

Recall that the term *metabolism* refers to all the chemical reactions of the body. When biologists and physiologists speak generally of metabolic rate, however, they are normally referring to the rate of energy metabolism. The standard measure of the rate of energy metabolism is called *basal metabolic rate* or *BMR*. The word "basal" derives from the same root as the word "basic."

Basal metabolic rate, then, refers to the rate of utilization of energy (how much energy is used) by an individual who has not eaten for 12 to 24 hours—this time interval being the standard for comparison between individuals. The units of BMR are usually given in calories per hour. This measure is arrived at by determining the amount of oxygen consumed per minute in normal activity.

CLINICAL CONSIDERATIONS

The Teeth and Gums

Dental caries. Caries (ka′re-ez) is a progressive disease of the teeth that destroys the enamel and causes *cavities*. The term "caries" is from the Latin word meaning dry rot. Dental caries is largely preventable by ingesting fluoride. When proper amounts of fluoride are ingested during dental development, a reduction in caries of about 60 per cent has been demonstrated.

Fluoridation of water is a widely accepted method of supplying dental fluoride.

Pyorrhea (pi″o-re′ah). This is an inflammatory process of the gums in which

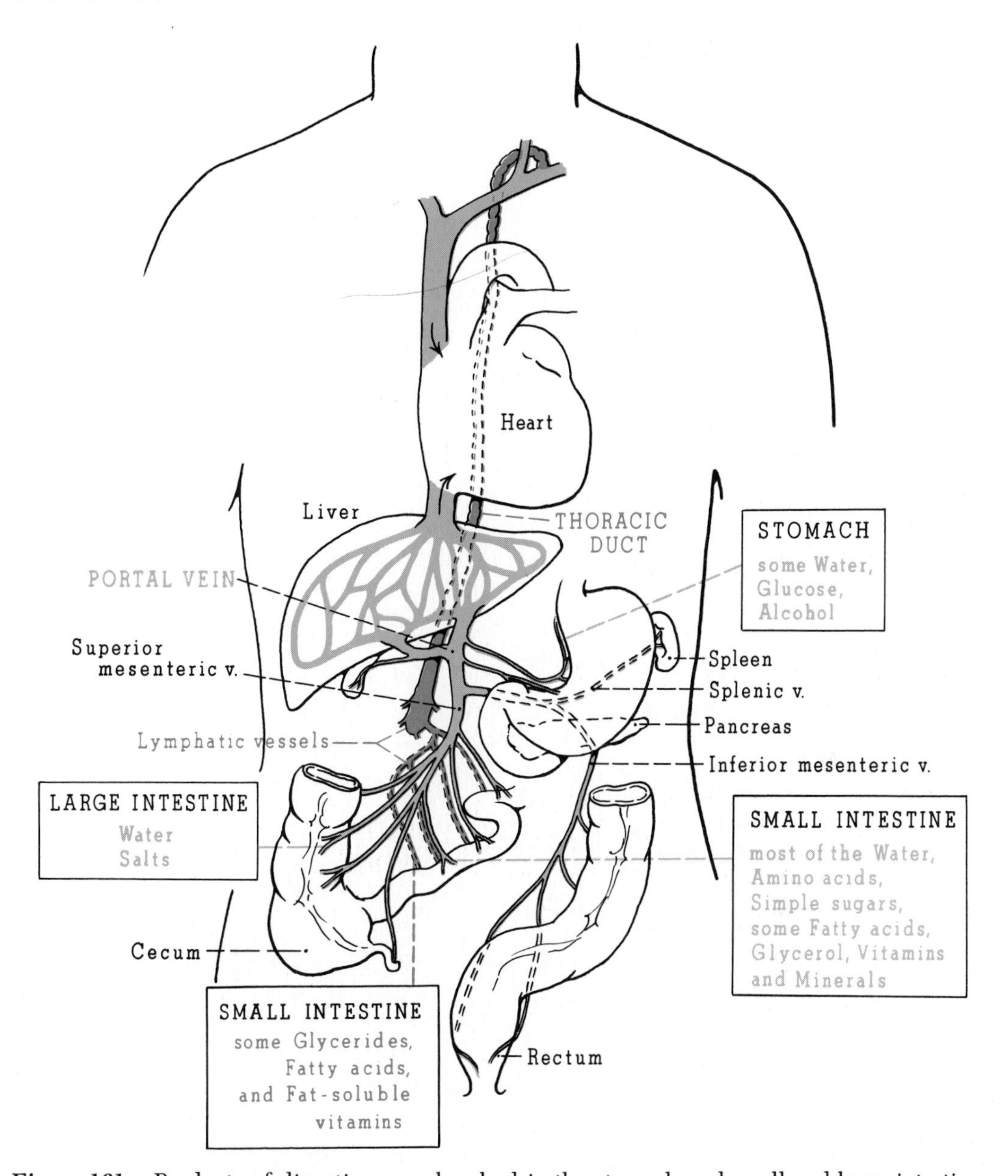

Figure 161. Products of digestion are absorbed in the stomach and small and large intestines.

the teeth become loose and fall out. Bacteria from dental infection can spread through the blood to the heart, causing inflammation of the lining membrane of the heart (bacterial endocarditis).

The Stomach and Duodenum

Ulcer. An ulcer is a sore or lesion on the surface of the skin of a mucous membrane. A peptic ulcer is an ulcer that occurs in either the stomach or duodenum. This relatively frequent condition results from an oversecretion of gastric juices, particularly hydrochloric acid, HCl. The major goal of treatment is to reduce this oversecretion of HCl through diet control and ingestion of antacids or by inhibiting nerve stimulation of the stomach. Occasionally, surgical removal of part of the stomach or duodenum is necessary.

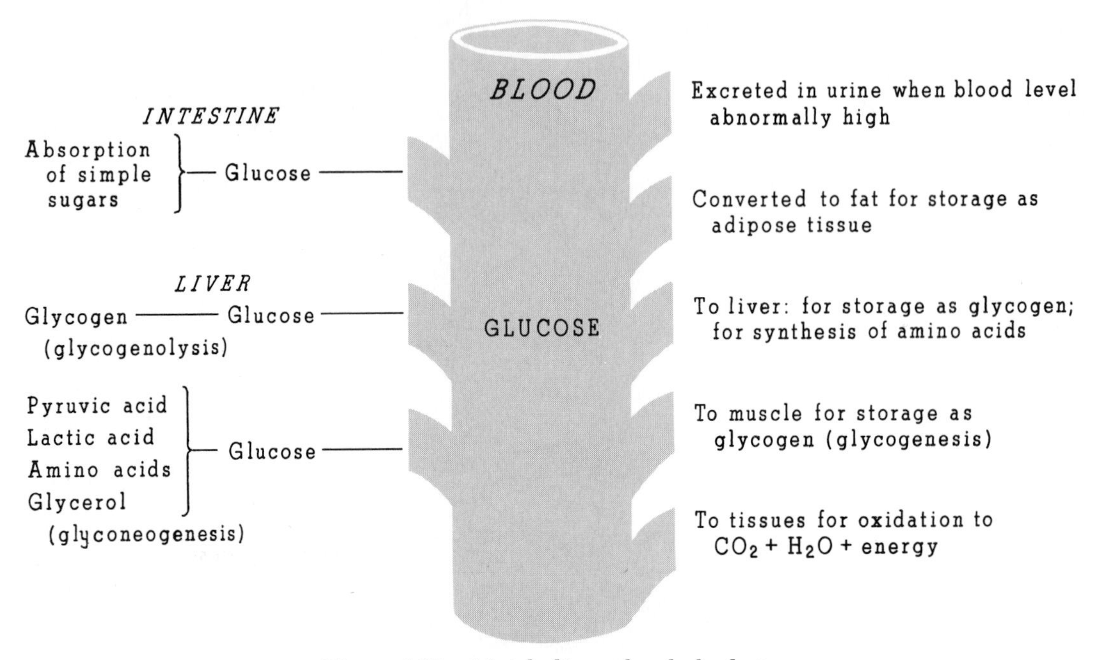

Figure 162. Metabolism of carbohydrate.

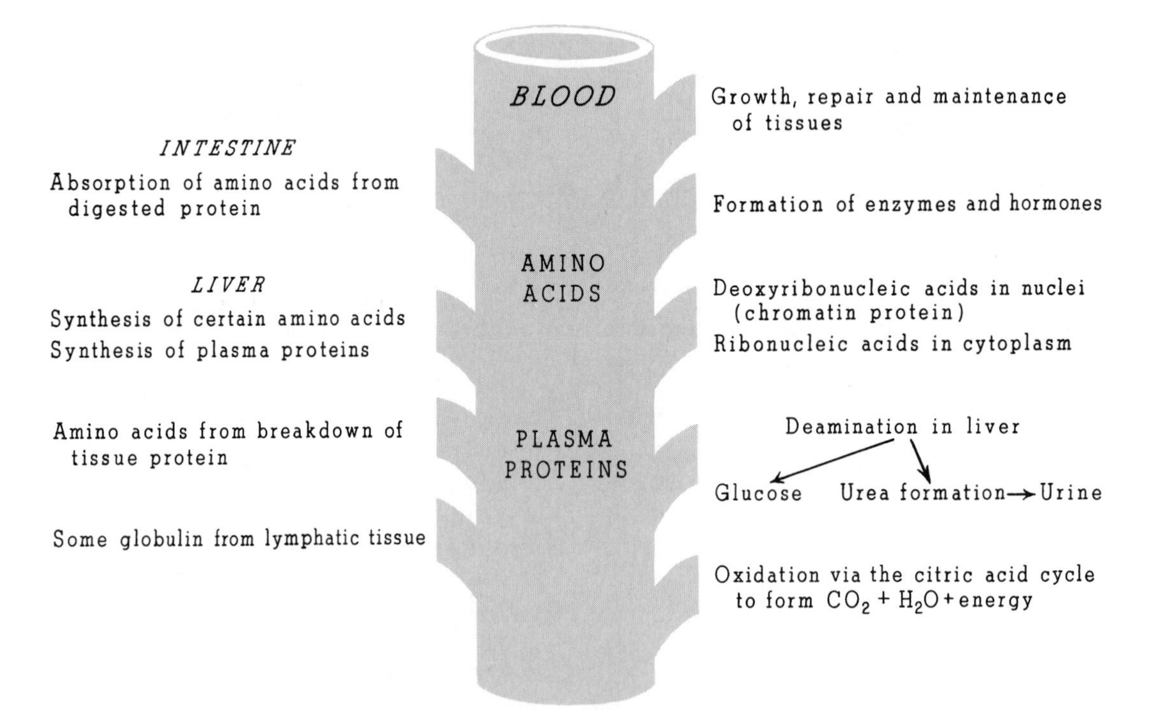

Figure 163. Metabolism of protein.

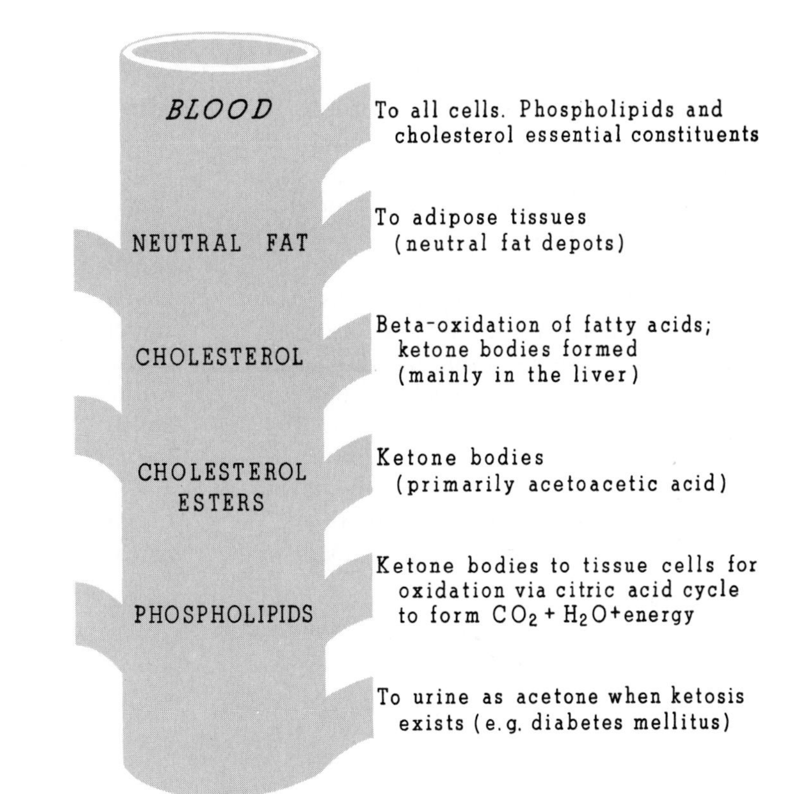

Figure 164. Metabolism of lipids.

Abnormal Absorption

Several conditions can lead to abnormal absorption of digested nutrients from the small intestine. A number of diseases and disorders of the small intestine itself will inhibit proper absorption.

Disorders of the accessory glands, such as the liver and pancreas, can cause inadequate absorption because of incomplete digestion.

A balanced diet is necessary for proper absorption to take place; the necessity of having vitamin D in the diet in order for calcium absorption to occur is an example of this. Some researchers have argued that arteriosclerosis, the cholesterol (lipid) disorder, can be prevented by proper diet. Modern nutrition research has turned up many cases where absorption of one nutrient is either increased or inhibited by the presence of some other substance in the diet.

The Liver and Gallbladder

Cirrhosis (sir-ro′sis). This is a disease of the liver leading to a progressive degeneration of liver cells. It is characterized anatomically by an increase in connective tissue (scar tissue) throughout the liver. In alcoholic cirrhosis, its most common form, the disease is preceded by a dietary deficiency. The alcoholic person tends to have a reduced intake of fat, protein, and carbohydrate, as well as of vitamins, especially B_{12}.

Cholecystitis (ko″le-sis-ti′tis). This is an inflammation of the gallbladder. The term derives from the Greek roots *chole,* meaning bile, and *kystis,* meaning bladder, + *itis,* meaning inflammation. *Cholelithiasis* (ko″le-li-thi′ah-sis), stones in the gallbladder, is associated with a chronic inflammation of the bladder wall, giving rise to a loss of normal capacity to absorb and concentrate. The term derives from the Greek roots *chole,* meaning bile, + *lithos,* meaning stone.

Jaundice (jawn′dis). A term derived from the French word for yellow, jaundice is a yellow discoloration of the skin, mucous membrane, and body fluids because of an excess of bile pigment. The most common type is obstructive jaundice, which is caused by internal blockage of the bile duct by gallstones or by a growth such as a tumor. Hepatic jaundice is a result of hepatitis, an inflammation of the liver, usually due to an acute infection.

LEARNING EXERCISES

1. Copy the drawing in Figure 146 and then, with the book closed, label all the parts of the digestive tract and the accessory organs.
2. Construct a table, listing the major substances absorbed in the stomach, small intestine, and large intestine.
3. Following Figures 162, 163, and 164 construct schematic diagrams showing the metabolism of carbohydrate, protein, and lipids. Complete accuracy is not as important as understanding the basic pathways.

CHAPTER 11

THE NERVOUS SYSTEM

THE NERVOUS SYSTEM: AN OVERVIEW

What Does the Nervous System Do?

The nervous system, in association with the endocrine system (Chapter 12), controls and coordinates the workings of the component parts of the body. In its most simple conception, the nervous system is a communication network that transmits information by electrical signals throughout the body.

The nervous system accomplishes its function by means of a vast network of nerves that carry signals into and out of the central nervous system (the brain and spinal cord). The brain, and to a lesser extent the spinal cord, serves to interpret and process incoming signals in order to produce appropriate outgoing signals. The higher centers of the brain (the cerebrum) are generally recognized as being responsible for thought processes and conscious activities.

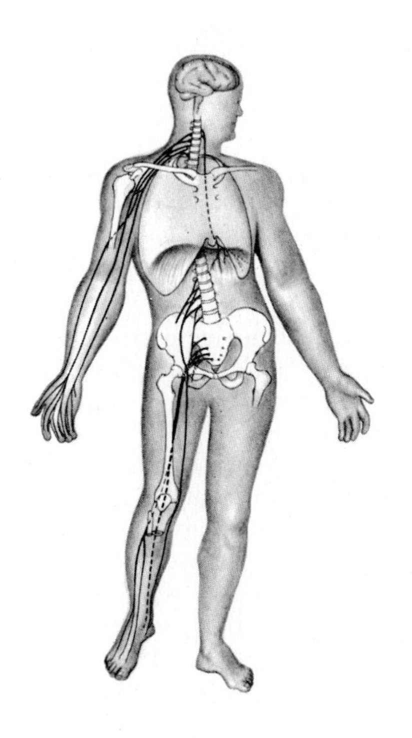

What Types of Cells Are Found in the Nervous System?

Two major kinds of cells make up the nervous system. The *neurons* (nerve cells) conduct and process electrical impulses (signals) from one area of the body to another. These are, therefore, the cells that perform the major work of the nervous system. The *neuroglia* (nu-rog′le-ah) are cells of a special kind of connective tissue. They do not transmit nerve impulses, but serve to support, repair, and perhaps nourish the neurons. The neuroglia, or glial (gli′al) cells (from the Latin word for glue) differ in size and shape according to their particular function.

THE NEURON

What Does a Neuron Look Like? How Does It Work?

Neurons are usually considered to have three main parts: the axon fibers, the dendrite fibers, and the main cell body.

The axon fibers, or *axons*, normally conduct impulses away from the main cell body. The dendrite fibers, or *dendrites*,

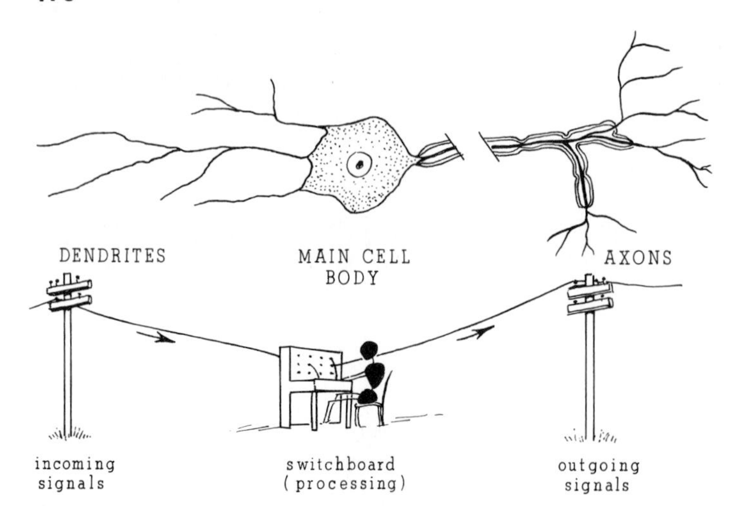

Figure 165. A typical neuron. Incoming signals travel along the dendrites, and outgoing signals travel along the axon. The main cell body processes the signals. The neuron is like a telephone communication system, except that in a neuron, signals normally travel in only one direction—in the dendrites and out the axon.

normally conduct impulses *toward* and into the main cell body. To remember which type of fiber conducts impulses which way, you should associate the "a" of an *axon* and *away* from.

These neuron fibers are actually part of the cell body, evidenced by the fact that they contain cytoplasm, continuous with the cytoplasm of the main cell body. In other words, the axons and dendrites are actually extensions of the main cell body. The axons and dendrites are often extremely long—up to two feet or more for neurons serving the feet. Anatomists often refer to these fibers simply as *nerves.* And this has become the proper definition of the word "nerve": that is, it refers to one of the long fibers of a neuron.

The main cell body of a neuron contains the nucleus of the cell. Current theory holds that the main cell body functions to change the nature and direction of signals passing through it. However, the mechanisms and reasons for such changes are not well understood. The main cell body can be thought of as like a telephone switchboard through which incoming signals are processed and directed to the appropriate outgoing fiber (Fig. 165).

How Are Nerve Fibers Insulated?

What Is the Function of Myelin?

Extending the comparison of the nervous system network to a telephone network, telephone wires are normally insulated (covered) with a nonconducting substance to prevent signals on one wire from interfering with signals on another. Similarly, nerves (long neuron fibers) are normally insulated with a nonconducting material called *myelin.* The myelin covering on nerves is made up of non-nervous Schwann cells, which wind around the nerve fibers, somewhat like several turns of thick tape around a telephone wire (Fig. 166).

Nerve cells, including their fiber extensions, all naturally appear gray. Schwann cells, however, appear distinctly white. As a result, some areas of the nervous system look gray and others white. Anatomists have come to refer to the different areas by use of the expressions *gray matter* and *white matter.*

The main cell bodies of the neurons are never covered with myelin; therefore, groups of main cell bodies are always gray matter. Furthermore, whenever gray matter is examined, it is most often composed of main cell bodies. White matter, on the other hand, is always made up of myelinated nerves (fibers).

What Do the Three Classes of Neurons Do?

There are actually three different classes, or categories, of neurons, distinguished on the basis of the duties they per-

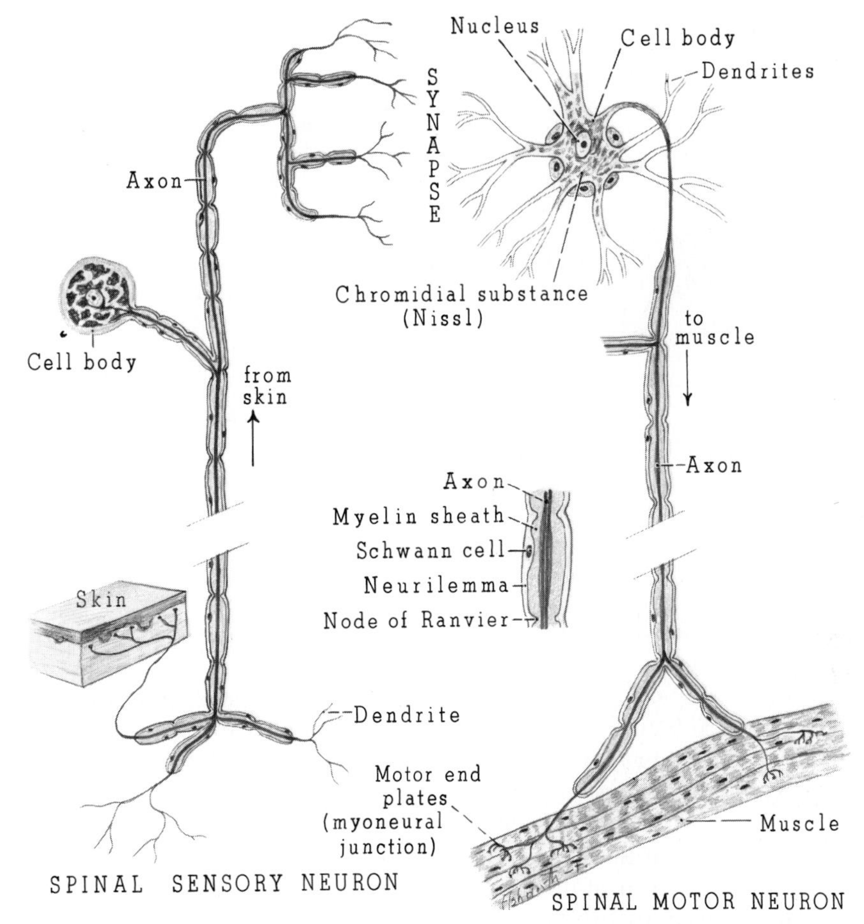

Figure 166. Motor and sensory neuron showing synapse in a two-neuron reflex arc.

form. *Sensory neurons* bring messages into the central nervous system. *Motor neurons* carry messages out of the central nervous system. *Interneurons* transmit messages between sensory and motor neurons, always within the confines of the central nervous system.

The sensory neurons carry all the sensory signals from all over the body into the central nervous system. These sensory signals inform the central nervous system of the internal workings of each part of the body *and* also of the external or environmental situation (as with the eyes and ears). We are only consciously aware of a small portion of these sensory signals as *sensations* or *perceptions*.

The motor neurons carry the signals that go out from the central nervous system to the various parts of the body. These motor signals activate or instruct each part of the body to accomplish its functions, either internal, such as digestion, or external, such as walking and talking.

The interneurons (sometimes called internuncial neurons) are located exclusively within the central nervous system. They carry messages between the sensory and motor neurons and are commonly connected to other interneurons. This network of interconnections of the interneurons in the central nervous system makes possible the coordination of complex behaviors.

What Is a Reflex Arc?

A *reflex arc* is a complete two-way communication unit. Since neurons carry messages in only one direction, a reflex arc

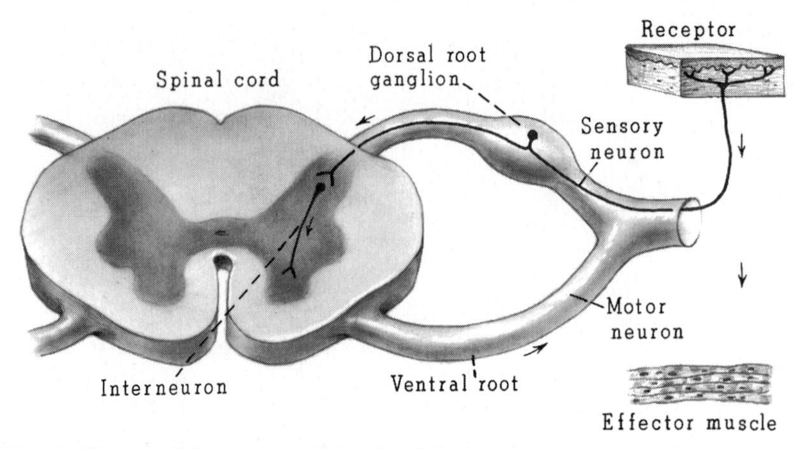

Figure 167. Reflex arc. The main cell body of the sensory neuron is located in the dorsal root ganglion. The main cell body of the motor neuron is in the gray matter of the cord. The interneuron is located *completely* inside the gray matter of the cord.

must be made up of at least two neurons, one sensory and one motor. In fact, most reflex arcs also involve one or more interneurons as well (Figs. 166 and 167).

As a rule, reflex arcs (properly called) control only the simplest or most primitive and automatic behaviors of the particular parts of the body. An example of a simple reflex action is the contraction of the pupil of the eye when exposed to bright light; this primitive response serves to protect the sensitive retina (back of the eye) from damage. The sensory neurons connected to the back of the eye bring the information that there is a bright light to the brain, and the motor neuron carries the activating signal back to the pupil, causing it to contract.

Most reflex behaviors that have been identified have a similar *primitive protective function* and they are fairly *automatic.*

As behaviors or responses to stimuli become increasingly complex, they are most often processed through the brain centers. Most everyday actions involve a smooth, integrated combination of involuntary (automatic) reflex actions and consciously controlled actions (involving the brain centers).

How are Signals Transmitted?

Neurons transmit impulses or signals by an electrical (or electrochemical) process. The process begins when the end of the dendrite is stimulated. The stimulus causes a rapid in and out exchange of sodium and potassium ions at the point of stimulation. This process is called "depolarization." It proceeds to adjacent areas of the dendrite fiber, and so on down the line, creating an electrical wave that sweeps along the length of the neuron, out to the tip of the axon (Fig. 168).

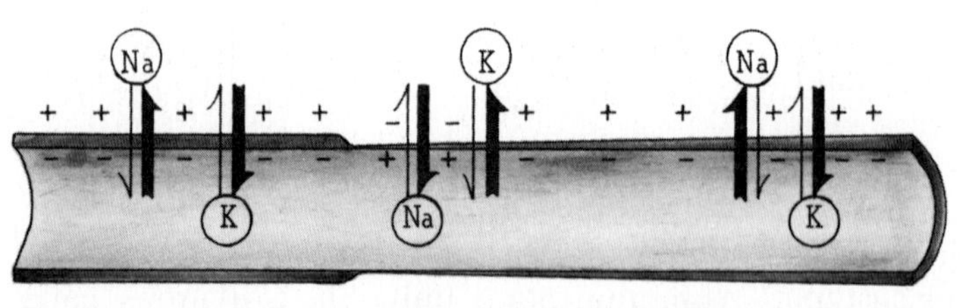

Figure 168. Conduction of nerve impulse.

The rapid in and out exchange of ions takes only a small fraction of a second, but the fiber cannot be stimulated to start another impulse until the ions return to their normal or "ready" position. This "return to normal" process is called "repolarization."

The larger the nerve fiber and the thicker the myelin sheath, the more rapid the rate of conduction of an impulse down the length of the fiber. The range is from 100 meters per second for the largest nerve fibers to about 0.5 meter per second for the smallest.

How Does an Impulse Get From One Neuron to Another?

The transmission of an impulse from one neuron to another occurs across a special neuron-to-neuron junction called the *neural synapse* (the point at which an impulse passes from one nerve cell to another) (Fig. 166).

Consider a sensory signal reaching the end of its axon in the central nervous system. What happens at that point is that the impulse stimulates the release of a special chemical, which is stored at the tip of the fiber. The chemical then becomes a messenger, travelling across the synapse to the dendrite of the next cell. On reaching the connecting neuron, the chemical messenger stimulates the next cell to transmit an electrical impulse, which then travels the length of that cell by the electrical process.

In a reflex arc, for example, the impulse begins in a sensory neuron and is transmitted electrically along the cell to the synapse. Here the chemical messenger is released, and quickly (almost instantaneously) crosses the junction, contacting the next neuron (either a motor neuron or interneuron). This contact stimulates a new (continued) electrical impulse in the contacted neuron and so on.

Since the chemical messenger is stored only at the ends of the axon fibers, the synapse is only a *one way junction.* This synaptic property is the real reason that nerve impulses normally travel in only one direction through a nerve cell. Although the nerve fibers are *able* to carry impulses in a reverse direction, this does not normally occur, since reverse impulses can never cross a synapse and, therefore, simply die out whenever they occur.

How Do Signals Start and Stop?

A normal sensory impulse, once started, does not necessarily lead to the production of a motor signal. Interneurons often stop the impulse. In a simple, direct, sensory-to-motor reflex arc, however, a sensory impulse will almost always lead to a motor signal activating the appropriate response; that is, the response to the stimulus is automatic.

At the end of each dendrite of a sensory neuron is a specialized microscopic ending called a *sensory receptor.* These receptors are sensitive to different types of stimuli. For instance, some are stimulated by cold and some by heat, some by light and some by sound; others are sensitive to specific concentrations of chemicals in the blood. One rather interesting type of sensory receptor is the *baroreceptor,* which is sensitive to the stretching of the walls of arteries caused by increased blood pressure. A sensory neuron transmits a signal when its sensory receptors are stimulated by their characteristic stimulus.

At the end of each axon of a motor neuron is a specialized microscopic ending called an *effector* (or effector ending). When an electrical motor impulse reaches these miniature organs they release a special chemical, which activates or deactivates the particular part of the body being served; for example, a muscle or digestive organ.

What Is the Language of the Nervous System?

For a long time scientists believed that neurons could each carry many types of messages, depending on how they were stimulated. Experiments have shown, however, that in most instances each neuron transmits only one type of signal. In this sense, neurons are unlike telephone lines. For instance, each sensory neuron or, more accurately, each sensory receptor is sensitive to only one type of stimulus, not several. Similarly, each motor neuron activates only one part and in only one way, not several; for instance, one part of one muscle at one strength.

As a result of this fact numerous neurons must be present in each area of sensitivity in order for the nervous system to be able to sense a variety of stimuli. And similarly, many different motor neurons must serve each area or organ in order for there to be a variety of possible motor responses for each component.

To take an example, one sensory neuron is needed in each area of the skin to transmit the message "cold" (another is needed for heat, another for pressure, and so on) and a different sensory neuron is needed to transmit the message "very cold." With motor neurons the situation is similar; each motor neuron serves a limited function, so that in many cases one motor neuron is needed to turn some part on and another is needed to turn it off (see Autonomic Nervous System below).

In other words, each neuron transmits only *one message.* So when it is cold, but not very cold, the first neuron, above, will fire but not the second. Each sensory neuron then has its own particular (characteristic) stimulus, by virtue of the type of sensory receptor endings it has. If that stimulus is present the neuron will be triggered; if that stimulus is not present, it will not fire.

Neurons cannot be stimulated partially or in degrees; they either fire and transmit their signal or they don't fire at all, and no signal is sent. They do not carry weaker or stronger signals—just one signal, just one message. This is known as the *all-or-none rule* governing the behavior of neurons. It might be remembered as the basic grammatical rule of the language of the nervous system.

These facts about the nature of neurons—about their language—help us to understand the basis for the organization of the network of millions of nerves throughout the body.

What Are the Two Main Divisions of the Nervous System?

The nervous system is structurally distinguished into two main parts: the *central nervous system* (CNS) (Fig. 169), which is composed of the spinal cord and brain, and the *peripheral nervous system* (PNS), which involves all the nerves and nerve centers of the outlying or peripheral parts of the body.

THE CENTRAL NERVOUS SYSTEM

What Does the Spinal Cord Look Like?

The spinal cord is enclosed in the bony *vertebral column* and extends from

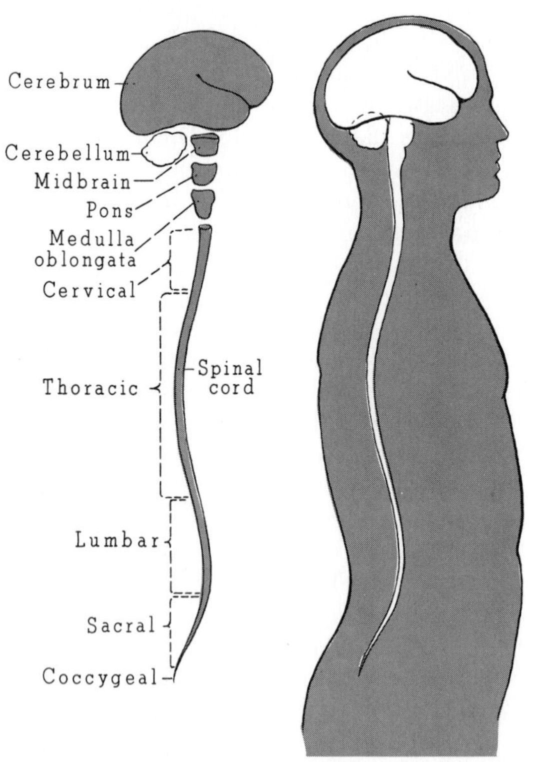

Figure 169. Diagram showing major anatomic divisions of the central nervous system.

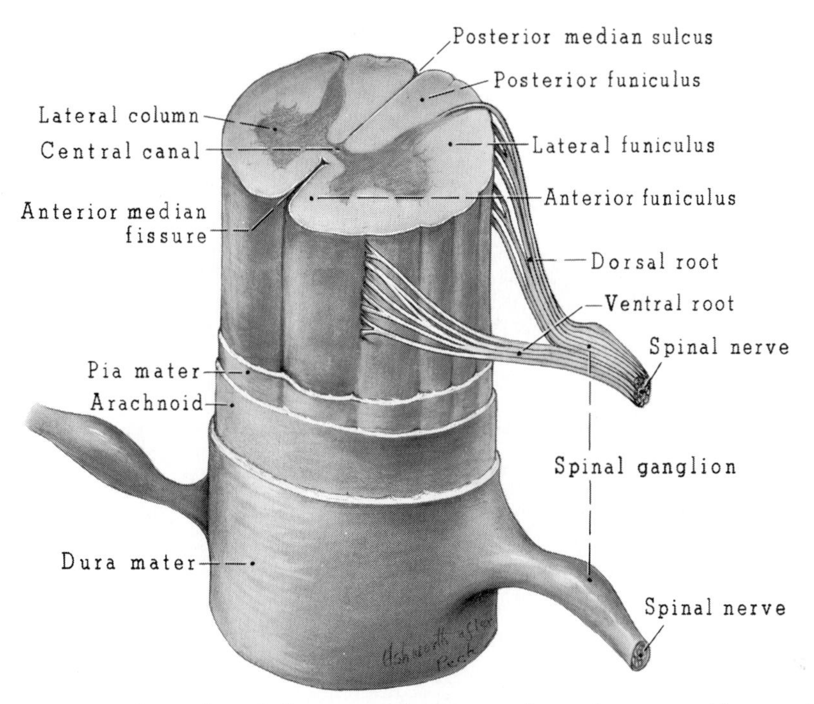

Figure 170. Section of spinal cord illustrating formation of spinal nerve and layers of meninges.

the lower brain, down the midline of the back, to the pelvic region. Inside the vertebral column the spinal cord is covered by three protective membranes called the *meninges* (me-nin'jez). The three layers of the meninges are the *dura mater* (lines the vertebrae), the *pia mater* (covers the cord), and the *arachnoid* (ah-rak'noid) *membrane,*which lies between them (Fig. 170). This middle layer is composed of a network of spaces filled with a specialized fluid called the *cerebrospinal fluid.* The

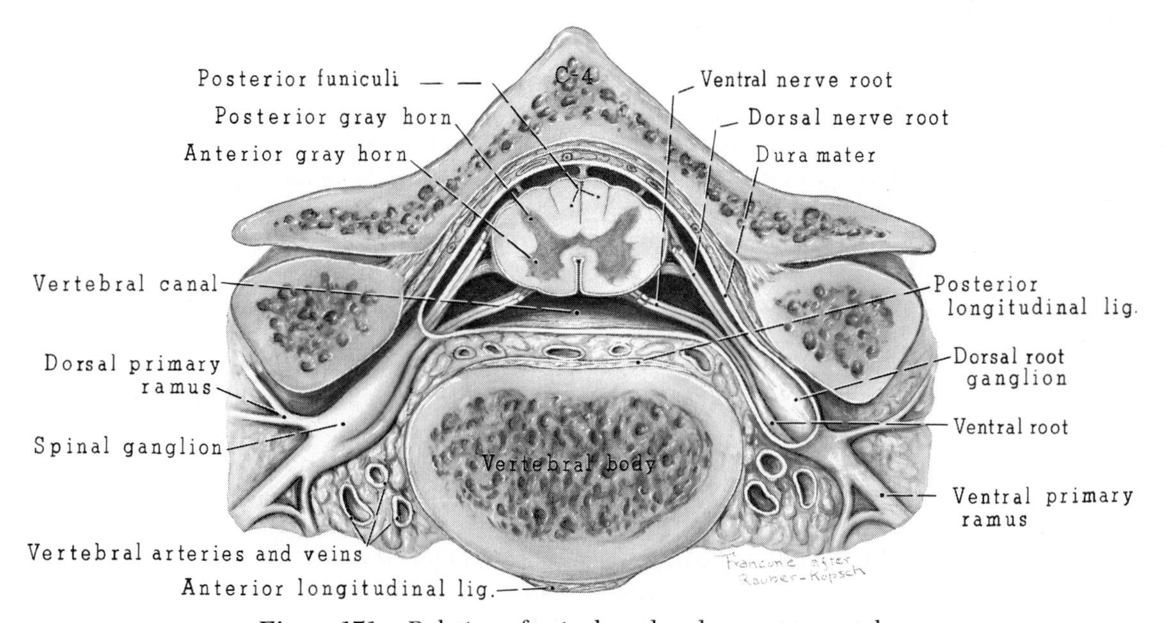

Figure 171. Relation of spinal cord and nerves to vertebra.

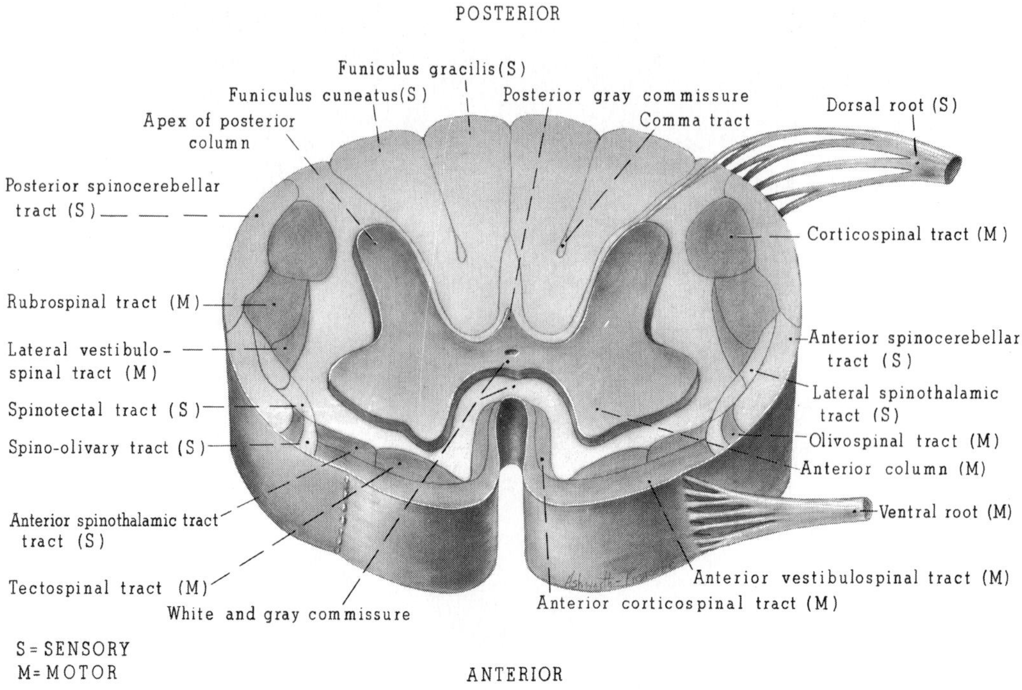

Figure 172. Major ascending and descending tracts of the spinal cord.

composition of the cerebrospinal fluid is very similar to that of the plasma portion of blood, but its function seems to be only protective, since no evidence exists that it participates in the operation of the nervous system. This fluid also surrounds and circulates through the hollows (ventricles) of the brain (Fig. 174).

A cross section of the spinal cord shows an inner region of gray matter in the shape of an H, surrounded by a region of white matter. The inner, gray-matter region is composed mainly of cell bodies of neurons. The outer, white-matter region is made up of bundles of myelinated axons known as *tracts,* which transmit impulses between the brain and the lower areas of the body (Figs. 171 and 172).

What Are the Two Functions of the Spinal Cord?

The activity of the spinal cord can best be understood by dividing it into two functions: the *reflex function* and the *brain communication* function. When a specific sensory impulse enters the spinal cord it can either be switched directly (or through an interneuron) to an outgoing motoneuron making a reflex arc or it can be switched to travel up the (white matter) nerve fiber tracts of the spinal cord to the brain. The *spinal cord reflexes* (primitive protective reflex behaviors below the head) result from the direct switching that is independent of the brain centers (Fig. 167).

The *brain communication* function of the spinal cord is performed by the long nerve fiber tracts (white matter) that run the length of the cord linking the brain centers with various parts of the body. The nerve tracts, which are from both sensory and motor neurons, transmit impulses directly to and from the brain and the sensory receptor and motor-effector sites (Fig. 173). In the later discussion of the peripheral nervous system we will talk about these nerve fibers as they enter (sensory) and leave (motor) the central nervous system to connect with the various receptors and effectors throughout the body. The nerve

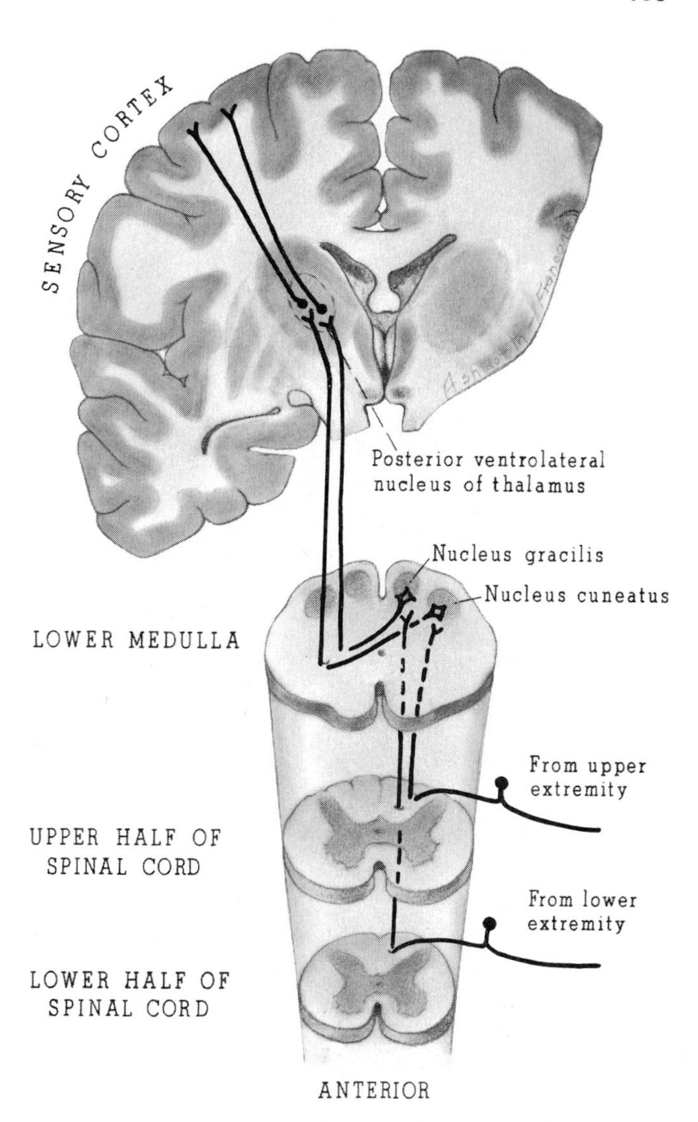

Figure 173. Brain communication with the lower body through the spinal cord (white matter).

fibers entering and leaving the cranial area are referred to as *cranial nerves*, and those to and from the spinal cord are referred to as *spinal nerves*.

These two functions of the spinal cord—reflex and brain communication—are easily seen to be distinct when an injury cuts the spinal cord all the way through. In such cases, the brain communication function *below the point of injury* is lost completely, resulting in a loss of sensation (anesthesia) and a loss of motor influence (paralysis). This is because the tracts—the direct communication lines—have been severed. However, the spinal cord reflexes continue to operate, so that all behaviors and responses controlled by spinal reflex arcs will persist. This occurs because the entire reflex arc (main cell bodies and nerve fibers) is located below the injury and *can* operate independently of the brain centers.

What Does the Brain Look Like?

As is the spinal cord, the brain is covered by a meninges, which is in fact continuous with the meninges of the spinal cord. The cerebrospinal fluid fills the spaces of the arachnoid layer of the meninges and also circulates through internal hollows or

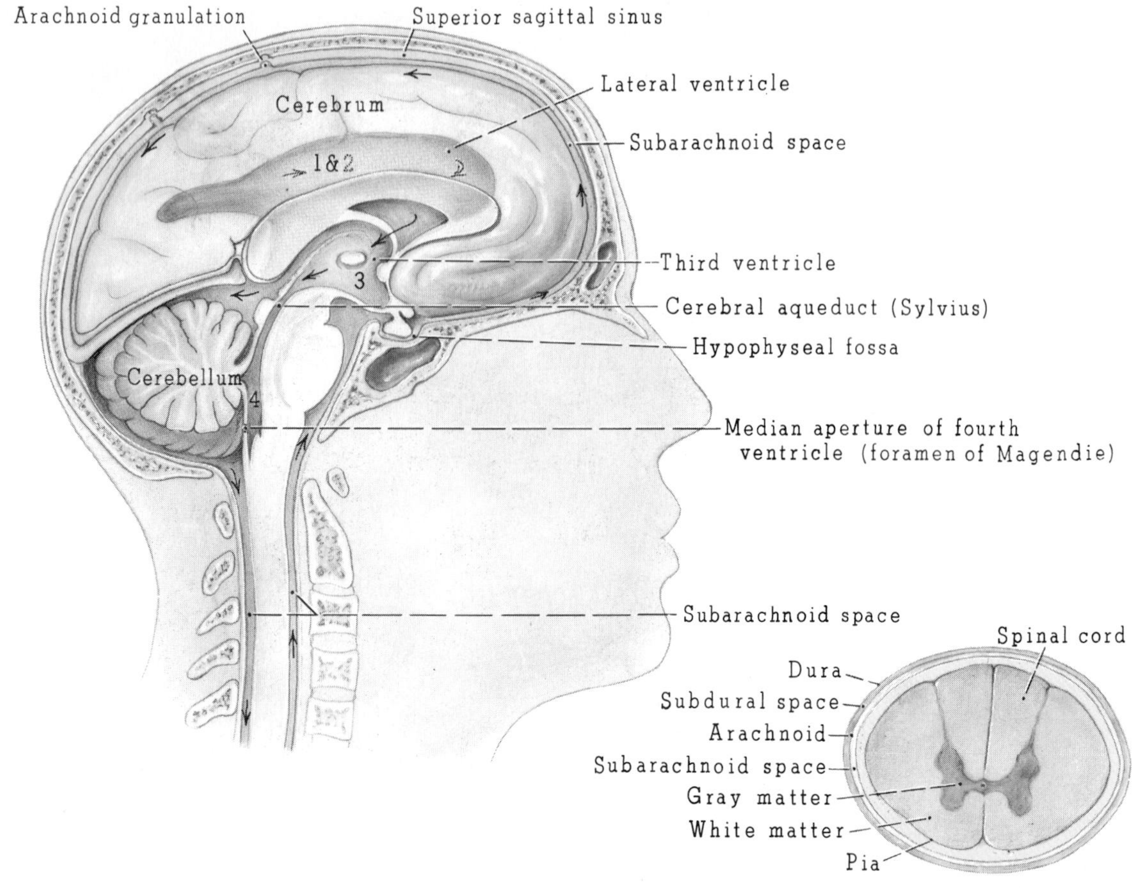

Figure 174. Circulation of cerebrospinal fluid in brain and spinal cord.

ventricles that lie deep inside the brain. These ventricles are connected by the *cerebral aqueduct* to the spinal cavity, which contains the spinal cord. The cerebrospinal fluid is continually being formed and reabsorbed by a sort of osmotic filtration process between the cavities and capillaries (Fig. 174). *Hydrocephalus*, or "water on the brain," results from blockages of this circulation such as by congenital (present at birth) abnormalities or tumors.

The brain itself can be structurally and functionally distinguished into several main parts, as follows (Figs. 175 and 176):

1. *The hindbrain,* includes the pons, medulla and cerebellum.

2. *The midbrain and interbrain,* includes the thalamus, hypothalamus, and important motor centers.

3. *The forebrain,* the largest portion, is also referred to as the cerebrum.

What Are the Structures and Functions of the Hindbrain?

Moving up the spinal cord to the hindbrain, one first encounters the *medulla* (medulla oblongata), which is continuous with the spinal cord. The medulla is very similar in many ways to the spinal cord, and has been called the bulb of the spinal cord. Internally, it contains several separate collections of the nerve cell bodies (gray matter), which are called *centers*. The outer portion of the medulla is made up of the continuation of the numerous myelinated (white matter) nerve fibers (both sensory and motor) that directly link the higher brain centers (through the spinal cord) to the various parts of the body.

The internal, gray-matter portions of the medulla contain several important re-

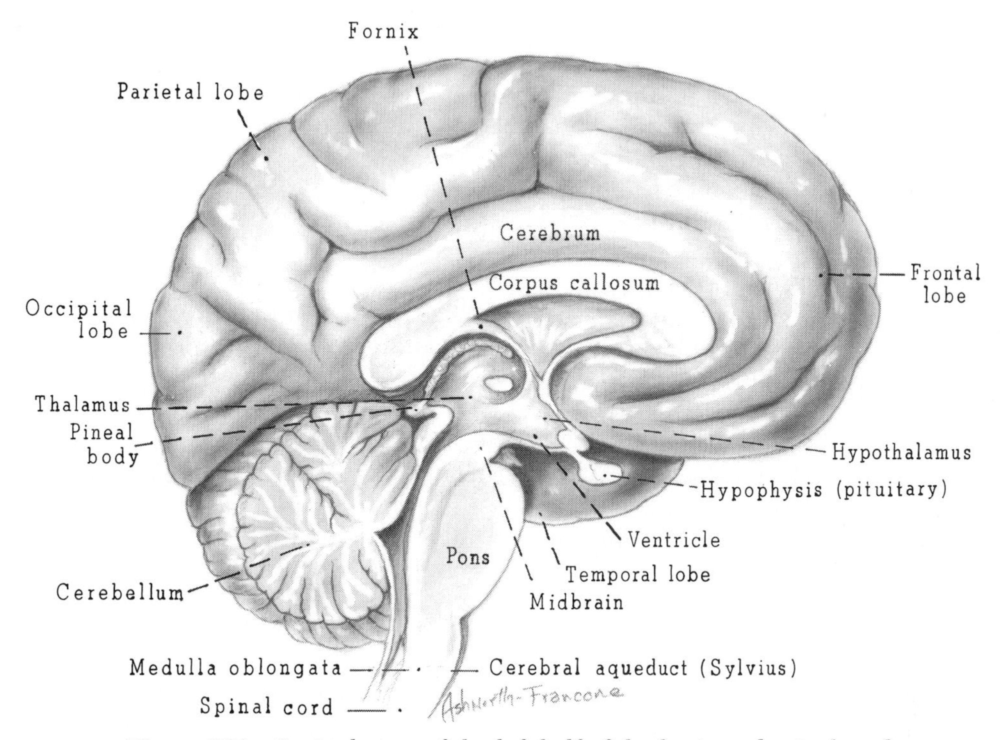

Figure 175. Sagittal view of the left half of the brain and spinal cord.

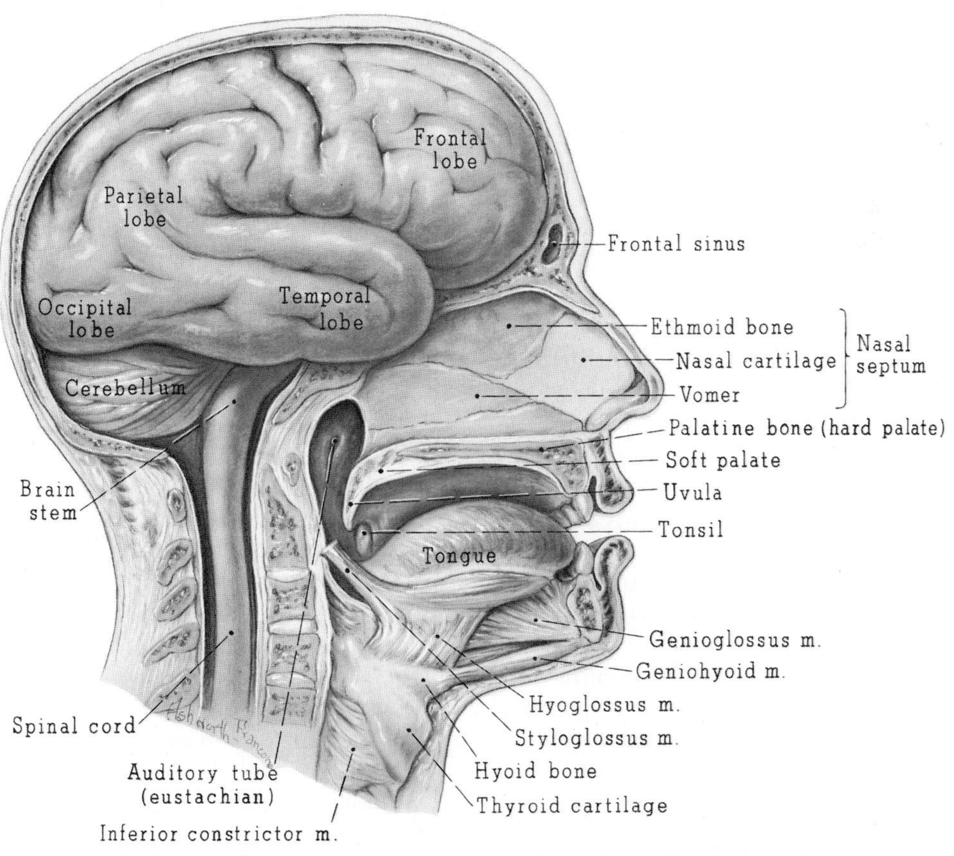

Figure 176. Sagittal section through the head (brain intact).

flex centers. Among these are the reflex centers controlling three vital functions. The *respiratory reflex center* controls the muscles of respiration in response to chemical and other stimuli. The *cardiac reflex center* has a role in controlling the rate of heartbeat. And the *vasomotor reflex center* (*vas* is Latin for vessel) activates constriction of the blood vessels and hence aids in maintaining proper blood pressure.

The *pons* is a bridgelike structure continuous with the medulla and lying anterior to the cerebellum. It is almost entirely composed of white matter linking the various parts of the brain to each other. The pons serves as a relay station from the medulla to the higher centers of the brain. In addition, several important reflex centers for cranial nerves are located in this area.

The *cerebellum* is a large, gray, oval body located posterior to the medulla and pons. The internal portion, called the *vermis*, is made up of white matter and is surrounded by two lateral *hemispheres* of gray matter; note that the medulla and spinal cord are of the opposite makeup—internal gray matter and external white matter. The determination of the functions of the cerebellum has come primarily from observing the problems that arise in patients with cerebellar injuries, and in experimental animals with all or part of the cerebellum removed. From these observations and studies it has been determined that the cerebellum functions in coordinating movements, such as in walking or more strenuous activity; in sustaining posture by maintaining muscle tone; and in controlling the balance and spatial orientation of the individual. A patient with cerebellar damage displays jerkiness of movement, lack of coordination, deterioration of muscle tone and posture, and has difficulty in maintaining balance.

Scattered throughout the area of the midbrain, pons and medulla are numerous large and small neurons that are related to each other by small processes (cellular extensions). These neurons and their fibers constitute the *reticular formation*, or the *reticular activating system* (RAS), which controls the overall degree of central nervous system activity, including the control of wakefulness and sleep, and at least part of our ability to direct our attention.

What Are the Structures and Functions of the Midbrain and Interbrain?

The area just above and forward of the pons is referred to as the *midbrain*. It is composed of several nuclear masses that are of great importance in motor coordination—relating and integrating motor signals from several parts of the brain. Of particular interest are three of these mass centers: the *superior colliculi* (ko-lik′u-li), which are involved in visual reflexes; and two posterior mass centers, the *inferior colliculi*, which are associated with hearing. When the superior part of the midbrain is injured or diseased, an abnormality of eye movements results, particularly a paralysis of upward gaze. The midbrain, pons, and medulla together are commonly referred to as the *brainstem*.

The *interbrain*, also referred to as the diencephalon, is located anterior to the midbrain. It is composed of two important structures, the thalamus and the hypothalamus. The *thalamus* (thal′ah-mus) is a relatively large mass of gray matter located in the medioposterior portion of each cerebral hemisphere, it operates to integrate and process sensory stimuli, suppressing some and magnifying others. The *hypothalamus*, a much smaller structure is located in the midline area below the thalamus. For its size, the hypothalamus is perhaps the most significant single structure so far considered. It has been found to be of central importance in a wide variety of functions, including control of the chief endocrine gland (the pituitary), which influences almost all secretions of the endocrine system; control of appetite, sleep, some emotions (such as fear and pleasure); and control of much of the activity of the autonomic (unconscious) nervous system, which activates and regulates the functioning of almost all of the internal (visceral) organs.

What Are the Structures and Functions of the Forebrain?

The forebrain, or *cerebrum,* is the largest part of the brain, and it controls all the "higher" functions and activities of the human body. It is divided into two hemispheres by the *longitudinal fissure.* These hemispheres are connected only in their lower middle portion, referred to as the *corpus callosum* (kah-lo′sum). The outer layer of the cerebral hemispheres, called the cerebral cortex, is composed of gray matter. The internal portion, called the cerebral medulla, is composed of white matter.

The nerve fibers (tracts) of the neurons contained in the cerebrum that communicate with the lower parts of the body "cross over" in the area of the medulla. As a result, the left hemisphere serves the right side of the body and the right hemisphere serves the left side.

The cerebral cortex of each hemisphere is divided into four *lobes.* Each lobe has been found to control certain categories of functions: (Fig. 177).

1. The *frontal lobe* contains the motor cortex, which controls the voluntary movements of all the muscles of the body. Also located in the frontal lobe are the *verbal speech* and *written speech* centers, which control the muscles of the tongue, soft palate, and larynx and the muscles of the hand and arm, respectively.

2. The *parietal lobe* contains the sensory reception area, and enables awareness of pain, touch, and temperature, as well as judgments of distances, sizes, and shapes.

3. The *temporal lobe* contains the auditory center, which enables awareness of distinct sounds, and also the auditory speech center, which functions in the understanding of spoken language.

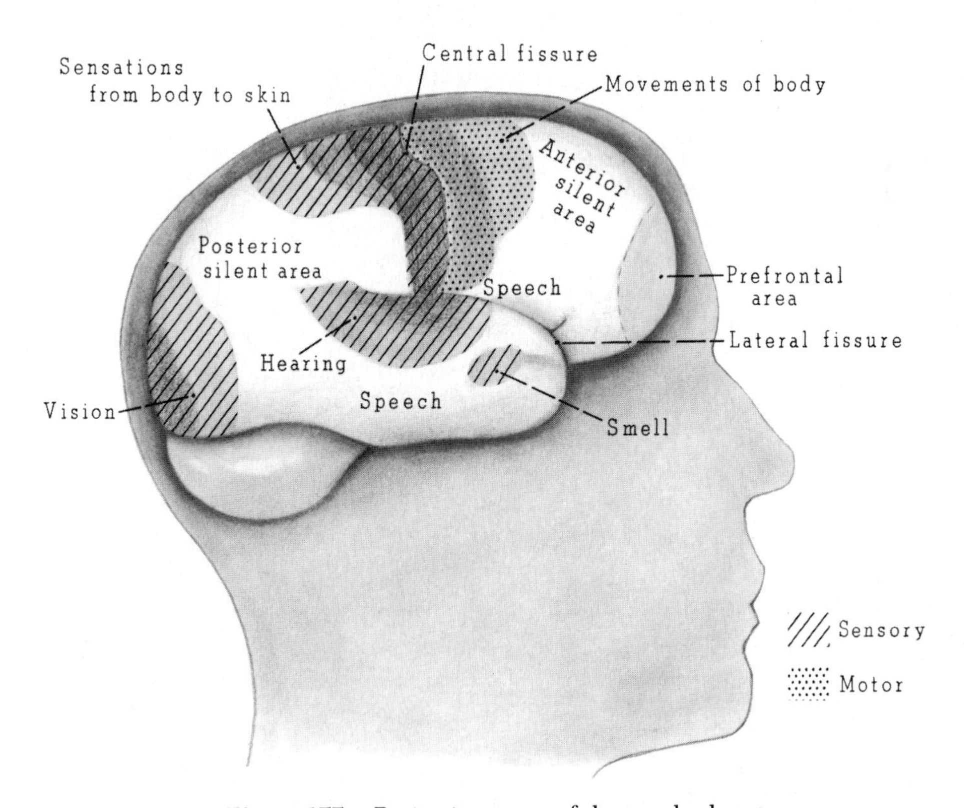

Figure 177. Projection areas of the cerebral cortex.

4. The *occipital* (ok-sip'i-tal) *lobe* contains the visual center, enabling awareness of distinct visual phenomena, and also the visual speech center, which functions in the understanding of written language (reading). The olfactory (smell) center is located in the cerebral core.

More generally, the cerebrum controls conscious mental processes, sensations, emotions, and voluntary movements.

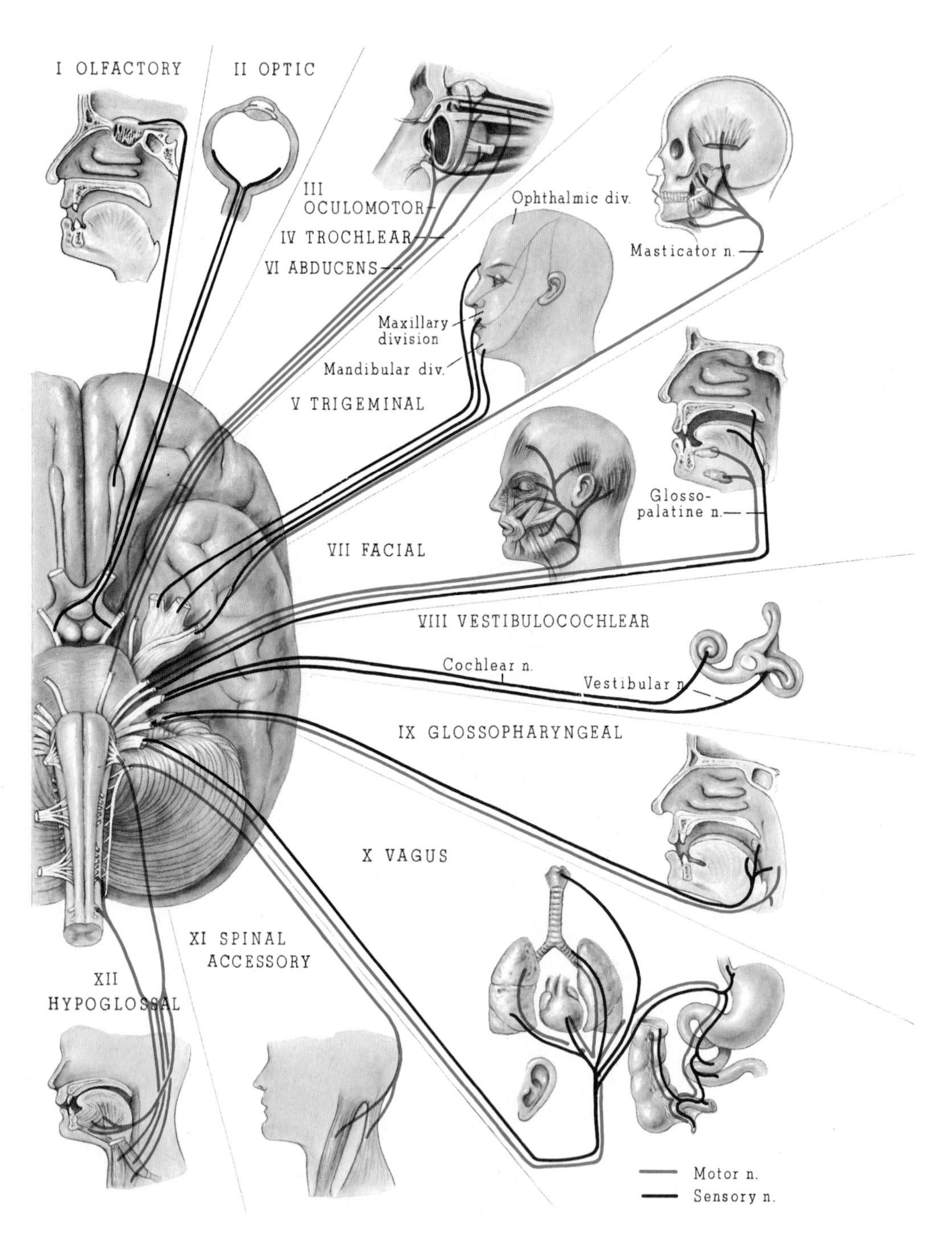

Figure 178. Distribution of cranial nerves. (After Netter.)

THE PERIPHERAL NERVOUS SYSTEM

What Are the Two Main Subdivisions of the Peripheral Nervous System?

The peripheral nervous system is made up of all the nerves and nerve centers of the outlying or peripheral parts of the body. Any part of the nervous system not contained in the central nervous system is a component of the peripheral nervous system. The two main subdivisions are the *cranial nerves* and the *spinal nerves.*

What Are the Cranial Nerves?

The cranial or cerebral nerves are 12 pairs of symmetrically arranged nerves attached to the brain (Figs. 178 and 179). Each leaves the skull through a foramen (small opening in the bone) at its base. The site where the fibers composing the nerve enter or leave the brain surface is usually termed the superficial origin of the nerve; and the more deeply placed group of neurons from which the fibers arise (motor) or around which they terminate (sensory) is called the nucleus of origin or *deep origin* of the nerves (Table 10). The sensory cranial nerve fibers bring in sensory impulses of special senses (such as smell, vision, and hearing) and general senses (such as pain, touch, temperature, deep muscle sense, pressure, and vibration). The motor cranial nerve fibers carry outgoing motor signals activating voluntary and involuntary muscles, organs and glands. Figures 180 to 183 detail several of the cranial nerves.

What Are the Spinal Nerves?

There are 31 pairs of nerves emerging from the spinal cord along almost its entire length (Fig. 184). Each pair is made up of a

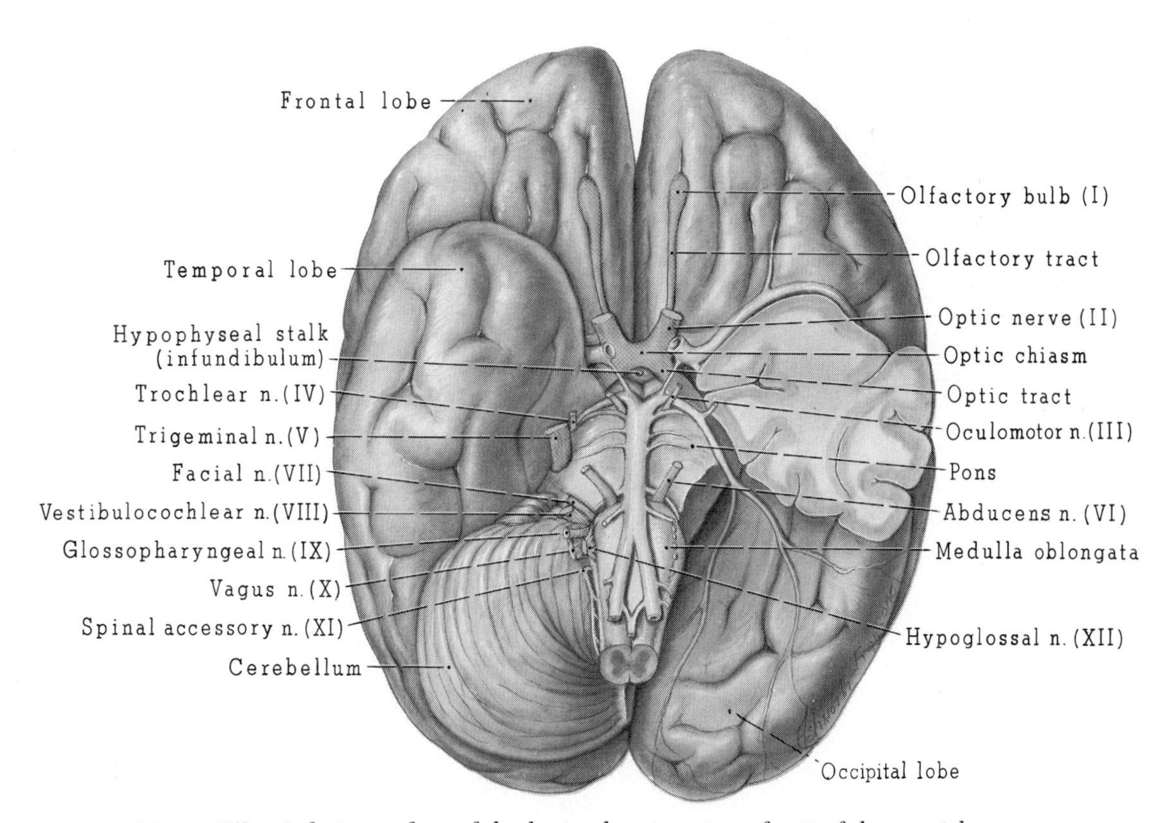

Figure 179. Inferior surface of the brain showing sites of exit of the cranial nerves.

Table 10. CRANIAL NERVES.

NUMBER	NAME	ORIGIN	EXIT FROM SKULL	FUNCTION
I	Olfactory	Cells of nasal mucosa	Cribriform plate of ethmoid	Sensory: olfactory (smell)
II	Optic	Ganglion cells in retina	Optic foramen	Sensory: vision
III	Oculomotor	Midbrain	Superior orbital fissure	Motor: external muscles of eyes except lateral rectus and superior oblique; levator palpebrae superioris. Parasympathetic: sphincter of pupil and ciliary muscle of lens
IV	Trochlear	Roof of midbrain	Superior orbital fissure	Motor: superior oblique muscle
V	Trigeminal	Lateral aspect of pons		
	Ophthalmic branch	Semilunar ganglion	Superior orbital fissure	Sensory: cornea; nasal mucous membrane: skin of face
	Maxillary branch	Semilunar ganglion	Foramen rotundum	Sensory: skin of face; oral cavity; anterior two-thirds of tongue; teeth
	Mandibular branch	Semilunar ganglion	Foramen ovale	Motor: muscles of mastication Sensory: skin of face
VI	Abducens	Lower margin of pons	Superior orbital fissure	Motor: lateral rectus muscle
VII	Facial	Lower margin of pons	Stylomastoid foramen	Parasympathetic: lacrimal, submandibular, and sublingual glands Motor: muscles of facial expression Sensory: taste, anterior two-thirds of tongue
VIII	Vestibulocochlear			
	Vestibular	Lower border of pons	Internal auditory meatus	Sensory: equilibrium
	Cochlear	Lower border of pons	Internal auditory meatus	Sensory: hearing
IX	Glossopharyngeal	Medulla oblongata	Jugular foramen	Motor: stylopharyngeus muscle Sensory: tongue (posterior one-third), taste, pharynx Branch to the carotid sinus
X	Vagus	Medulla oblongata	Jugular foramen	Sensory: external meatus, pharynx, and larynx Motor: pharynx and larynx Parasympathetic: thoracic and abdominal viscera
XI	Accessory	Medulla oblongata	Jugular foramen	Motor: trapezius and sternocleidomastoid muscles
XII	Hypoglossal	Anterior lateral sulcus between olive and pyramid	Hypoglossal canal	Motor: muscles of tongue

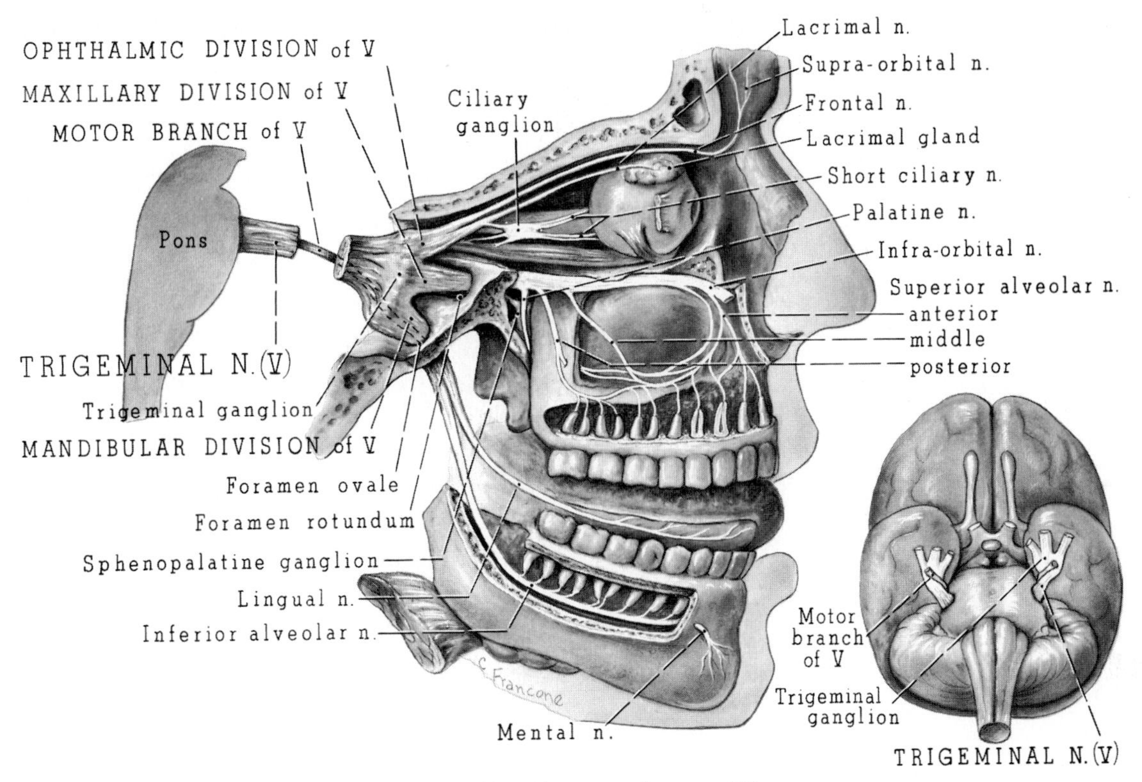

Figure 180. Trigeminal nerve (V).

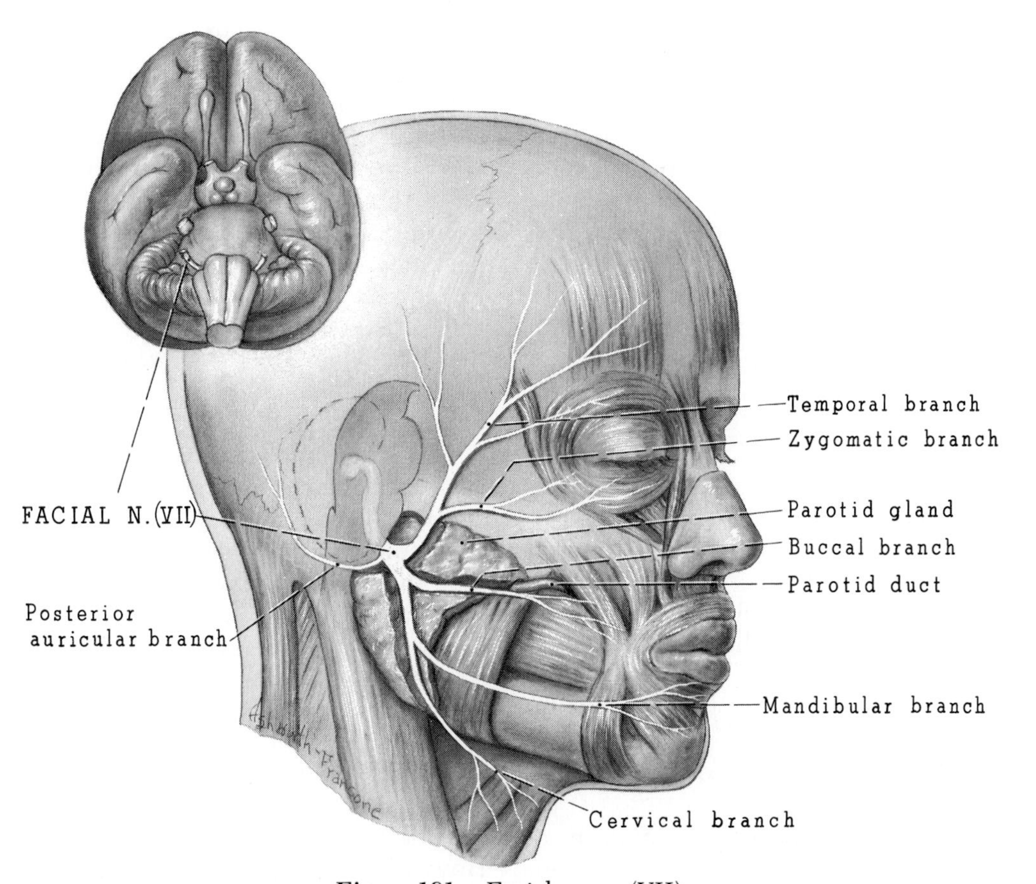

Figure 181. Facial nerve (VII).

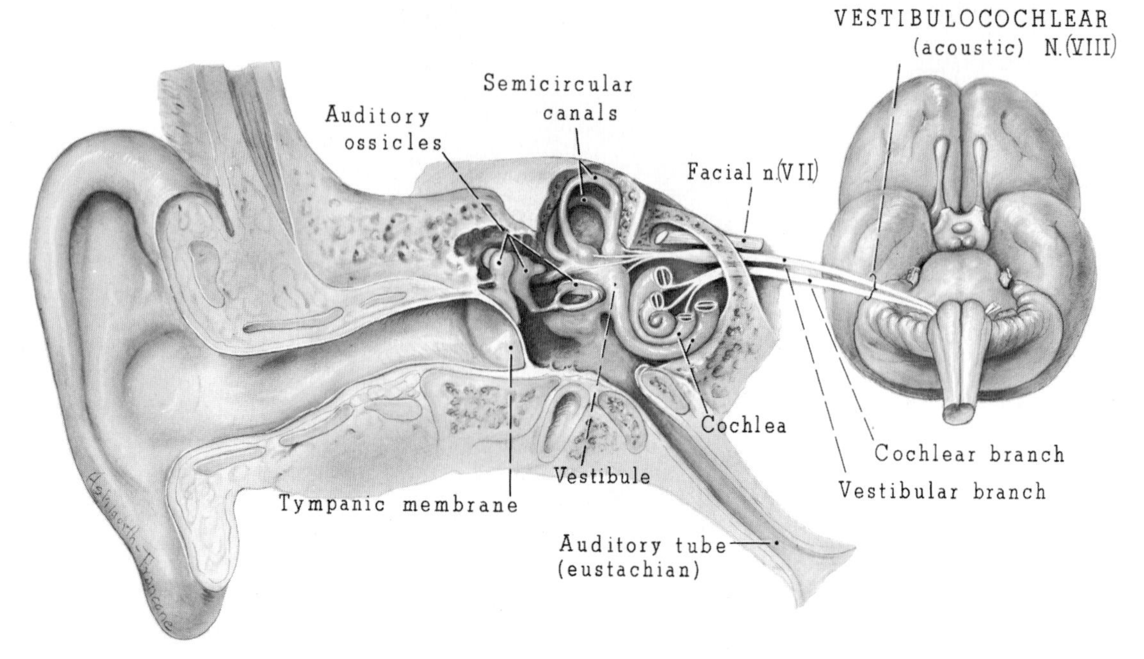

Figure 182. Vestibulocochlear nerve (VIII).

dorsal root and a *ventral root.* The ventral root nerve carries motor signals, and the dorsal root nerve carries sensory signals. Along each dorsal root are found concentrations of sensory nerve cell main bodies; these collections are called dorsal root ganglia. The ganglia are gray matter, since main cell bodies are never covered by myelin, even when located outside the central nervous system. The main cell bodies for the ventral (motor) roots are located inside the spinal column, forming the internal gray matter of the spinal cord (Fig. 185).

Each spinal nerve branches into a small, posterior division and a larger, anterior, division a short distance after emerging from the spine. The anterior branches interlace to form networks called *plexuses* (plek'sus-eez), from which branches then re-emerge to innervate the motor effector organs (Figs. 186 to 188).

What Is the Autonomic Nervous System?

The autonomic nervous system is a functionally specialized subsystem of the overall nervous system. It generally controls bodily functions that are considered to be involuntary, acting on the internal (visceral) *effectors* such as smooth and cardiac muscle, exocrine glands (sweat and salivary), and some endocrine glands (Fig. 189). Anatomically and functionally, the autonomic nervous system is the motor pathway linking the control centers of the brain with the internal organs and secretory cells. The corresponding sensory portion of the nervous system serving these areas is not considered to be part of the autonomic system.

The autonomic nervous system, then, is strictly a motor system (Table 11).

Unlike the motor portion of the voluntary (somatic) division of the nervous system, the autonomic system displays in part an organizational network of motor ganglia. These *ganglia* are synaptic junctions and relay centers in the divided motor signal transmission system of the autonomic system. On the spinal side there are motor neurons with their main bodies inside the spinal cord (as with the voluntary system), and these are called preganglionic neurons. However, unlike the voluntary system,

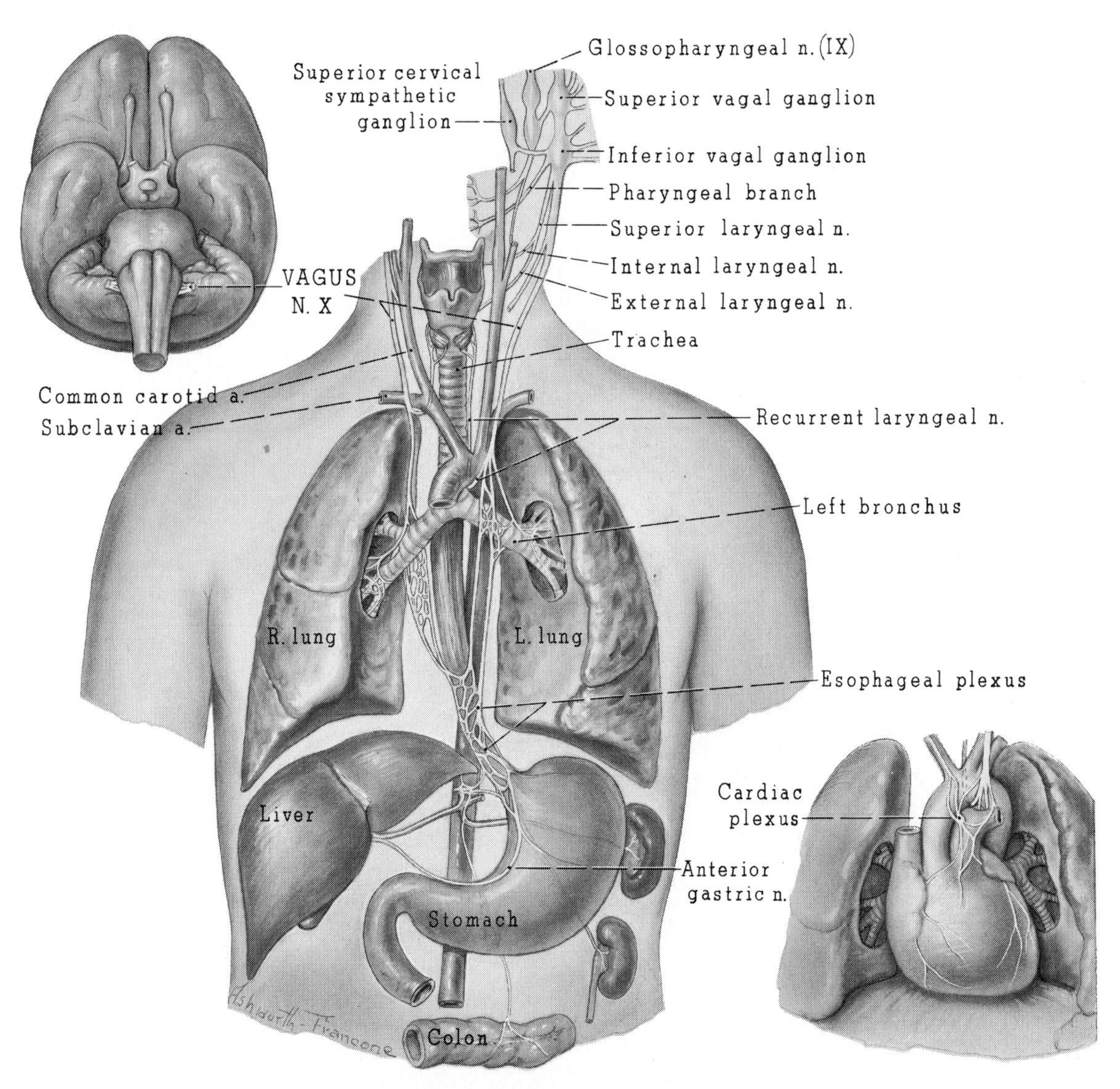

Figure 183. Vagus nerve (X).

these preganglionic neurons do not reach the effectors, instead they synapse (connect) with the postganglionic neurons, whose main cell bodies go to make up the bulk of the ganglion. It is the postganglionic neuron nerve fiber that finally carries the motor signal from the ganglion to the effector organ. Recall that in the somatic (voluntary) nervous system the motor neuron cell bodies are located inside the spinal cord, with fibers extending directly out to the effector (such as skeletal muscles).

What Are the Sympathetic and Parasympathetic Subsystems?

The overall function of the autonomic nervous system seems to be to maintain homeostasis (since it controls the most important homeostatic effector organs) of the internal environment, particularly with respect to temperature and composition. The autonomic nervous system can be further distinguished into the sympathetic and parasympathetic nervous systems. These two systems both operate in the organs

Text continued on page 201.

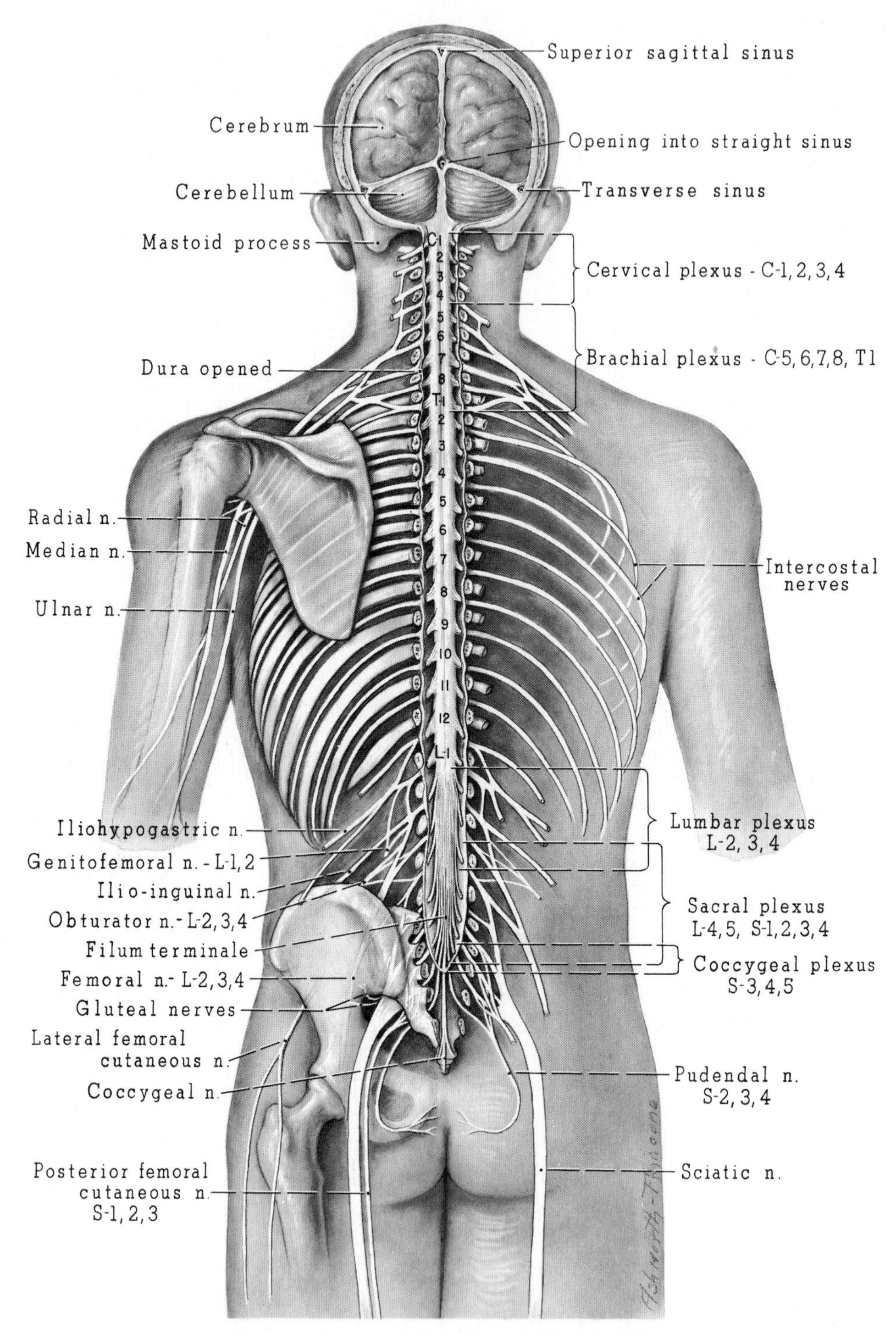

Figure 184. Spinal cord and nerves emerging from it.

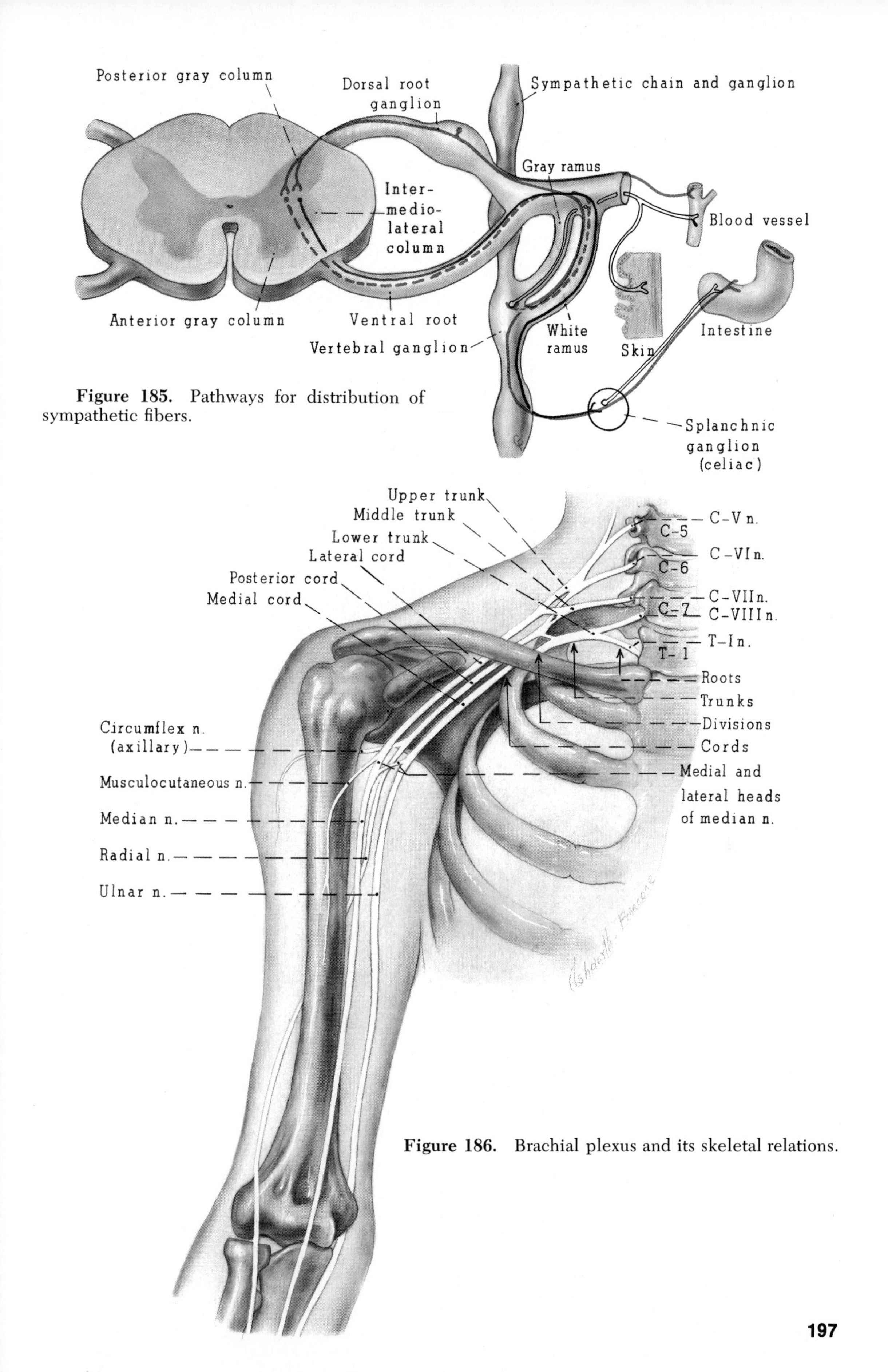

Figure 185. Pathways for distribution of sympathetic fibers.

Figure 186. Brachial plexus and its skeletal relations.

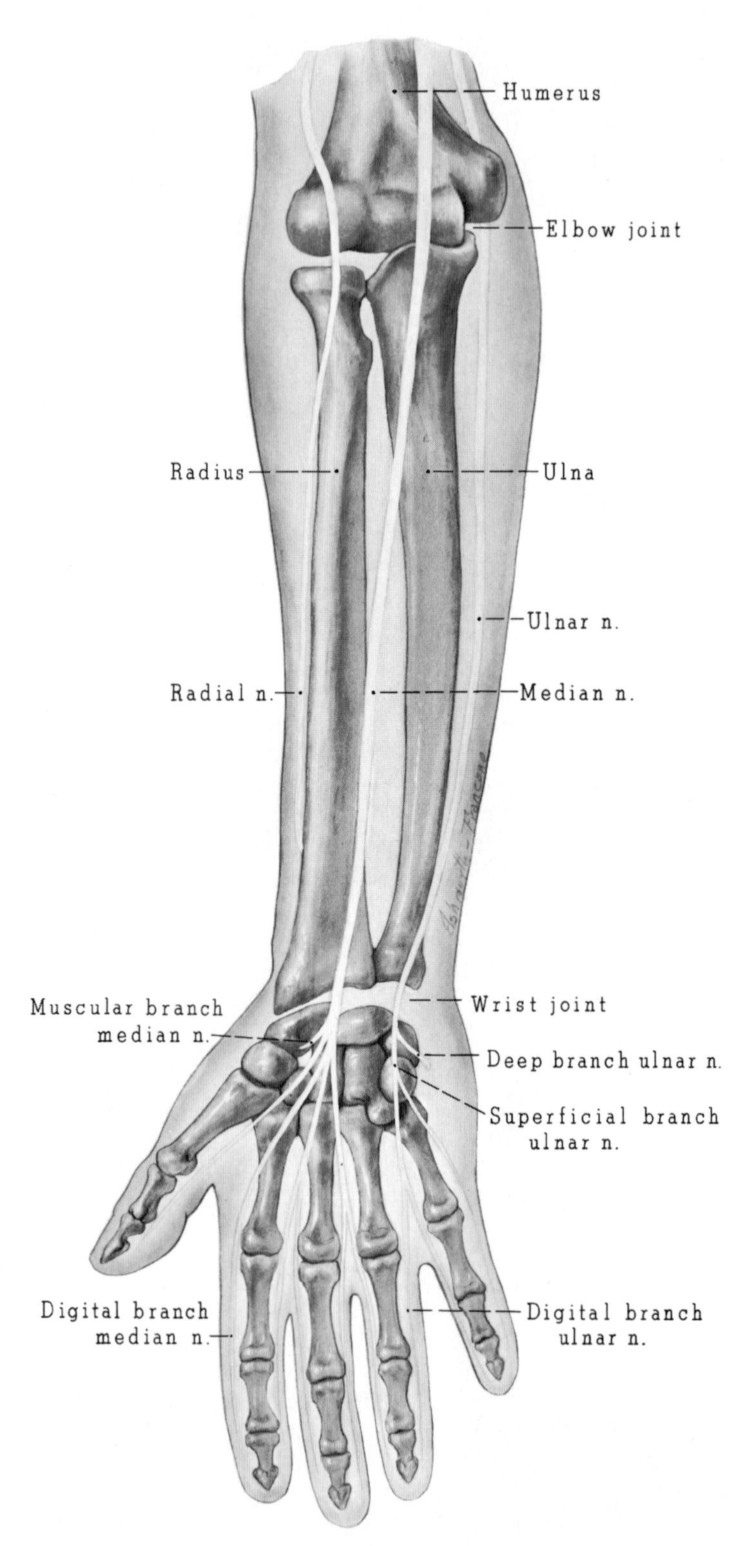

Figure 187. Nerves of right forearm and hand (palmar view).

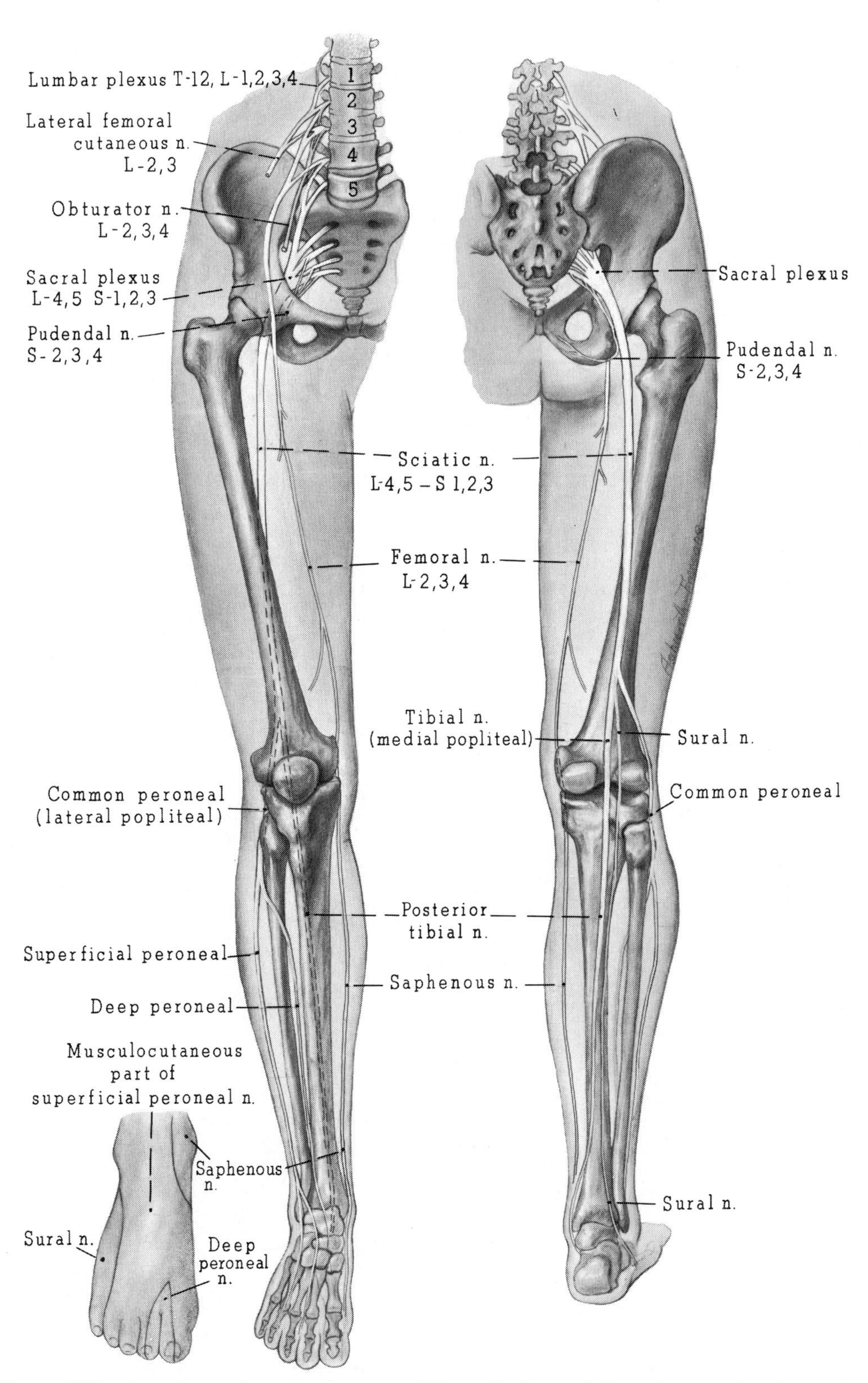

Figure 188. Anterior and posterior view of the right leg and foot, showing lumbar and sacral plexuses and the regions supplied. Insert shows areas of the foot supplied by the nerves.

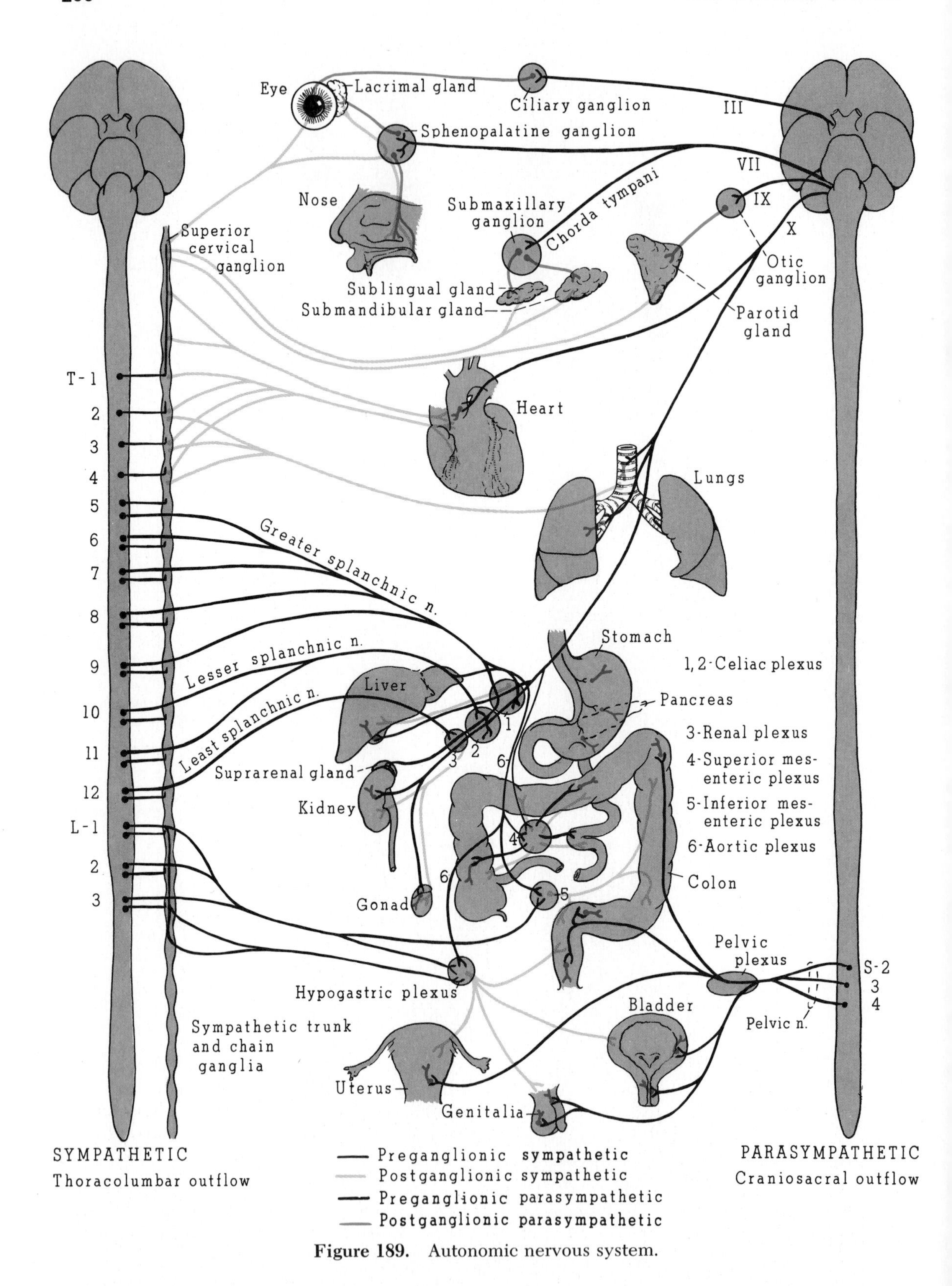

Figure 189. Autonomic nervous system.

Table 11. FUNCTIONS OF THE AUTONOMIC NERVOUS SYSTEM

ORGAN	SYMPATHETIC STIMULATION	PARASYMPATHETIC STIMULATION
Eye	Accommodates for far vision	Accommodates for near vision
Iris	Dilates pupil	Constricts pupil
Ciliary muscle	Lens flattens (inhibits)	Lens bulges (stimulates)
Glands		
Lacrimal	Vasoconstriction	Stimulation of secretion high in enzyme content
Sweat	Copious sweating	Secretion of tears
Heart		
S.A. node	Increases rate	Decreases rate
Muscle	Increases force of contraction	
Lungs		
Bronchi	Dilation	Constriction
Stomach		
Sphincter	Contraction	Inhibition
Glands	Inhibition	Secretion
Intestine		
Wall	Inhibition	Increases tone of musculature
Anal sphincter	Contraction	Decreases tone of musculature
Pancreas	Diminishes enzyme secretion	Stimulates secretion of pancreatic enzymes
Suprarenal gland		
Medulla	Secretion	No known effect
Kidney	Decreases output	No known effect
Urinary bladder		
Detrusor	Inhibition	Excitation
Trigone (sphincter)	Excitation	Inhibition
Penis	Ejaculation	Erection
Arterioles in abdomen and skin	Constriction	
Arrector muscles of hair follicles	Contraction	

controlled by the autonomic system. The two sets of nerves from these subsystems have opposing functions—one stimulates and the other inhibits the activity of a given organ. Thus, the heart rate is slowed by the parasympathetic and accelerated by the sympathetic division. In general, the parasympathetic system is concerned with restorative processes, and the sympathetic with processes involving energy expenditure. The sympathetic system is sometimes called the "fight or flight" system, since stimulation through the sympathetic division alone does those things which would prepare one to either fight or flee.

The sympathetic nerves arise from all the thoracic segments and from the first few lumbar segments of the spinal cord. The parasympathetic nerves arise from the third, seventh, ninth, and tenth cranial nerves, and from the second, third, and fourth sacral segments of the spinal cord.

SPECIAL SENSES

How Are the Sensory Receptors Organized?

At the beginnings of the dendrites of the sensory neurons are specialized receptors, microscopic sense organs. Literally millions of these sensory receptors are scattered in almost every part of the body. Some receptors are designed to respond to special types of external stimuli in a limited area; examples are the receptors for vision, hearing, taste, and smell. Other receptors, such as those sensitive to touch, pressure, temperature, and positions of the body, are more generally distributed. When stimulated, a receptor organ produces impulses in the sensory neuron, which transmits the impulses to the spinal cord or brain.

The sensation of pain does not seem to arise from special receptor organs, but rather from *free nerve endings;* that is, branchings of the dendrite fibers without any specialized end organs. These pain sensation endings are the most widely distributed of any type of sensory receptor. They are found in the skin, muscles, and joints, and, to a lesser extent, in most internal organs (including the blood vessels and viscera).

Internal sensations such as hunger and thirst seem to be more complex, arising from a variety of conditions that stimulate receptors in the stomach, intestines, and throat, as well as receptors that monitor the level of various nutrients in the blood. Hunger, in particular, involves several types of sensory stimulation, taken individually or together.

The two major receptor structures considered next operate in conjunction with the functioning of gross anatomical sensory organs that aid in gathering sensory information from the enviornment as the attention directs.

What Does the Eye Look Like? How Does It Work?

The lining of the eyeball is made up of three partially specialized membrane layers: the sclera, the choroid, and the retina. The *sclera* (skle′rah) is the tough outer layer composed of connective tissue. What is commonly referred to as the "white" of the eye is part of the frontal portion of the sclera (Fig. 190). In the center of the frontal surface of the sclera is a specialized transparent area known as the *cornea.* Lying over the cornea is a limited protective mucous membrane referred to as the *conjunctiva* (kon-junk-ti′vah), which aids in keeping particles from irritating the corneal surface.

The *choroid* layer is also specialized in the frontal area of the eyeball, being made

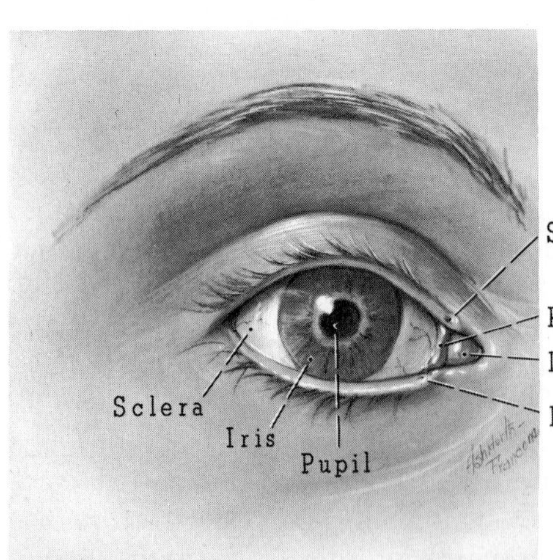

Figure 190. External appearance of the eye and surrounding structures.

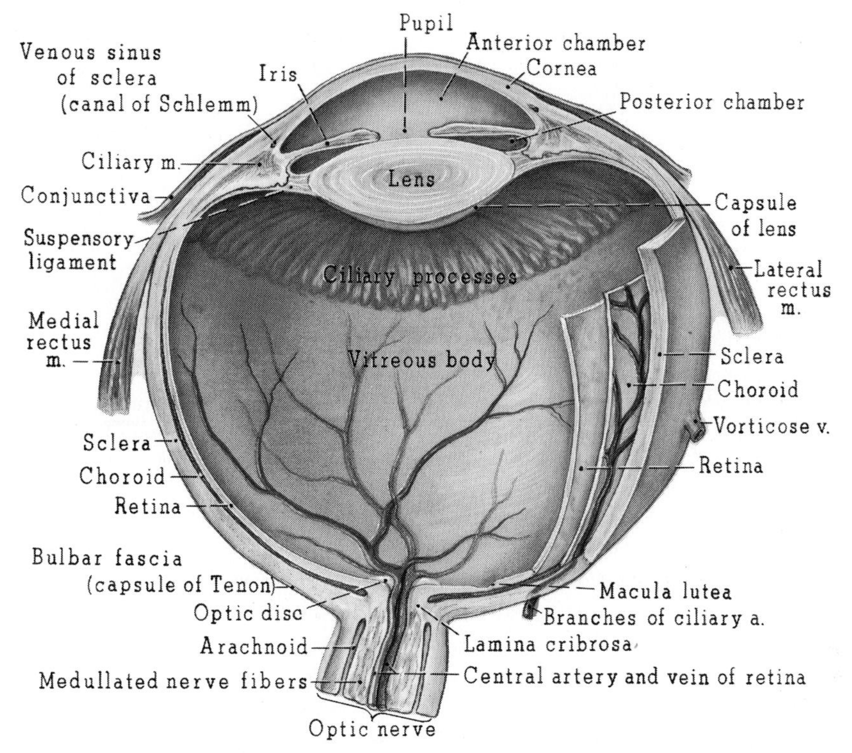

Figure 191. Mid-sagittal section through the eyeball showing layers of retina and blood supply. (After Lederle.)

up of two involuntary muscles, the iris and the ciliary muscle. The *iris* is a colored, doughnut-shaped muscle. The open area in the middle of the iris is called the *pupil* (Figs. 190 and 191). In normal bright daylight the iris is relaxed and the pupil opening is small. In dim light the spokelike muscle fibers of the iris contract, widening the pupil opening and thus allowing more light to enter the interior of the eyeball.

The *ciliary muscle* is attached to the *lens* of the eye; both of these structures lie behind the iris and pupil opening. The lens is held in place by a ligament attached to the ciliary muscle. When one focuses on distant objects, the ciliary muscle relaxes

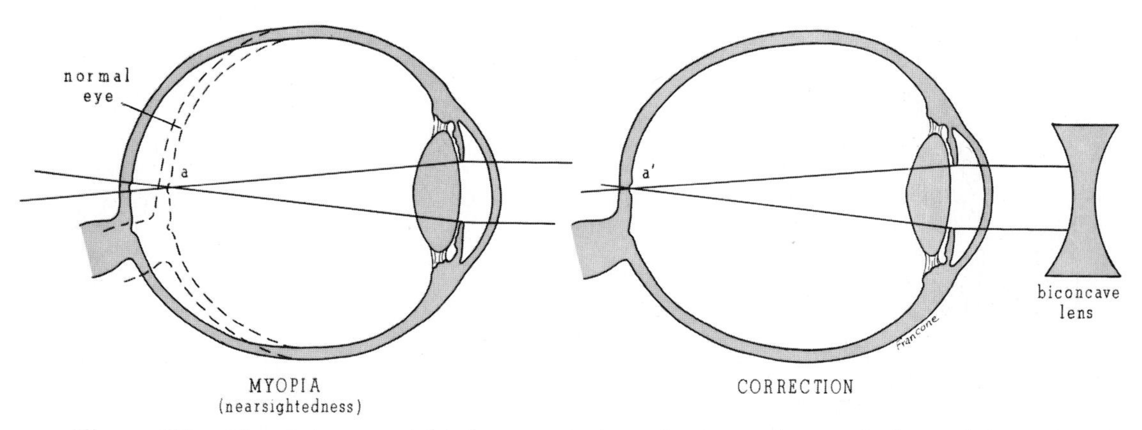

Figure 192. Myopia or nearsightedness; note how the image focuses in front of the retina. A biconcave lens is used as a corrective device for this condition. *a* indicates incorrect point of focus; *a'* indicates focus after correction.

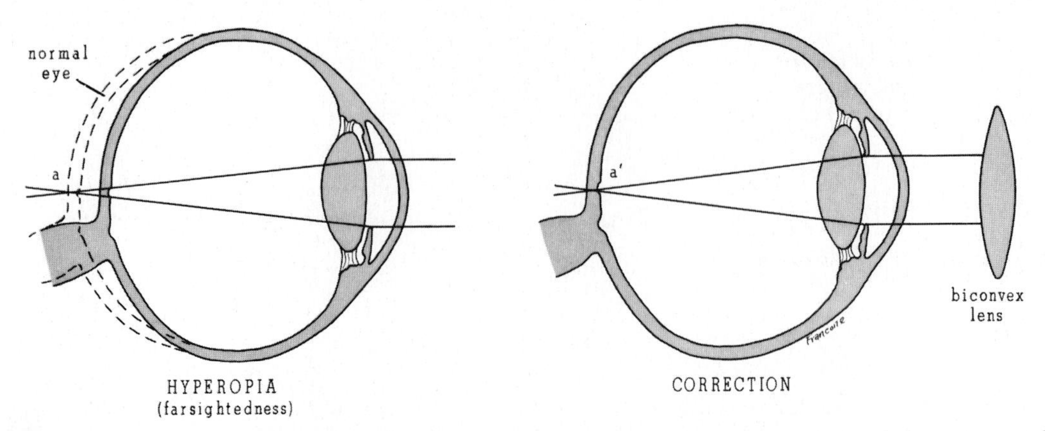

Figure 193. Hyperopia or farsightedness; note how the image focuses behind the retina. A biconvex lens is used as a corrective device for this condition. *a* indicates incorrect point of focus; *a'* indicates focus after correction.

and the curve of the lens is flattened. When one is looking at close objects, as when reading, the ciliary muscle contracts, causing the lens to bulge into a more pronounced curvature. These changes in the curvature of the lens operate to focus light images on the internal surface (retina) of the back of the eyeball.

The interior lining of the eyeball is referred to as the *retina.* The portion of the retina directly across the chamber behind the lens is the only area sensitive to light. This light-sensitive area contains two types of microscopic light receptors, the rods and the cones, which derive their names from their shapes. The *cones* are stimulated by fairly bright light, and are of central importance in daylight and in most color vision. The *rods* operate in dim light and are necessary for good night vision. The rods tend to be located away from the point of central focus. This is demonstrated by the fact that, if one looks slightly away from a dimly lit object or a dim light source (such as a star at night), the object is seen more clearly. Figures 192, 193, and 194 show some common abnormalities in vision, and how they are corrected.

What Does the Ear Look Like? How Does It Work?

The ear consists of three portions: an external, a middle and an inner ear. The

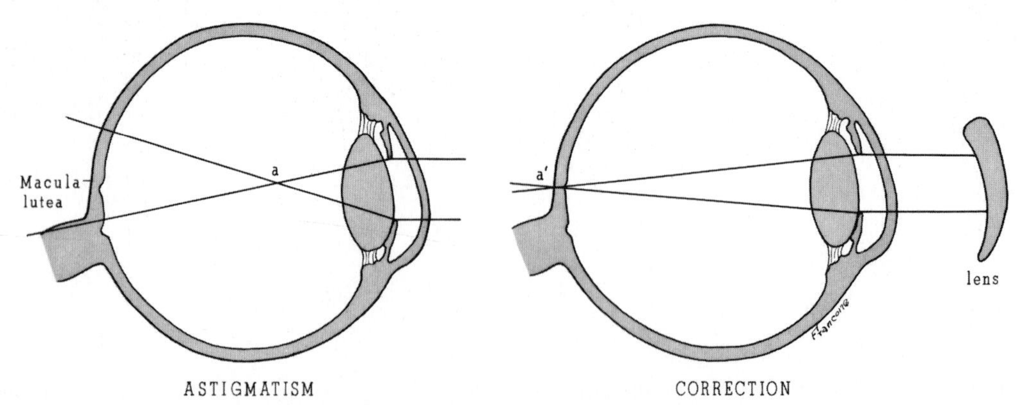

Figure 194. Astigmatism: uneven focusing of the image resulting from distortion of the curvature of the lens or cornea. *a* indicates incorrect point of focus; *a'* indicates focus after correction.

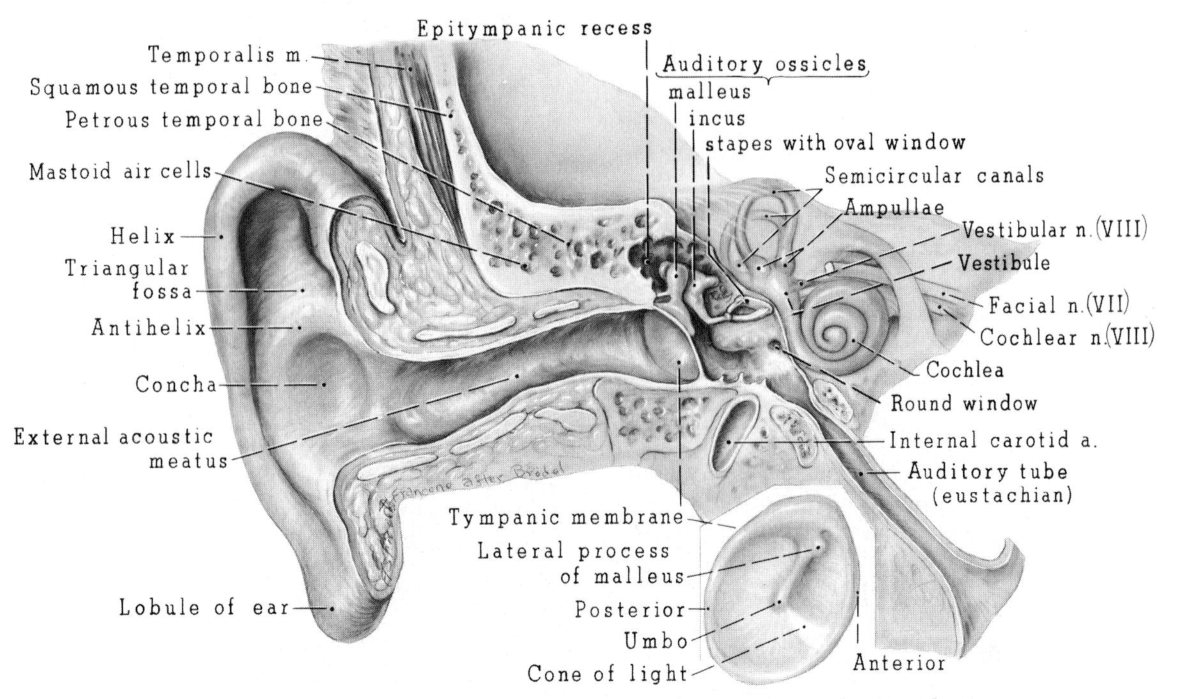

Figure 195. Frontal section through the outer, middle, and internal ear.

external ear has two parts, the ear flap (auricle or pinna) and the ear canal (external acoustic meatus). The ear flap functions to collect sounds, which then travel along the ear canal to the eardrum (Figs. 195 and 196).

The *middle ear* (tympanic cavity) is a tiny hollow in the temporal bone. The tym-

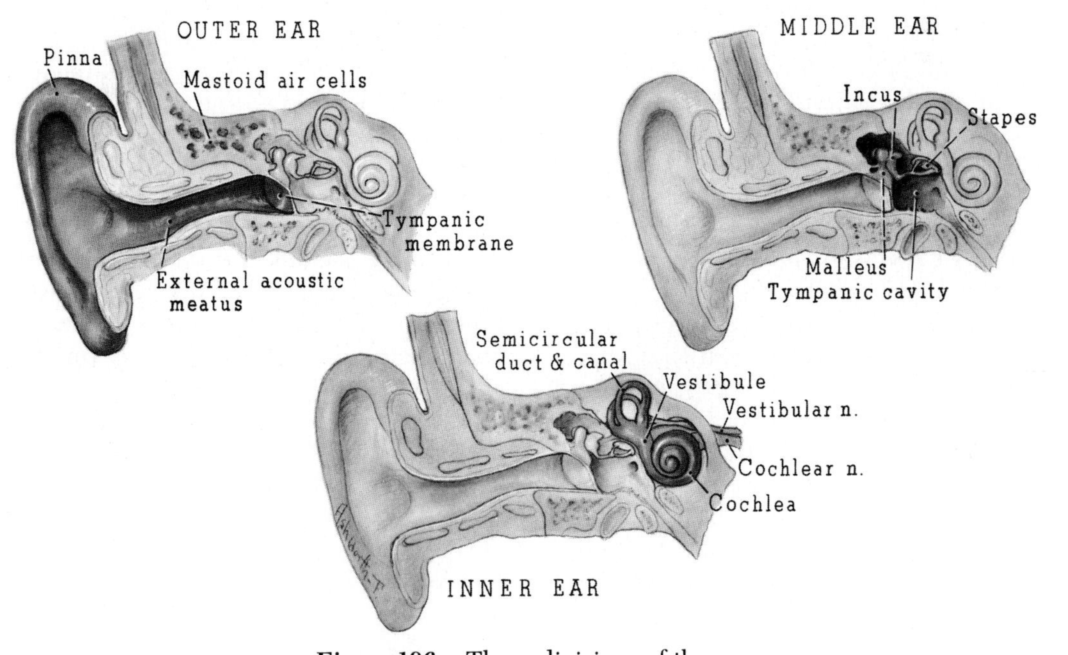

Figure 196. Three divisions of the ear.

panic membrane, or eardrum, separates the middle ear from the external ear canal. Within the middle ear cavity are three very small bones: the malleus (hammer), the incus (anvil), and the stapes (stirrup). The malleus (mal'ee-us) is attached to the eardrum and to the incus. In sequence, the other end of the incus is connected to one end of the stapes, and the stapes then terminates at the oval window (the opening to the inner ear). Sound vibrations striking the eardrum are translated to the inner ear by the mechanical movement of the three small bones of the middle ear.

The middle ear cavity has multiple openings: the opening covered by the tympanic membrane; the auditory, or eustachian, tube (connecting with the throat); the openings into the mastoid cavity (sinuses); and the openings into the inner ear. The eustachian tube is frequently the pathway for the spread of infections, which may begin with a sore throat and lead to middle ear infection (otitis media) or infection of the mastoid spaces (mastoiditis).

The *inner ear* consists of bony and membranous labyrinths. The bony labyrinth, composed of a series of canals hollowed out of the temporal bone, is filled with a fluid called perilymph. The membranous labyrinth lies within the bony labyrinth and is filled with a fluid called endolymph. Each labyrinth has three parts: the *vestibule* and *semicircular canals,* which contain the sensory receptors for balance, and the *cochlea* (kok'le-ah) (like a snail shell), which contains the sensory receptors for hearing. The portion of the cochlea that contains the hearing receptors is referred to as the organ of Corti.

Summary. When sound is produced, the atmosphere is disturbed by sound waves radiating from the source. As sound waves impinge on the eardrum, the membrane vibrates at the same frequency as the source creating the sound. Sound vibrations are carried from the tympanic membrane by the small bones of the middle ear to the inner ear to be transformed into nerve impulses.

CLINICAL CONSIDERATIONS

What Are the Results of Injury to the Spinal Cord?

The spinal cord may be either partially injured (as by blunt trauma) or it may be severed completely (cut through). *Partial injuries* produce weakness, abnormal sensations, and decreased ability to sense pain and temperature. If the spinal cord is *severed completely,* communication between the higher centers and the sensory and motor neurons is lost below the point of transection. This results in a total loss of sensation and a loss of conscious motor control (paralysis) below the point of transection.

What Is Multiple Sclerosis?

Multiple sclerosis (skle-ro'sis) is a disease of the central nervous system that is characterized by a patchy demyelinization (loss of the myelin covering) in many (multiple) areas, resulting in a variety of symptoms involving both sensory and motor systems. These symptoms include virtually all the dysfunctions (abnormal functionings) of the nervous system. Multiple sclerosis is a chronic disease, and the symptoms will sometimes disappear for months or years at a time.

What Is Hydrocephalus?

Hydrocephalus, or "water on the brain," is a condition that occurs when blockage of circulation of cerebrospinal fluid increases the pressure on the brain and spinal cord. The signs and symptoms of hydrocephalus may be evident at the time of birth. The head enlarges and the veins of the scalp dilate and become prominent (Fig. 197).

Surgical treatment consists of diverting the cerebrospinal fluid from one compartment into another in the normal fluid

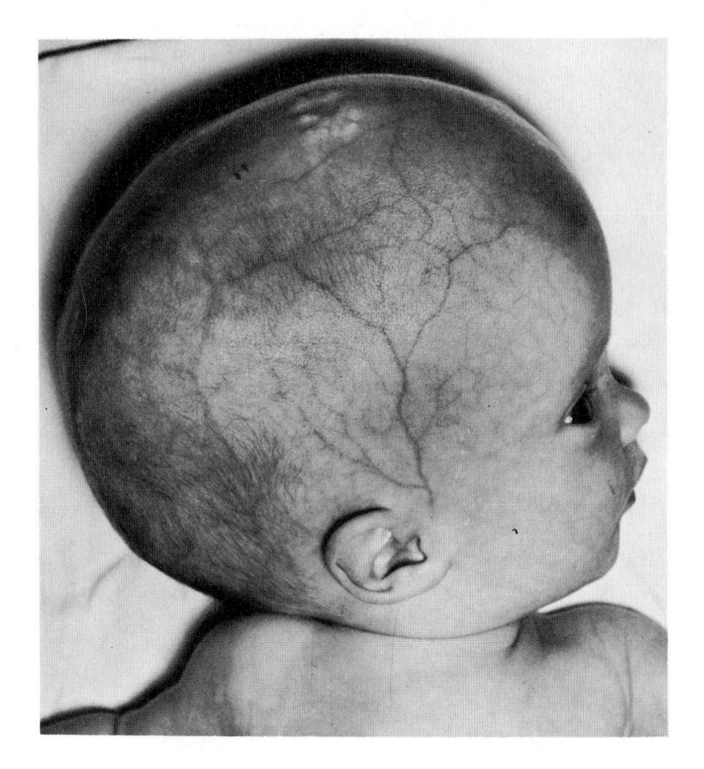

Figure 197. Child, age 4 months, with hydrocephalus.

pathways, or from the cerebrospinal fluid compartments to some other area of the body, where it can be absorbed. One diverting technique is described in Figure 198.

What Is Meningitis?

Meningitis (men″in-ji′tis) is an infection of the meninges. The diagnosis depends on the history of the infection, the so-called meningeal signs (such as stiffness of the neck), and any abnormalities of the spinal fluid. In the small infants, manifestations of mild meningitis are sometimes masked for days, with the symptoms suggesting a cold (upper respiratory infection).

What Are Convulsive Seizures?

Convulsive disorders are more commonly known as *epilepsy.* An attack is characterized by only momentary suspension of consciousness is called *petit mal* (French for "little illness"). An attack in which immediate loss of consciousness and a violent generalized convulsion occur is called *grand mal.*

What Is a Concussion?

Concussion is defined as a temporary loss of consciousness or paralysis of nervous function as a result of a violent blow or shock. Even when consciousness is not lost, in a few days symptoms may arise including headache, dizziness, loss of self-confidence, nervousness, fatigue, inability to sleep, and depression. No account has been given of the mechanism of these symptoms.

Other symptoms may occur that are an indication of some process in addition to concussion. Symptoms include delayed traumatic collapse, epilepsy, coma, or acute drowsiness, confusion, or headaches. The basis of such symptoms can be several, ranging from contusion (a bruising) to laceration (a tearing wound), to subdural hemorrhage (bleeding beneath the dura).

Interestingly, many patients who suffer

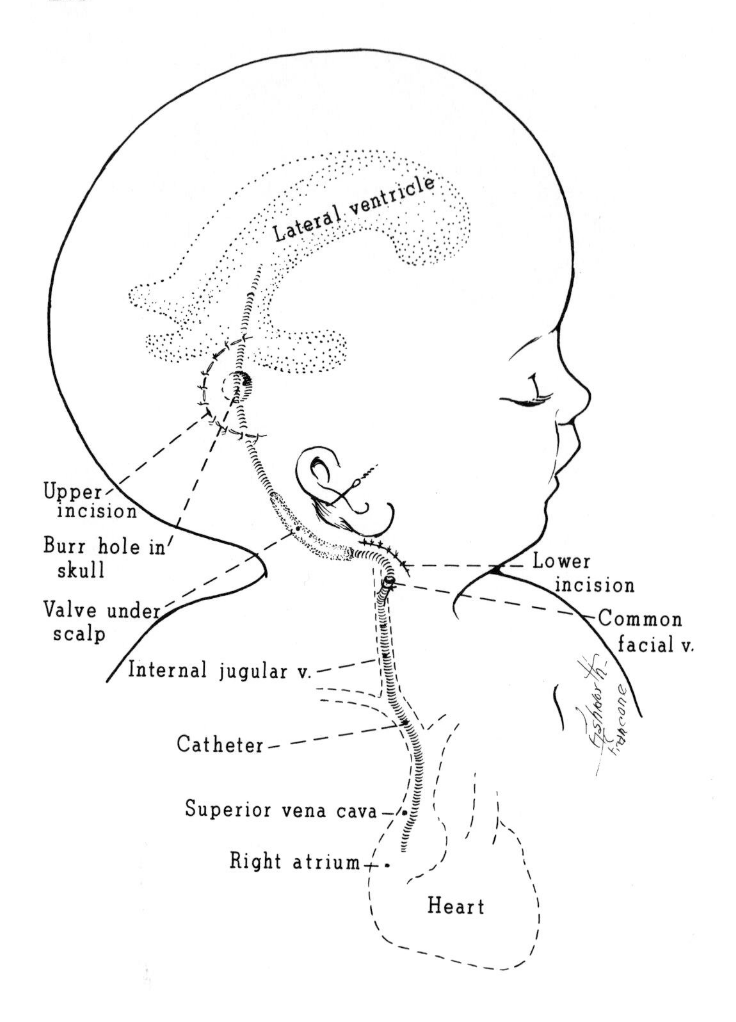

Figure 198. Operative procedure for hydrocephalus in which a catheter drains the ventricular system into the right atrium.

actual skull fractures do not have serious or prolonged disorders of cerebral function. On the other hand, autopsy in fatal head injuries may reveal an intact skull in 20 to 30 per cent of cases.

What Is Parkinson's Disease?

Parkinson's disease is characterized by involuntary tremors (shaking), decreased strength in movement, and rigidity; intellectual capabilities are not affected. Because of the characteristic shaking or tremors, it is commonly referred to as "shaking palsy."

What Is a Stroke?

Strokes, more formally referred to as cerebral vascular accidents, are specific disorders caused by a disease process in a blood vessel in the brain most commonly either a blood clot or hemorrhage. In most cases the onset is abrupt and the development rapid, and symptoms reach a peak within seconds, minutes, or hours.

The specific symptoms of the resulting disorders vary depending on the part of the brain affected. Symptoms may be few and relatively unimportant or numerous and eventually fatal. Frequently symptoms include paralysis, difficulty in speaking, and inability to write. Partial or complete recovery may occur over a period of hours to months.

LEARNING EXERCISES

1. Draw a diagram of a three-neuron reflex arc, with a sensory neuron, or interneuron, and a motor neuron. By use of arrows, indicate the direction (one-way) of transmission.
2. Following Figure 175, draw an outline of a sagittal section of the brain, and label all parts about which you have read.
3. Make a table, listing the primary functions of each of the major parts of the brain and spinal cord.
4. Construct a diagram of the external eye, and label the major portions. Do the same with a sectional drawing of the internal eye.
5. Following Figure 196, construct diagrams of the external, middle, and internal ear, and label the major parts.

CHAPTER 12

THE ENDOCRINE SYSTEM

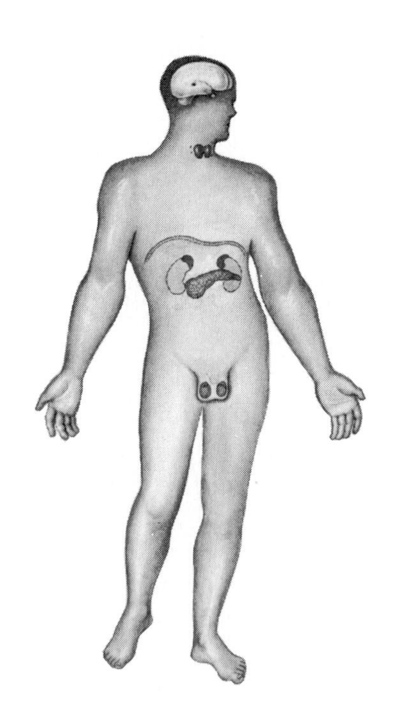

THE ENDOCRINE SYSTEM: AN OVERVIEW

What Does the Endocrine System Do?

There are actually two general systems in the body, performing the overall functions of *control* and *communication.* The nervous system is responsible for rapid, short-interval control, and accomplishes this through the use of fast-traveling nerve impulses. The *endocrine* (en′do-krin) *system,* on the other hand, is responsible for more general and longer-lasting control of bodily states. It accomplishes this by the use of *hormones* (chemical stimulants), which are secreted into and circulated by the blood stream. Nervous system controls act over seconds or fractions of seconds; endocrine system controls act over minutes, hours, and days.

In comparing the two systems, we find that the nervous system is responsible for the moment-to-moment control and coordination of the skeletal muscles in the performance of movements, and the glands of the endocrine system secrete the hormones that control growth and stimulate sexual development during puberty. The endocrine glands also secrete hormones that regulate the amount of glucose in the blood, and hormones that will excite the whole body, as when one is preparing to fight or flee in some situation.

What Structures Make up the Endocrine System?

The endocrine system is made up of *eight endocrine glands.* A gland is any organ that produces and secretes special substances. Endocrine glands are those that secrete hormones directly into the blood. Another class of glands, the exocrine, discharge their secretions (usually enzymes) by means of ducts, and are not part of the endocrine system. The glands of the digestive system and the skin are examples of *exocrine glands.* A few glandular organs, such as the pancreas, are made up of both types of glands operating independently. (The term, endocrine, is from the Greek *endo,* within + *krine,* to separate or secrete. The term exocrine is similar, *exo,* outside + *krine,* to secrete).

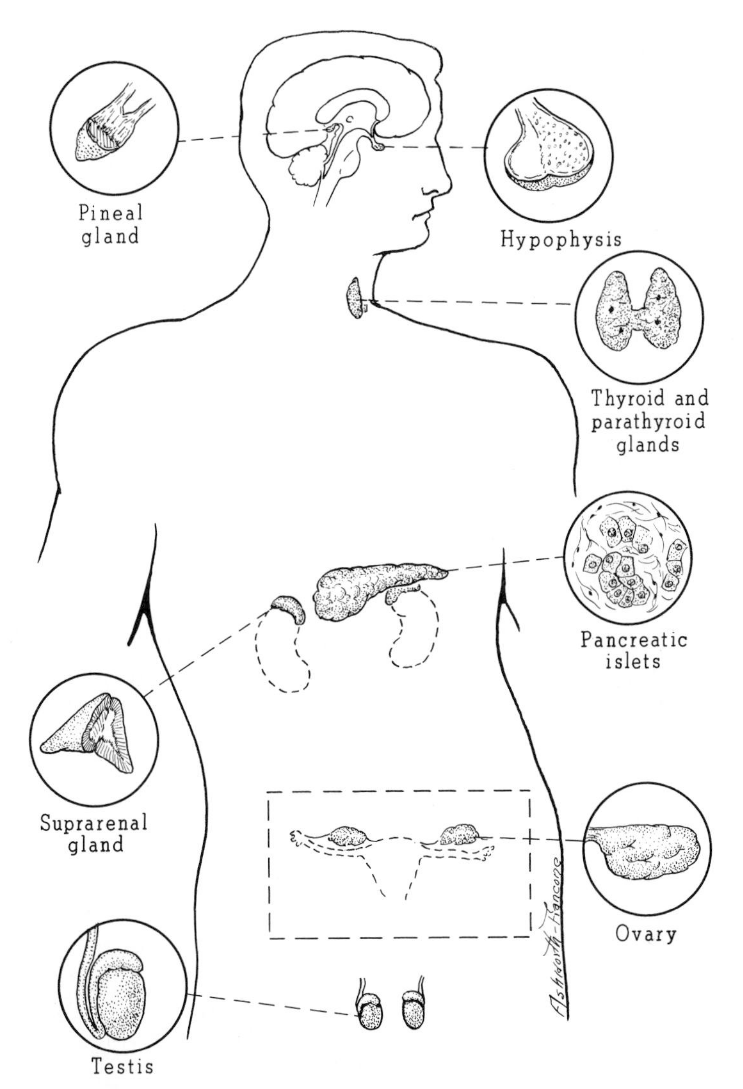

Figure 199. Eight glands of internal secretion produce substances which work in harmony with the nervous system to control and coordinate all activities of the body.

As can be seen in Figure 199, the glands of the endocrine system are located throughout the body. They are considered to make up *one system,* not only because of their similar type of function and similar means of active influence, but also because of the many important interrelationships. These interrelationships will become apparent as we proceed.

What Is a Hormone?

A *hormone* is a chemical substance (or stimulant) secreted by one part of the body (an endocrine gland) that controls or helps to control a function elsewhere in the body. The term "hormone" is derived from the Greek word *hormaein,* meaning to activate, or set in motion.

Hormones pass (are secreted) directly into the blood through the capillaries running through each of the endocrine glands, and are taken to other parts of the body through the blood stream.

Some hormones act *generally,* stimulating a specific activity over the whole body. For instance, insulin stimulates the absorption of glucose from the blood by the cells throughout the body. Other hormones directly affect only a *limited* area—these areas are called *target organs* for the hormone. For example, ADH (antidiuretic hormone) affects the renal tubules to increase their rate of water reabsorption. The direct effect of ADH on other tissues in the body is negligible.

An excess or deficiency in the secretion of a particular hormone may result in a specific disease state. For instance, diabetes mellitus is a condition in which insulin secretion is deficient.

THE PITUITARY GLAND

What Is the Pituitary Gland?

The *pituitary* (pi-tu′i-tar″e) *gland* is a small, pea-sized organ located below the hypothalamus in the midbrain. Another name for the pituitary is the *hypophysis* (hi-pof′i-sis), from the Greek word meaning an undergrowth—so named because the gland seems to be growing on a stalk underneath the hypothalamus (Fig. 200). The pituitary, or hypophysis, is really two distinct and independently operating endocrine glands. The front half of the organ is called the *anterior lobe,* also called the *adenohypophysis;* the back half is called the *posterior lobe,* also called the *neurohypophysis;* "neuro-" because it is more directly associated with the nervous tissue of the hypothalamus above.

What Does the Anterior Lobe of the Pituitary Gland Do?

The anterior lobe of the pituitary gland has been found to secrete at least *six hormones* of major importance. Five of these are commonly referred to by their abbreviated names—GH, ACTH, TSH, FSH, and

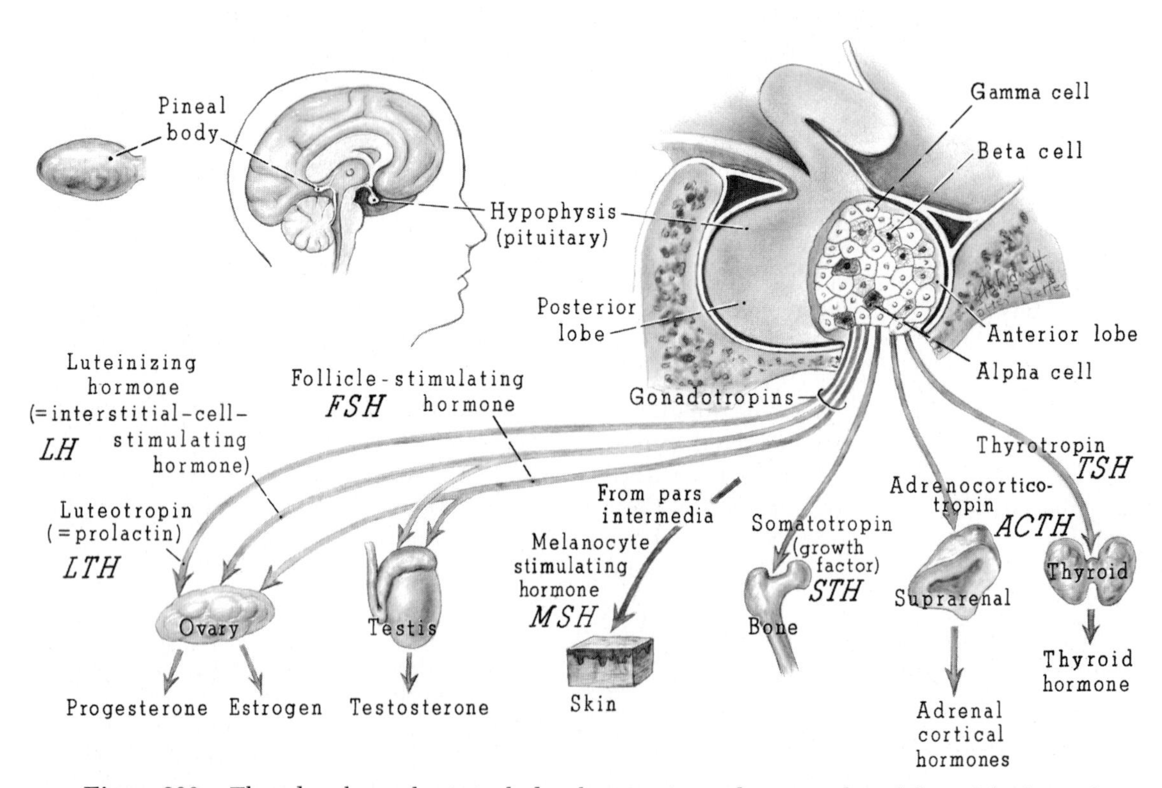

Figure 200. The adenohypophysis includes the anterior and intermediate lobes of the hypophysis and produces several hormones controlling the activity of a number of endocrine glands.

LH. All these letters may seem a little overwhelming at first, but notice that each abbreviation ends with an H, and this always stands for "hormone." In the beginning you should take the time to recall, out loud, the full name whenever you encounter the sequence of letters—this will help you to learn the full names in a short time.

The anterior lobe (adenohypophysis) has often been called the "master" endocrine gland, or the "leader of the endocrine orchestra," since four of its hormones act on other endocrine glands to control both their growth and rate of secretion. In other words, four hormones from the anterior pituitary have other endocrine glands as their "target organs." These include adrenocorticotrophic hormone (ACTH), which acts on the adrenal cortex, and thyroid-stimulating hormone (TSH), which acts on the thyroid gland. Two hormones influence the ovary: follicle-stimulating hormone (FSH), which stimulates growth of the follicles of the ovaries, and luteinizing hormone (LH), which acts on one ovarian follicle each month (the one releasing the ovum) to stimulate its development into the corpus luteum.

ACTH. *Adrenocorticotrophic* (ad-re″-no-kor″te-ko-trof′ik) *hormone* regulates the growth and activity of the cortex (outer portion) of the adrenal glands—the adrenal cortex—as indicated by the name. The suffix "*-trophic*" comes from the Greek word meaning to help to grow or act. The literal translation of "adrenocorticotrophic hormone," then, is "the hormone that helps the adrenal cortex to grow and act." ACTH stimulates the adrenal cortex (itself an endocrine gland, considered below) to grow and secrete cortisol (hydrocortisone) and similar substances.

TSH. *Thyroid-stimulating hormone* controls the growth of the thyroid gland and the secretion of the thyroid hormones.

FSH and LH. *Follicle-stimulating hormone* and *luteinizing hormone* are called *gonadotropic* (gon″ad-do-trop′ik) hormones, since they control the development and level of activity of the ovaries (female gonads) in the female, and the testes (male gonads) in the male. FSH in the female stimulates the growth of the graafian follicles to the point of maturity each month, until one ruptures and releases its ovum. FSH also stimulates the ovarian follicle cells to produce and secrete the female sex hormones, the estrogens and progesterone. In the male, FSH influences the development of spermatozoa.

Luteinizing hormone in the female also stimulates the growth process of the follicles. At the point of maturity, there is an additional sudden increase in LH secretion from the anterior pituitary, which causes *ovulation* (rupture of the follicle and release of the mature ovum). After ovulation, LH influences the ruptured follicle to develop into the *corpus luteum.* This latter process is called *luteinization* (lu″te-in″i-za′shun), and is the function that earned LH its name, luteinizing hormone. In the male, LH stimulates the testes to produce the basic male sex hormone, *testosterone* (tes-tos′ter-on).

GH and Prolactin. The two hormones secreted by the anterior lobe of the pituitary gland that do not work directly on other endocrine glands are growth hormone (GH) and prolactin, or lactogenic hormone.

The mechanisms by which *growth hormone* (also referred to as *somatropin;* abbreviated STH) accelerates growth are incompletely understood. Generally speaking, it alters the metabolism of cells, influencing them toward increased reproduction. One key discovery, however, is that GH promotes the movement of amino acids (protein building blocks) into the cells. This seems to stimulate the cells to increase production of proteins—a major step in the reproductive and growth sequences. GH is also known to affect carbohydrate and fat metabolism, increasing the use of fat and decreasing the cellular uptake of glucose. Underproduction of growth hormone in adult life leads to Simmonds' disease (Fig. 201). Overproduction of growth hormone in adolescence, leads to gigantism, and in adult life, to acromegaly (Fig. 202).

Prolactin promotes the growth of

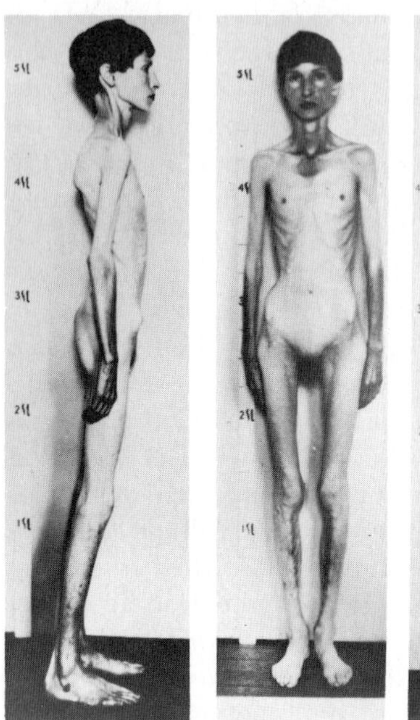
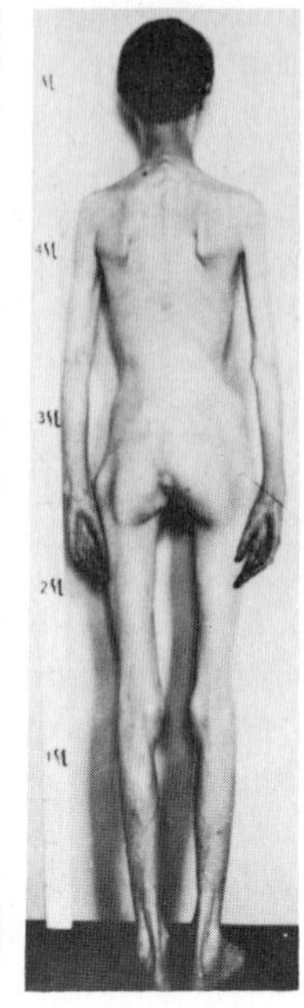

Figure 201. Simmonds' disease. (Courtesy of Escamilla and Lisser: California & West. Med., *48*:343, 1938.)

breast tissue during pregnancy, and operates in the maintenance of *lactation* (milk production) after the birth of the child. More specifically, the suckling reflex stimulates the secretion of prolactin, and prolactin, in turn, stimulates the production of more milk. Prolactin's other name, *lactogenic* hormone means milk-producing hormone, (*lact-* is Latin for milk; *genic* means to make).

What Does the Posterior Lobe of the Pituitary Gland Do?

The posterior lobe of the pituitary gland secretes two hormones—*oxytocin* (ok″se-to′sin) and *antidiuretic* hormone (ADH).

Oxytocin. This hormone stimulates the uterus to contract at the time of childbirth, and stimulates the *release* of breast milk from the mammary glands into the ducts for suckling. Oxytocin, (from the Greek *oxys,* swift, + *tokos,* childbirth), acts on the smooth muscle of the pregnant uterus to maintain labor and to decrease hemorrhage after delivery (by forcing stronger uterine contraction). Another function of oxytocin is to influence the glandular cells of the lactating breast to release the formed milk into the ducts. Suckling by the infant, in turn, stimulates additional secretion of oxytocin.

ADH. *Antidiuretic hormone* stimulates cells in the kidney to draw (or reabsorb) water out of the forming urine and replace it into the blood stream. The effect is to decrease the amount of water loss from the body into the urine. When the body is very low on water (dehydrated), ADH stimulates the body to retain water by preventing further loss of water in urine formation. When there is an excess of body water, ADH secretion drops to almost nothing, and as a result urine (water) output greatly increases until a proper water balance in the body is reached. Then ADH secretion returns to a level that stimulates normal production of urine. The term "diuretic" comes from the Greek words *dia,* meaning intensive, and *uresis,* meaning urination. So "diuretic" means intensive urination and "antidiuretic" means nonintensive or low production of urine. *Antidiuretic hormone, then, is the hormone that decreases the rate of formation of urine in order to save water.*

Diabetes insipidus is a disease caused by an abnormal decrease in the production of ADH. The result is excessive output of urine—up to 20 liters a day. The term "diabetes" comes from the Greek word for siphon, it is intended to indicate excessive urination. The term "insipidus" is Latin and means "tasteless," which distinguishes diabetes insipidus from diabetes mellitus, in which the urine tastes sweet, since it contains glucose (a sugar). Diabetes mellitus is a disorder in which insulin production by the pancreas is impaired; it will be discussed later in the chapter. Although both types of diabetes have the common symptom of excessive urination, the underlying disorders are quite different.

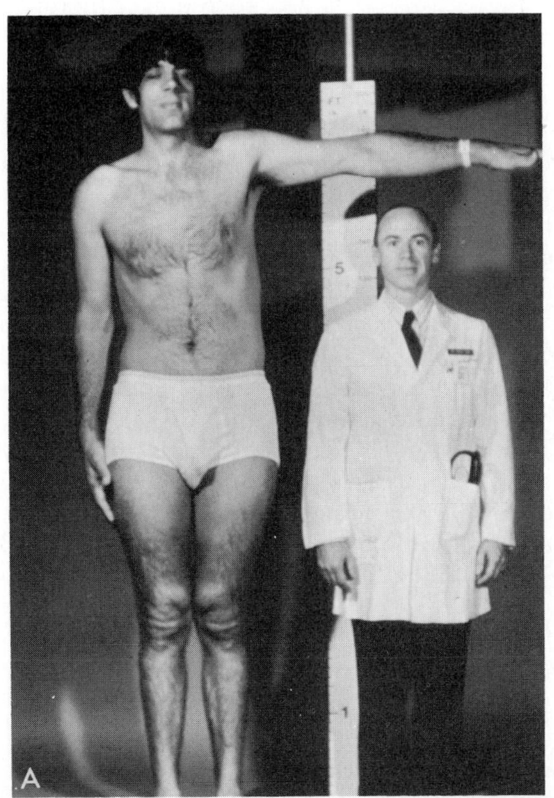

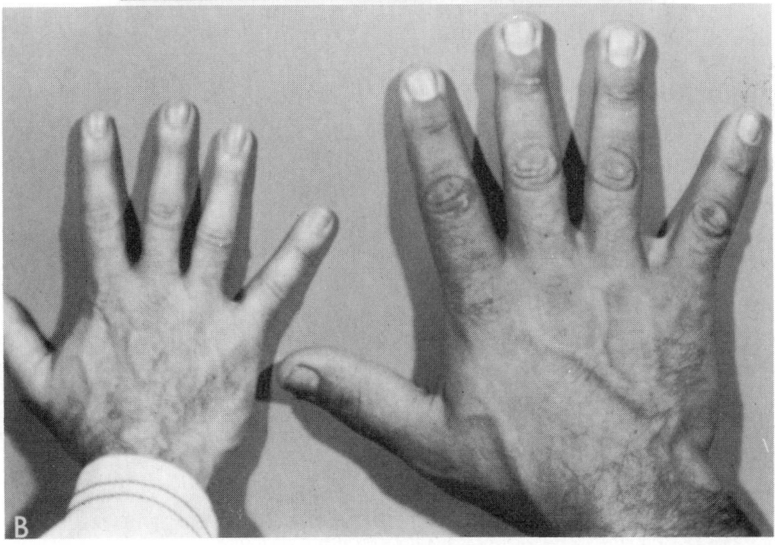

Figure 202. Gigantism with acromegaly in a male aged 28. *A*, Height approximately 7 feet, 6 inches. *B*, Hands of same individual as compared to normal sized hand.

THE THYROID GLAND

What Does the Thyroid Gland Do?

The thyroid gland is located in the neck, in front of the trachea, with its major portion lying on either side of the trachea (windpipe) (Fig. 203). The thyroid gland produces and secretes two hormones, *thyroid hormone,* also called *thyroxin,* and *calcitonin,* a recently discovered hormone.

The growth and level of activity (secretion) of thyroid hormone is controlled by the anterior pituitary gland secretion of thyroid-stimulating hormone (TSH).

Thyroid hormone (thyroxin) stimulates all the cells of the body to be more active. It has been called the "metabolism activating hormone." When secretion of thyroid hormone increases, the entire energy metabolism (and consequently all cellular activity) of the body speeds up. When thyroid hormone secretion decreases, all cellular (and therefore, bodily) activity slows down. In abnormal thyroid conditions, excessive secretion of thyroid hormone can cause the patient to be nervous, jumpy, and *hyperactive. Exophthalmos* (ek″sof-thal′mos) (Fig. 204) is indicative of hyperthyroidism. If an abnormally low amount is secreted, the person tends to slow down considerably, and to have little energy; in adults, this is referred to as *myxedema* (mik-se-de′mah) (Fig. 205).

One of the chemicals necessary for production of thyroid hormone is *iodine.* If iodine intake is too low, the thyroid gland enlarges in an attempt to prevent any further loss of iodine from the body. This is one cause of *goiter,* an enlargement of the thyroid gland that causes the front of the neck to have a puffy appearance. The term "goiter" is from the Latin word for throat.

Calcitonin (kal″si-to-nin) is a recently

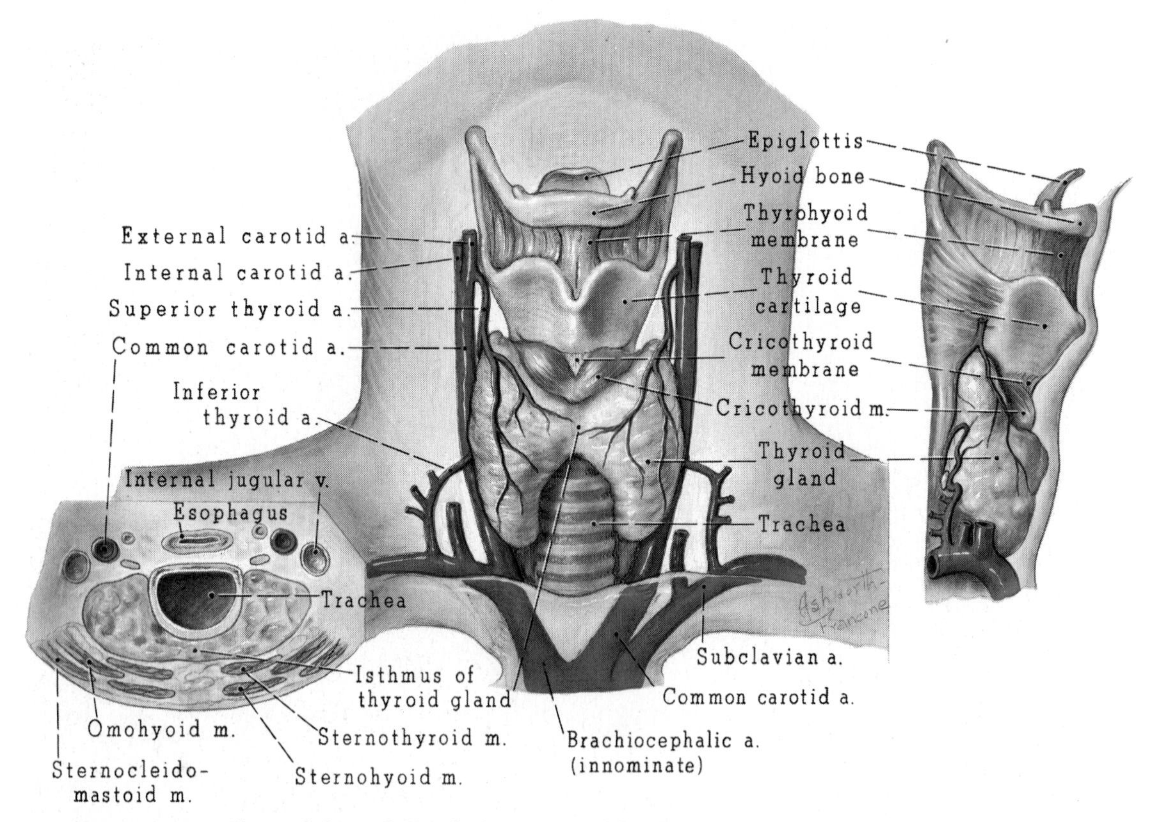

Figure 203. Plate of thyroid gland showing its blood supply and relations to trachea; in cross section, anterior, and right lateral views.

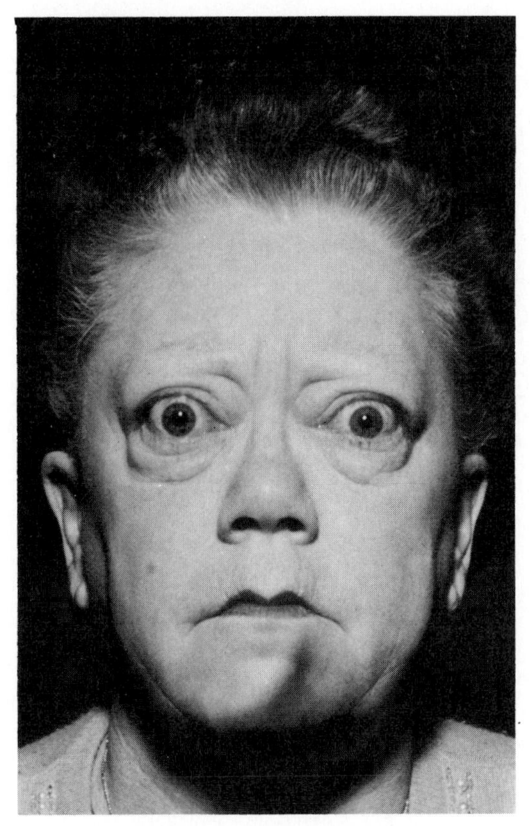

Figure 204. Exophthalmos. Note startled appearance and loss of eyebrows.

discovered hormone that is associated with the thyroid gland. There is some question, however, as to whether it is actually produced by thyroid gland cells, and its secretion does not seem to be controlled by TSH. Calcitonin is secreted when the calcium level in the blood gets too high. Somehow, calcitonin works to decrease, and bring back toward normal, the concentration of calcium in the blood. It has an effect opposite to that of parathyroid hormone, from the parathyroid gland.

THE PARATHYROID GLANDS

What Do the Parathyroid Glands Do?

There are four small parathyroid glands, located on the back side of the thyroid gland (Fig. 206). They secrete only one hormone, called *parathyroid hormone*, or *parathormone* for short.

Parathyroid hormine is secreted when the calcium concentration in the blood becomes too low. It has the effect of increasing the blood calcium concentration by promoting the reabsorption of solid calcium deposits from the bones into the blood. Calcitonin has the opposite effect of promoting deposition.

Parathyroid hormone has also been shown to stimulate the uptake (absorption) of ingested calcium from the intestines into the blood. Absorption of calcium from the intestines is even more immediately dependent on the presence of vitamin D.

Parathyroid gland disorders can be very serious, since nerve and muscle cells are extremely sensitive to the concentration of calcium in the blood. With too much calcium in the blood—from excessive secretion of parathormone—brain and heart

Figure 205. Myxedema. Note thick lips, baggy eyes, loss of hair, and dry skin.

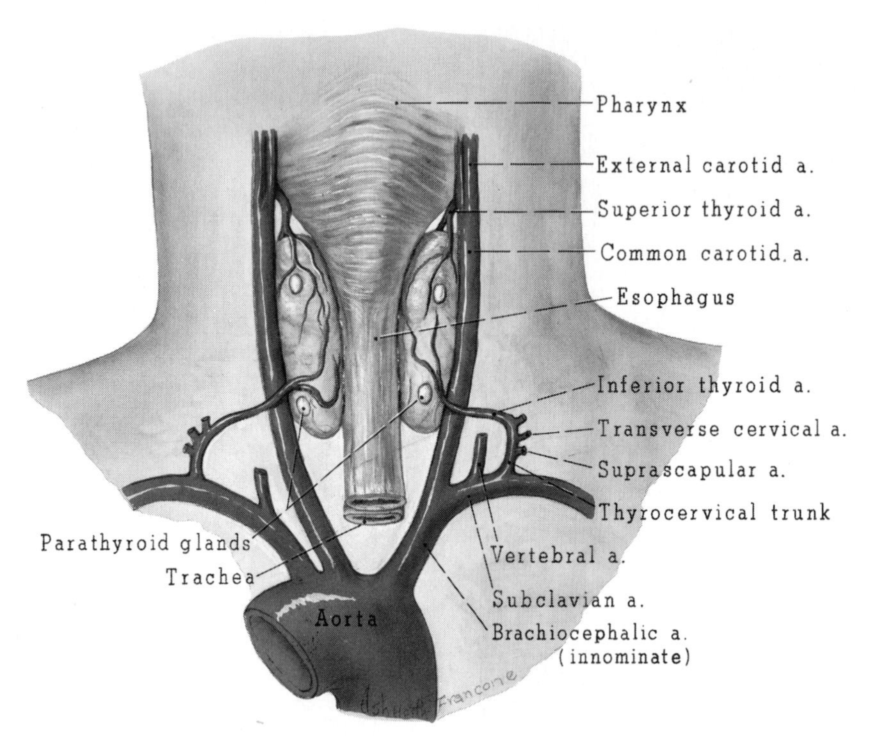

Figure 206. Posterior view of the neck and thyroid gland, showing the approximate location of the parathyroid glands.

cell activity becomes *depressed* and a person will become less responsive; and eventually the heart will stop. With too little calcium—from too little parathormone secretion—the nerve and muscle cells become *hypersensitive,* or over-responsive, resulting in severe jerkiness, muscle spasms, and cramps. This condition is known as *tetany* (tet'ah-ne). The term "tetany" derives from the Greek word meaning convulsive tension.

THE ADRENAL GLANDS

There are two *adrenal glands,* one above each kidney (Fig. 207). They have another name, *suprarenal* (literally, above + kidneys) glands. Each adrenal gland has a cortex, or outer portion, and a medulla, or inner portion. The cortex and medulla of each adrenal gland are actually different in both embryonic origin and function. They are in reality two separate and distinct glands. Does this two-glands-in-one arrangement remind you of another endocrine gland?

What Does the Adrenal Cortex Do?

The *adrenal cortex* secretes three general types of substances: *mineralocorticoids,* represented principally by *aldosterone; glucocorticoids,* represented chiefly by *cortisol* (also called hydrocortisone); and *androgens,* represented by *testosterone.*

The mineralocorticoids (min"er-al-o-kor'ti-koids) help to control the amounts of several *mineral salts* in the body fluids. Aldosterone is the most active mineralocorticoid. It stimulates the kidney to retain (or reabsorb from the forming urine) sodium. It also has the opposite effect for potassium. A decrease in normal aldosterone secretion has the reverse effect of increasing sodium loss and retaining potassium.

The *glucocorticoids,* (gloo"ko-kor'ti-koids), chiefly the hormone *cortisol,* have an effect similar to thyroid hormone, that is,

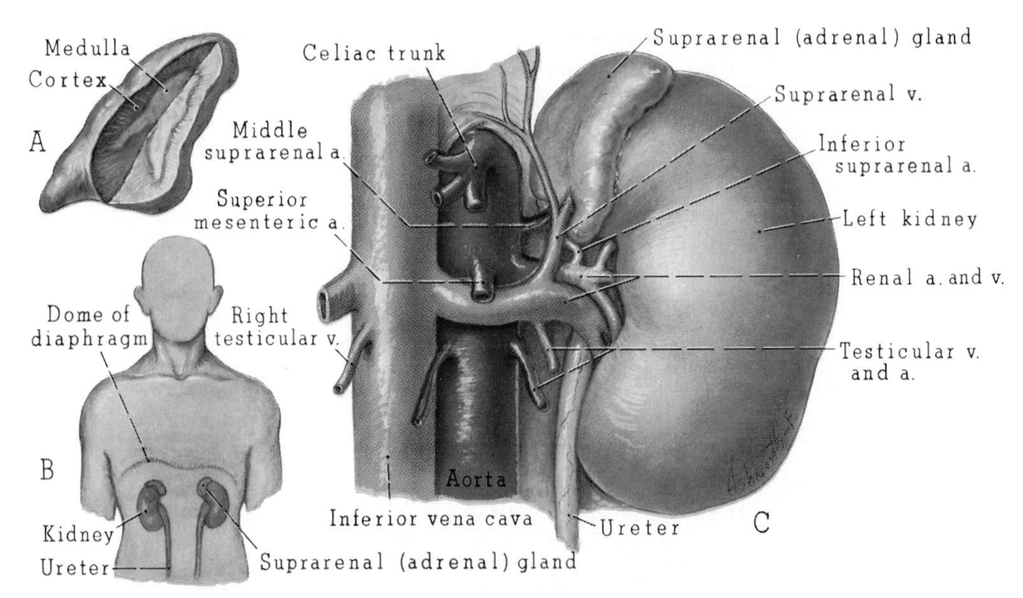

Figure 207. *A,* Suprarenal gland sectioned to show the medulla. *B,* Anatomic position of kidney and suprarenal glands. *C,* Anterior aspect of left kidney, showing adrenal gland and vascular supply.

they stimulate the energy metabolism of the body. Normal amounts of the glucocorticoids are necessary for normal cellular metabolism. Equally important is the role of increased glucocorticoid secretion during periods of *stress.* During such periods, cortisol secretion leads to a breakdown of proteins into amino acids. These amino acids then circulate through the blood to the liver, where they undergo *gluconeogenesis* (gloo″ko-ne″o-jen′e-sis)—the process of making new glucose out of amino acids. This results in an immediate new supply of glucose to the blood for the quick energy needed to cope with stressful situations.

Surgery, hemorrhage, infections, severe burns, and intense emotions are examples of extreme stimuli that produce stress. Glucocorticoids are sometimes called the stress hormones. Persons with poorly functioning adrenal cortexes have a lower survival rate through periods of stress.

Androgens produce masculinization. The most important androgen is testosterone, which is secreted primarily by the testes. Adrenal androgens are of minor importance, except when an adrenal tumor develops, in which case excessive quantities of androgenic hormones are produced. This can cause a child or even an adult female to take on an adult, masculine (male) appearance, including growth of the clitoris to resemble a penis, growth of a beard, a change in the voice quality to bass, and increased muscular development.

What Does the Adrenal Medulla Do?

The adrenal medulla produces and secretes *epinephrine* (ep″i-nef′rin) (also known as adrenaline) and *norepinephrine* (also known as noradrenaline), with constant proportions of each for each species. The human adrenal medulla secretes ten times as much epinephrine as norepinephrine.

The adrenal medulla functions in conjunction with the sympathetic nervous system (part of the autonomic nervous system). The medullary hormones are released in response to sympathetic nervous system stimulation, which typically occurs in situations in which a person is prepared to fight or flee. Epinephrine operates to supercharge the body very quickly. It increases heart rate, blood pressure, and blood glucose levels. In times of stress and excitement the adrenal medulla will quickly infuse (literally squirt) large amounts of epinephrine into the blood.

THE PANCREAS

What Do the Pancreatic Islets Do?

The *pancreatic* (pan″kre-at′ik) *islets* are small clumps of specialized cells scattered throughout the pancreas (Fig. 208). The term "islet" (i′let) means little island. These islets were discovered by a German pathologist, Paul Langerhans, and are sometimes referred to as the *islets of Langerhans.* They operate independently of the exocrine portion of the pancreas, which secretes digestive juices via a duct into the duodenum during digestion (Fig. 154).

The pancreatic islets contain two types of cells: beta cells, which produce and secrete *insulin,* and alpha cells, which produce and secrete *glucagon.*

Insulin regulates the cellular uptake of glucose from the blood, and as a consequence serves to regulate the concentration or amount of glucose present in the blood. Insulin is secreted most heavily after a meal, when glucose is being absorbed from the small intestine into the blood. This stimulates the cells to take up the glucose from the blood and insures that an excessive buildup of glucose in the blood does not occur. Insulin secretion decreases when the concentration of glucose in the blood decreases (between meals). In *diabetes mellitus,* the beta cells of the pancreatic islets do not secrete enough insulin (or none at all) and as a result the blood glucose level rises

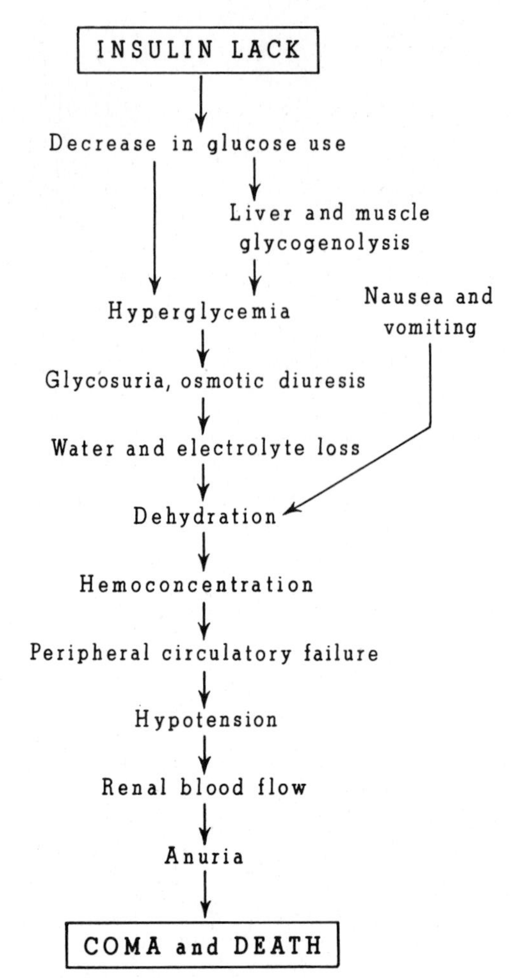

Figure 209. Progressive symptoms resulting from lack of insulin. (From Tepperman: *Metabolic and Endocrine Physiology.* Chicago, Year Book Medical Publishers, 1962.)

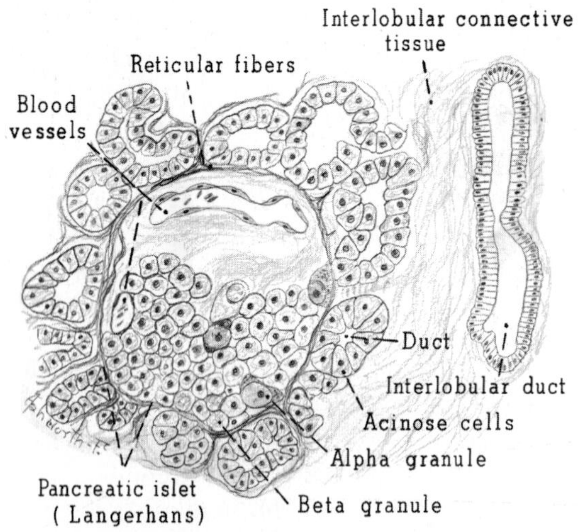

Figure 208. Microscopic section of pancreas showing pancreatic islets (of Langerhans).

dangerously; this leads to mental confusion, coma, and eventually death (Fig. 209). The kidneys attempt to rid the blood of the excess glucose by producing large amounts of urine. The term "diabetes" means excessive urine. The term "mellitus" comes from the Latin word, meaning honey-like; this is appropriate because the glucose in the urine gives it a sweet taste. (Tasting the urine was the method used by the ancients to diagnose diabetes mellitus.) Glucose is not normally passed into the urine by the kidney. Diabetes insipidus, which results from deficient antidiuretic hormone (ADH) secretion by the posterior pituitary, also results in excessive urination, but the urine does not taste sweet—thus the appropriateness of the term "insipidus," which means tasteless.

Glucagon has an effect on the blood glucose concentration that is the opposite of insulins. Glucagon works to increase the amount of glucose in the blood. Secretion of glucagon is stimulated by a decrease from normal glucose levels. Glucagon accomplishes its task by acting on liver cells to promote the breakdown of *glycogen* and release of the resulting glucose. Recall that glycogen is a large storage molecule made up of many glucose molecules.

The major stimulant, then, for the release of either insulin or glucagon is a change in the circulating blood glucose concentration. Elevation of blood glucose results in insulin secretion; similarly, a fall in blood glucose stimulates secretion of glucagon.

THE PINEAL GLAND

What Does the Pineal Gland Do?

The *pineal* (pin′e-al) *gland* is located in the midbrain area. Its precise function is still somewhat of a mystery. One suggestion is that it influences the activity of the anterior pituitary gland. Most recently, investigators have suggested that the pineal gland may secrete a hormone called *melatonin* (mel″ah-to′nin), which somehow serves as a "biological clock" for regulating the activities of the ovaries. At this time, however, these suggestions are merely interesting speculation.

THE GONADS (SEX GLANDS)

What Is the Endocrine Function of the Ovaries?

The ovaries are two medium-sized, ovoid glands located in the pelvic portion of the female abdomen and attached to a supporting broad ligament. The outer layer of the ovary consists of a specialized germinal epithelium that produces *ova.* Two types of hormones are produced and secreted by the ovaries, the *estrogens* and *progesterone.* These are female sex hormones. The structure and function of these two endocrine glands will be discussed more fully in Chapter 13, The Reproductive System.

What Is the Endocrine Function of the Testes?

The testes are two medium-sized, ovoid glands suspended by the spermatic cord and surrounded and supported, below the pelvic region, by the *scrotum.* Two major types of specialized tissue are found in testicular substance. Tubules containing germinal epithelium function in the formation of spermatozoa. *Interstitial* (in″ter-stish′al) *cells* produce and secrete *testosterone,* the "masculinizing hormone." The structure and function of the testes will be discussed more fully in Chapter 12.

CLINICAL CONSIDERATIONS

Anterior Lobe of Pituitary Gland

Since this is the source of several hormones that directly control other endocrine glands, abnormal increases or decreases in the production of these hormones will inevitably affect these target endocrine glands.

Overproduction of anterior pituitary hormones in adolescence leads to *gigantism,* and in adult life, *acromegaly* (Fig. 202). In gigantism, the target endocrine glands as well as the whole body increase in size; metabolism is increased, and there are disturbances in glucose utilization. In acromegaly (ak″ro-meg′ah-lee) similar increases in organ sizes occur, with metabolic disturbances as well as noticeable growth and thickening of the skeletal framework. In both disorders, the oversecretion of growth hormone is thought to produce the most serious clinical symptoms.

Underproduction of anterior pituitary hormones, commonly known as *Simmonds' disease,* can occur at any age but is most frequent in adults. In children, it results in *dwarfism* or *Froehlich's syndrome* (obesity and lack of sexual development). In adults, the clinical syndrome is represented by insufficiency of the secondary target organs, particularly the adrenals, thyroid, and gonads. Patients gain weight, have metabolic disturbances, and are easily fatigued. It is remarkable that most patients survive many years, with the average duration of the disease 30 to 40 years.

Posterior Lobe of Pituitary Gland

Oversecretion of posterior pituitary hormones, principally antidiuretic hormone (ADH), is rare and is almost always associated with a cancer of the gland. It results in excessive retention of water and a condition of water intoxication.

Diabetes insipidus is a disease caused by an abnormal decrease in the production of ADH. The result is excessive output of urine – up to 20 liters a day.

Thyroid Gland

Hyperthyroidism, an excessive secretion of thyroxin, produces a marked increase in metabolic rate that is reflected by an increase in pulse rate, temperature, and blood pressure, accompanied by extreme nervousness, irritability, weight loss, and fatigability. *Exophthalmos* (Fig. 204) is indicative of hyperthyroidism. Skeletal decalcification has been noted which may be due to overproduction of the only recently discovered hormone, calcitonin.

Hypothyroidism, insufficiency of thyroid hormone, produce *myxedema* – puffiness and edema of face with a slowing of all metabolic activities, and weight gain.

Parathyroid Glands

Oversecretion of parathormone causes an abnormal increase in the calcium concentration of the blood. Since nerve and muscle cells are very sensitive to calcium ion concentration, the brain and heart activity soon become *depressed* and the patient becomes less responsive.

Undersecretion of parathormone results in an abnormal decrease in blood calcium. The nerve and muscle cells become *hypersensitive,* or over-responsive, resulting in severe jerkiness, muscle spasms, and cramps. This condition is known as tetany. The term "tetany" derives from the Greek word meaning convulsive tension.

Adrenal Cortex

A common type of oversecretion of adrenal cortical hormones is known as *Cushing's disease,* and involves a complex of effects. Cushing's disease is characterized by development of a humped back, puffiness of the face, masculinizing effects, hypertension, and increased concentration of blood glucose (Fig. 210).

Hypersecretion of individual hormones produces more selective effects. *Aldosteronism,* involving overproduction of the sodium retaining hormone aldosterone, results in water retention (edema), hypertension, excessive loss of potassium, and a tendency toward excessive concentrations of sodium. Oversecretion of cortisol, which

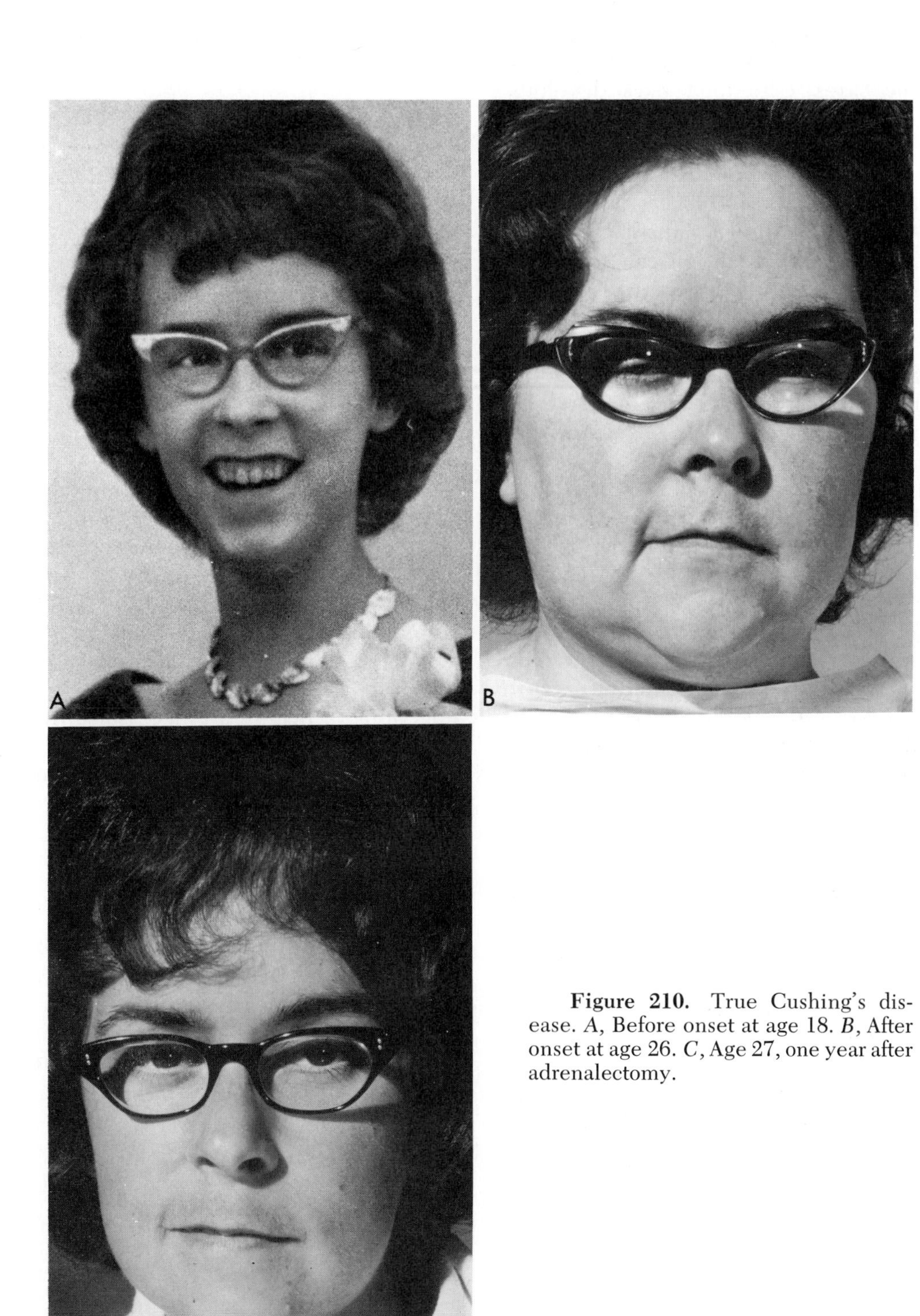

Figure 210. True Cushing's disease. *A*, Before onset at age 18. *B*, After onset at age 26. *C*, Age 27, one year after adrenalectomy.

is rare by itself, produces disturbances of protein, lipid, and glucose metabolism; in extreme cases a humped back develops. Adrenogenital syndrome results from oversecretion of the adrenal sex hormones (androgens), leading to intense masculinizing effects throughout the body.

Addison's disease, or insufficient production of the adrenal cortical hormones, has several consequences, corresponding to the lack of each of the specific hormones. Lack of aldosterone secretion allows large amounts of sodium ions (and water) to be lost into the urine. Loss of cortisol secretion makes it impossible for a person with Addison's disease to control blood glucose levels; mobilization of both proteins and fats is also reduced. Generally speaking, the most detrimental effect is the lack of responsiveness to the energy needs of the body. This leads to serious complications during periods of stress, such as diseases, surgery, or emotional crisis. Even a mild respiratory infection can sometimes cause death.

Adrenal Medulla

Hypersecretion of the adrenal medulla is rare, and is almost always a result of cancer of the gland. The main consequences of oversecretion of the hormones (epinephrine and norepinephrine) is hypertension.

The adrenal medulla does not seem to be essential to life or normal functioning; as a consequence, no disorders of undersecretion have been described.

Pancreas

The primary disease of the endocrine portion of the pancreas is *diabetes mellitus*. When an insufficient amount of insulin is secreted, the blood glucose level rises markedly. As a result, the kidneys attempt to rid the blood of the excess glucose by producing large quantities of urine. Glucose in the urine is a major clinical sign of diabetes mellitus, since glucose does not normally pass into the urine.

Occasionally the insulin-producing cells of the pancreatic islets have not been destroyed but are merely inactive, causing a mild form of diabetes. In such cases, the patient can be maintained on islet-cell stimulating drugs, which can be taken orally. In the more common cases of diabetes in which the cells are destroyed, injections of insulin are necessary. Insulin cannot be taken orally, since it is broken down in the digestive system before reaching the blood.

LEARNING EXERCISES

1. Draw an outline of the body, and locate and label each endocrine gland.
2. For each endocrine gland, list the hormones it produces and secretes, and briefly explain the target organ, and/or general function of each.

CHAPTER 13

THE REPRODUCTIVE SYSTEM

The function of the reproductive systems is to bring about the formation of a new member or members of the species. In one-cell organisms reproduction is accomplished through simple mitotic division, resulting in two "new" daughter cells. Higher species, such as the human, have evolved into a two-sex (male and female) system of reproduction.

Formation of a human offspring occurs through the growth of a single egg cell (from the female) fertilized by a single sperm (from the male) (Fig. 211). This fertilized egg grows and matures within the reproductive system by means of billions of individual cell divisions to produce the offspring at birth.

All the structures of the male and female reproductive systems are designed to bring about *fertilization,* and the subsequent *development* and *nourishment* of the new offspring (Fig. 212).

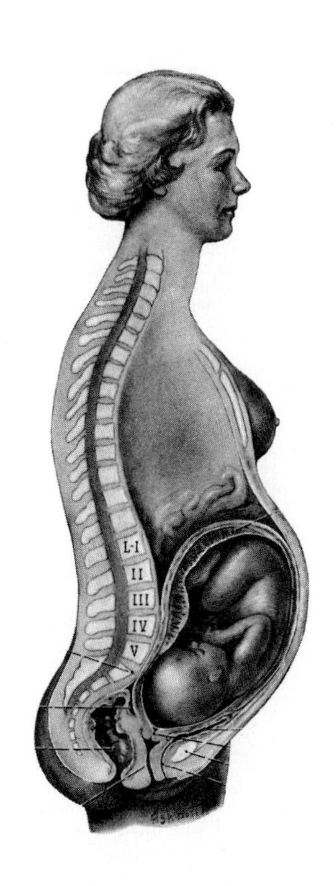

THE MALE REPRODUCTIVE SYSTEM

What Are the Scrotum and Penis?

The scrotum and penis are the visible external male organs of reproduction. The testes are also external but are contained in the scrotum (Figs. 213 and 214).

The *scrotum* is a pouch that hangs behind the penis. It is a continuation of the abdominal wall and is divided into two sacs by a septum (Latin for wall). Each sac contains one of the testes, with its epididymis or connecting tube leading up into the abdominal cavity.

The *penis* contains the male *urethra,* which functions to carry both urine from the bladder and *semen* (Latin for seed) from the ejaculatory duct. The penis is composed primarily of erectile tissue surrounding many small compartments, or spaces, that are normally collapsed (Fig. 215). During periods of sexual stimulation, the arteries that supply the penis dilate, and a large quantity of blood under pressure enters the erectile-tissue compartment. This causes the penis to become fixed and erect, and facilitates its penetration into the female vagina during inter-

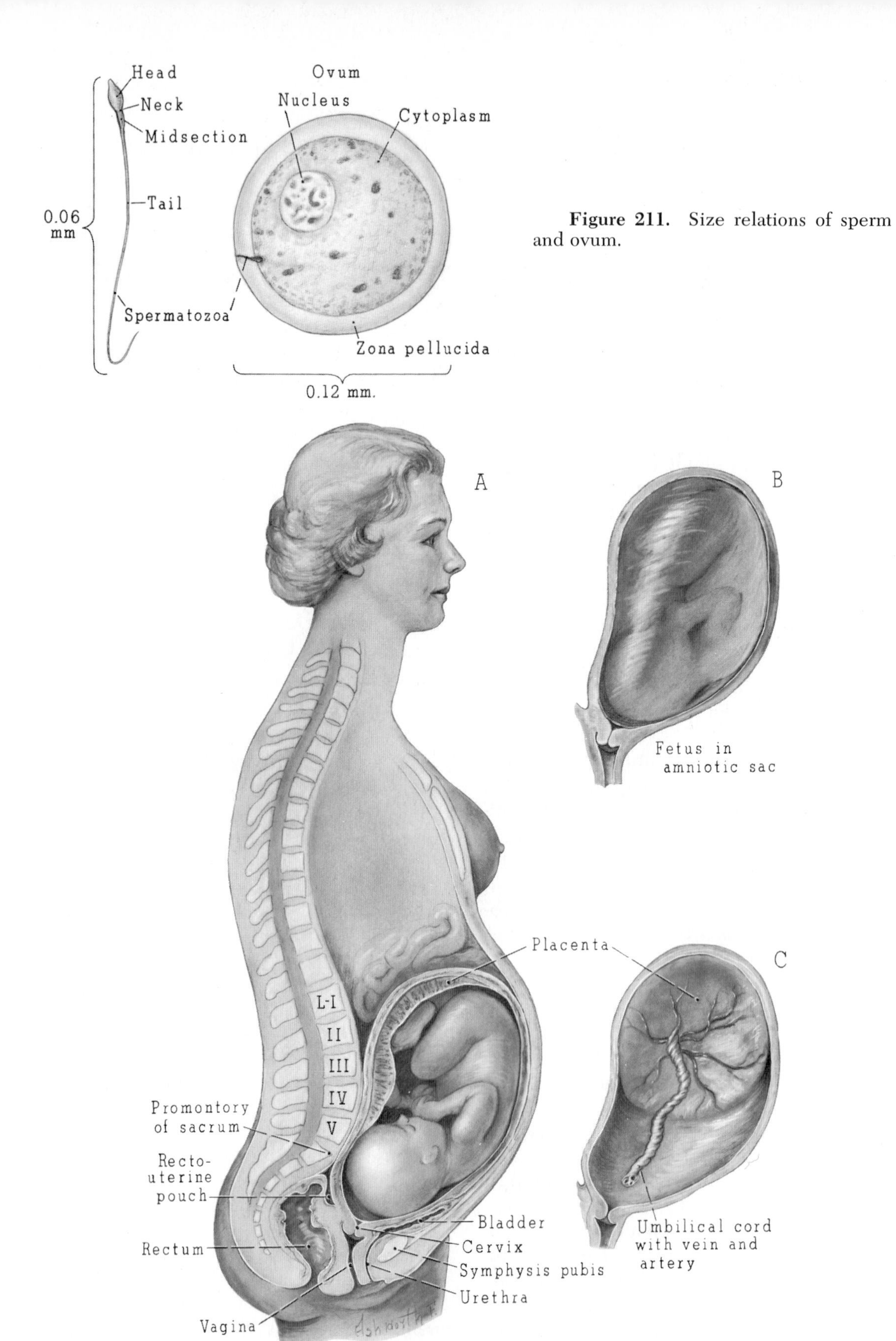

Figure 211. Size relations of sperm and ovum.

Figure 212. *A*, Mid-sagittal section of a pregnant woman showing fetal position. *B*, Amniotic sac with fetus. *C*, Placenta in uterus with fetus removed.

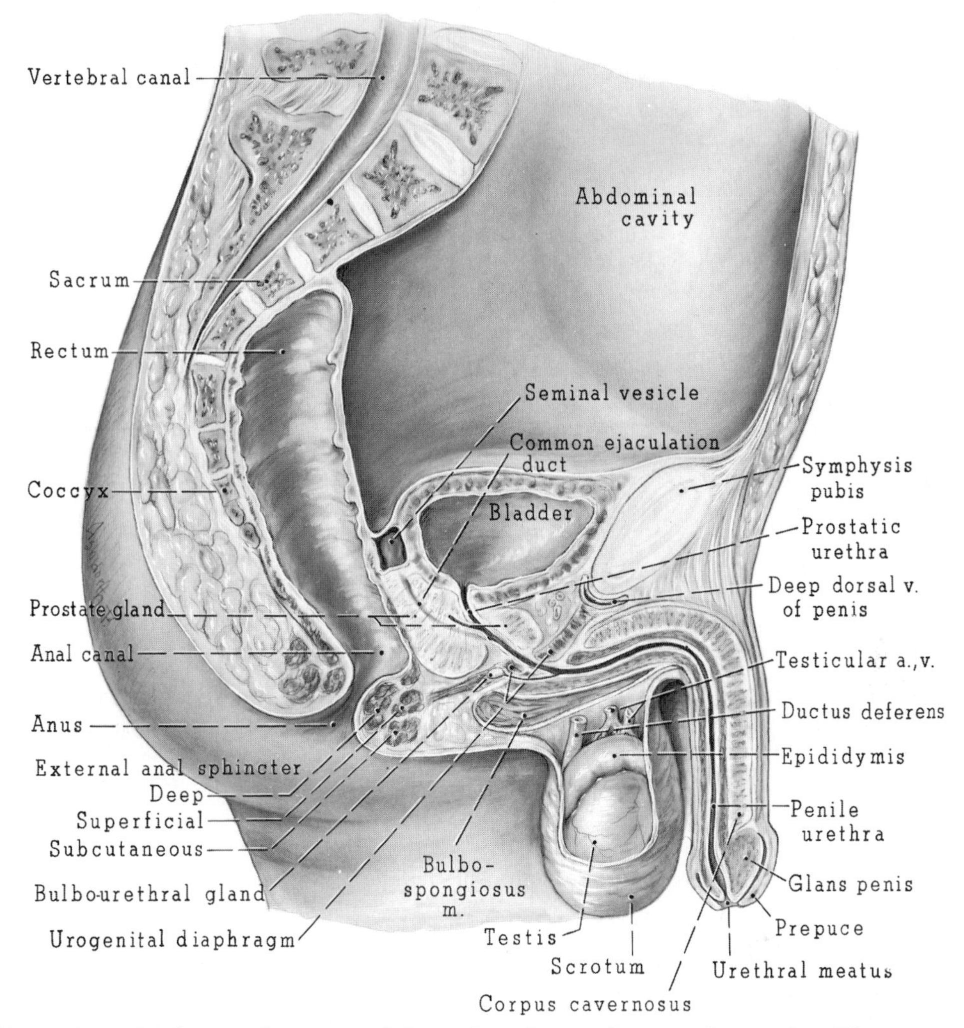

Figure 213. Mid-sagittal section of the male pelvis and external genitalia. (The course of the ductus deferens is shown in Figure 217.)

course. During intercourse, semen passes into the vagina, setting the stage for fertilization of the female egg.

The end of the penis is covered by a loose skin which is folded inward and then backward upon itself; this is called the *prepuce* (from the Latin *praeputium,* meaning foreskin). In current practice, this foreskin is usually removed in newborn boys by a simple surgical procedure known as *circumcision.* Removal of the foreskin serves to keep this area cleaner (Fig. 216).

What Are the Internal Male Organs of Reproduction?

The internal organs of reproduction in the male can be divided into *three groups.*

First, there are the male *gonads,* or *testes.* These function to produce sperm and secrete the male sex hormone, *testosterone.*

The second group consists of a series of ducts, including the *epididymis* (ep″i-did′i-mis), *ductus deferens,* and *urethra.*

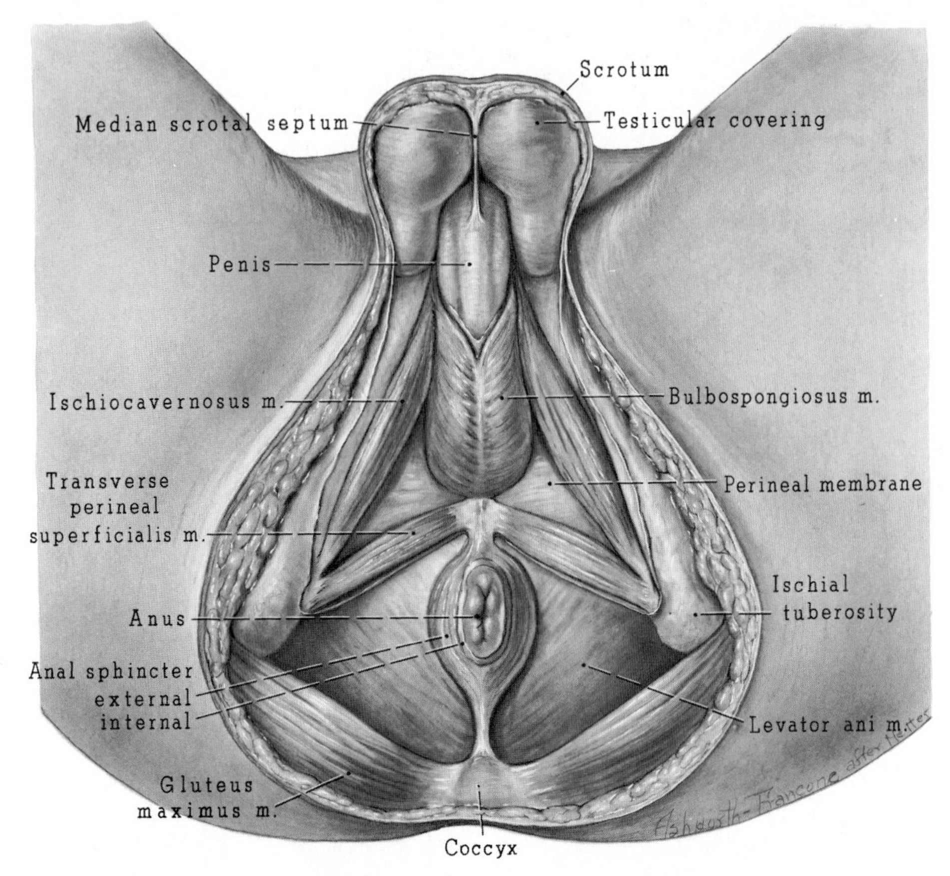

Figure 214. Male perineum with skin and superficial fascia removed.

These carry the sperm and semen from the testes and accessory organs into the vagina during intercourse.

The third group of internal organs is the accessory glands; the *seminal vesicles,* the *prostate,* and the *bulbourethral* (Cowper's) *glands.* These glands secrete fluid that carries the sperm through the penile urethra into the female vagina during intercourse.

What Are the Functions of the Testes?

The male *gonads* (Latin for seed) are called the *testes* or *testicles* (singular testis or testicle). They correspond to the ovaries in the female. The two testes are the organs that produce the male reproductive cells—the *sperm* (Greek for seed), or *spermatozoa* (Greek, *sperma* = seed + zoon = animal). The testes also contain glandular tissue that secretes the male sex hormone, *testosterone,* into the blood. The testes, therefore, are endocrine glands as well (Fig. 217).

Each testis is an oval organ about 2 inches in length and located in the pouch-like scrotum. Each testis is divided into about 250 wedge-shaped lobes. Each lobe contains one to three narrow, coiled tubes known as *seminiferous tubules* (the term "seminiferous" derives from the Latin words *"semen,"* meaning seed, and *"fero,"* meaning to carry). Male reproductive cells, known as *sperm* or *spermatozoa,* are found within these tubules at different stages of development. If uncoiled, a tubule would measure about 2 feet in length. Any one of the millions of sperm cells formed by each testis may join with a female reproductive

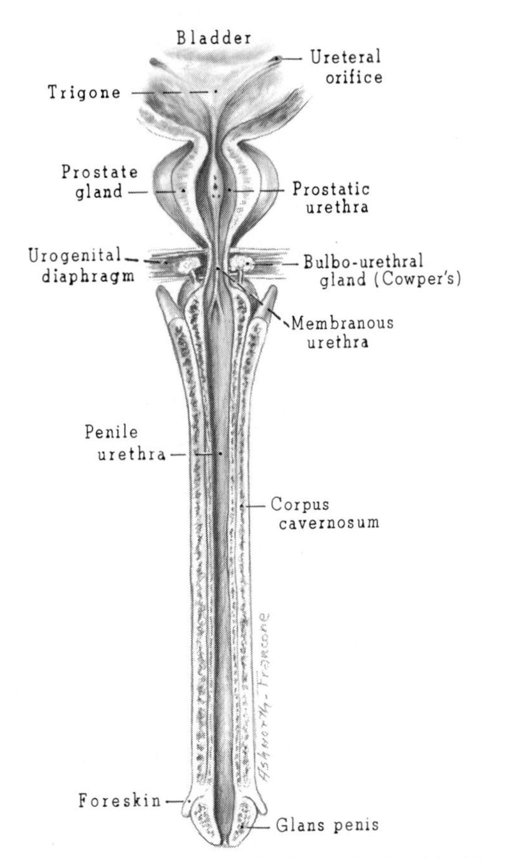

Figure 215. Section through the bladder, prostate gland and penis.

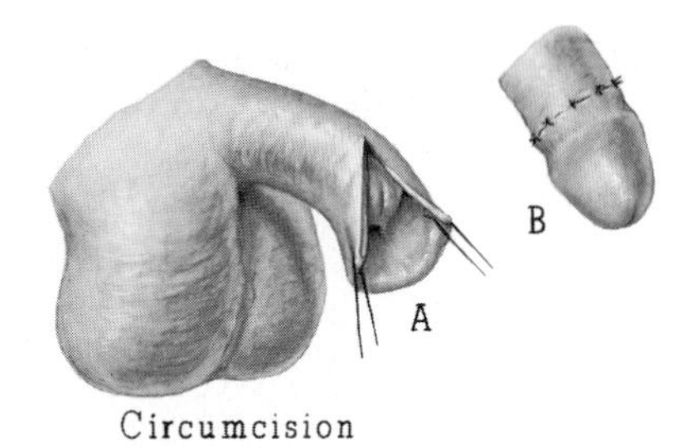

Figure 216. In *circumcision* the prepuce is removed. *A* shows incision in the prepuce. Closure of the wound after removal of the prepuce is shown in *B*.

cell (the egg or ovum) to eventually become a new human being.

Scattered among the tubules are the *interstitial cells,* which produce and secrete testosterone. These cells perform the endocrine activities of the testes. The increase in secretion of testosterone during puberty produces dramatic changes, transforming a little boy into a man. Testosterone lowers the pitch of the voice, increases the muscular development, promotes beard growth, and influences changes in the size and shape of the bones. Testosterone is known as the "masculinizing" hormone. The secretion of testosterone by the interstitial cells is controlled by the anterior pituitary gland secretions of FSH, which is referred to as interstitial cell stimulating hormone, ICSH, when discussing the male. FSH and ICSH are the same substance.

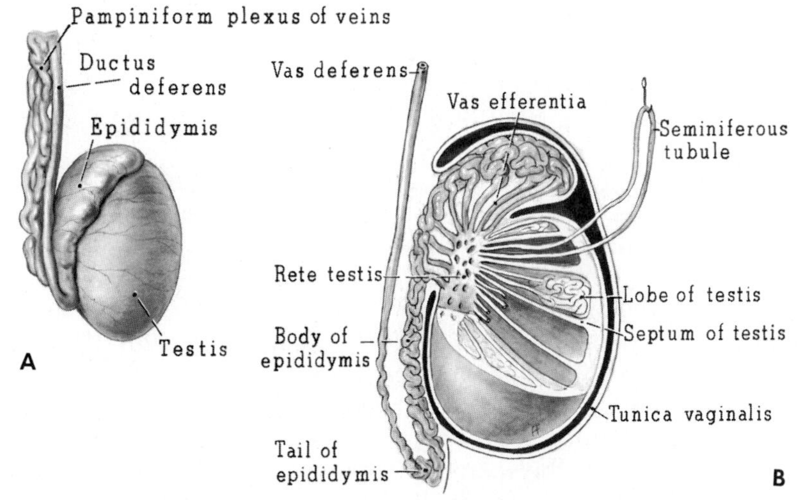

Figure 217. *A* shows the male testis. *B* is a diagram of a section of the male testis, showing detail of a seminiferous tubule.

What Are the Functions of the Epididymis, Ductus Deferens, and Urethra?

The *epididymis* (Greek, *epi* = upon + *didymos* = the two or twins) extends from the upper end of each testis downward along the posterior side for about 1½ inches. About 16 feet of tube are coiled within this short distance. The epididymis is the first part of the duct system leading from the testes.

The *ductus deferens* (Latin *ductus* = duct, and *defero* = carry down) can be considered a continuation of the epididymis and has been described as "the excretory duct of the testis." It is also called the *vas deferens* (Latin, *vas* = vessel). It extends from the testis about 18 inches, up into the abdomen, where it contacts, and forms a common duct (the ejaculatory duct) with, one of the seminal vesicles. The ejaculatory duct opens into the *urethra,* which carries both semen and urine out of the body.

A *vasectomy* (removal of a segment of the vas deferens) is a minor surgical operation wherein the ductus deferens is either tied off or severed or sewn closed. This achieves artificial sterility by stopping the normal passage of sperm.

What Are the Functions of the Seminal Vesicles, Prostate, and Cowper's Glands?

There are two *seminal vesicles.* They are membranous pouches lying behind the urinary bladder near its base, each consisting of a single tube coiled upon itself (Fig. 218). The seminal vesicles secrete a fluid that is one of the components of semen. *Semen,* which is the thick, whitish fluid that carries the sperm, leaves through the urethra, and enters the vagina during intercourse. The seminal vesicle fluid serves to give the sperm motility (ability to move). The tube of each seminal vesicle ends in a straight, narrow duct joining the ductus deferens to form the ejaculatory duct. The ejaculatory duct actually ejects the spermatozoa-containing fluid from the seminal vesicle into the urethra.

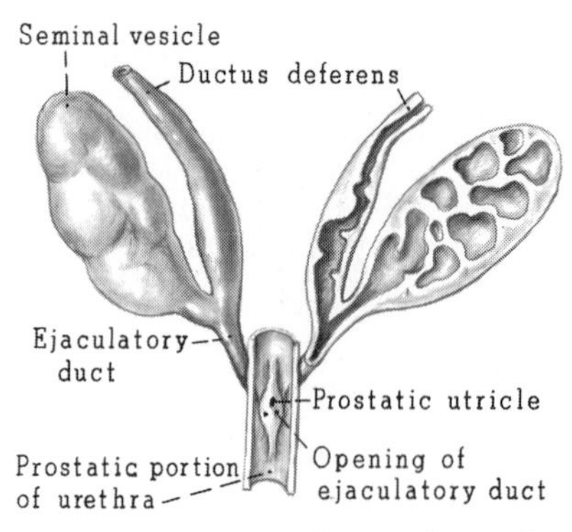

Figure 218. Seminal vesicle and related parts. On the left the vesicle and duct are intact; the right side is sectioned to show internal detail.

The *prostate gland* is a cone-shaped body about the size of a chestnut lying under the urinary bladder. It surrounds the first inch of the urethra and secretes an alkaline (high pH, low acidity) fluid, which also aids the motility of the sperm cells. In older men a progressive enlargement of the prostate commonly obstructs the urethra and interferes with the passage of urine. This condition calls for the surgical removal of part of the prostate gland. The prostate is also a frequent site of cancer in elderly men.

The *bulbourethral glands,* also called *Cowper's glands,* are two glandular bodies about the size of a pea located below the prostate on either side of the urethra. These also secrete alkaline fluid as a part of the semen.

THE FEMALE REPRODUCTIVE SYSTEM

What Are the External Reproductive Structures of the Female?

The external female reproductive organs are collectively known as the *vulva* (Figs. 219 and 220). "Vulva" is a Latin word meaning a wrapper, covering, or seed covering. The structures of the vulva are as follows. The *labia majora* (major lips) are two longitudinal, rounded folds of skin that are similar in structure to the scrotum in the male. Two smaller folds of skin, the

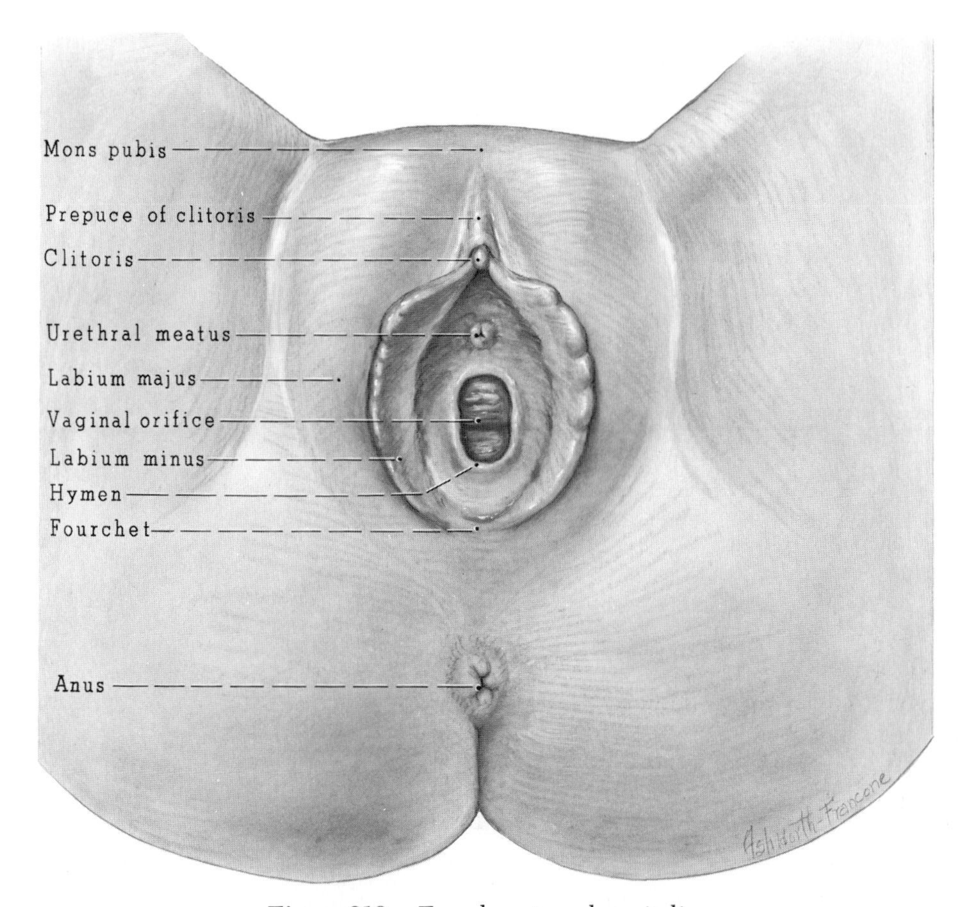

Figure 219. Female external genitalia.

labia minora lie between the labia majora. The *clitoris* is a small, rounded projection of erectile tissue, nerves, and blood vessels; it occupies the apex that is formed by the anterior meeting of the labia minora. The clitoris is partially hooded by a *prepuce.*

The *vestibule* (from the Latin word for outer chamber) of the vagina lies between the labia minora. Situated within the cleft of the vestibule are the *hymen,* the *vaginal orifice* (opening), the *urethral orifice* and the openings of the *vestibular glands.*

The hymen (Greek for membrane) is a thin fold of vascularized mucous membrane separating the vagina from the vestibule. It may be entirely absent or it may cover the vaginal orifice partly or completely. Anatomically, neither its absence nor presence can be considered a criterion of virginity.

INTERNAL FEMALE ORGANS OF REPRODUCTION (Figs. 221 and 222)

What Are the Ovaries?

The female *gonads* are called *ovaries* (singular ovary; from the Latin word *ovum,* meaning egg). They correspond to the testes in the male. The ovaries are the organs that produce the female reproductive cells—the *ova* (plural of ovum). The cells of the ovaries also contain glandular tissue, which secretes the female sex hormones, the *estrogens* and *progesterone.*

The *ovaries* are two oval-shaped structures about 1½ inches in length. They are located in the upper part of the pelvic cavity, one on each side of the uterus, and are anchored to the uterus by the *ovarian ligament.*

The inner structure of the ovary con-

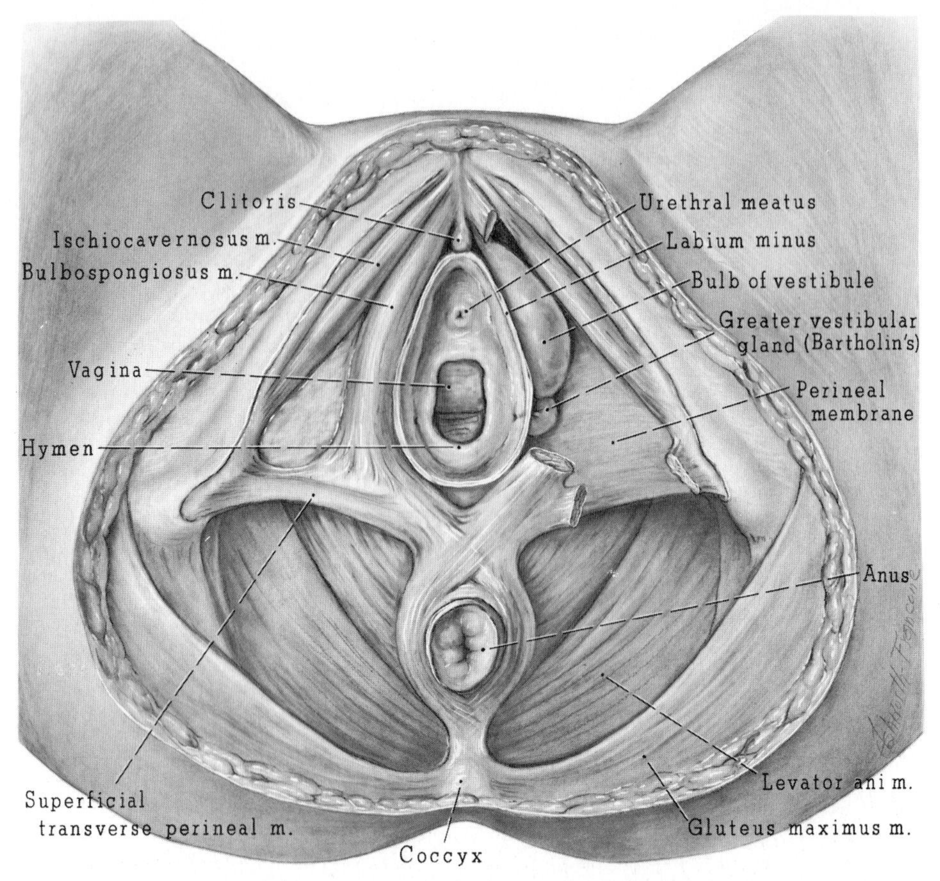

Figure 220. Female perineum with skin and superficial fascia removed.

sists of a meshwork of several thousand sacs, too small to be seen without a microscope. These are the *graafian* (graf'e-an) *follicles* (sacs); they are found at different stages of development within each ovary. "Graafian" derives from DeGraaf, the Dutch anatomist who first identified the follicles some 300 years ago. The term "follicle" derives from the Latin word meaning a small sac. Within each follicle is an ovum which matures as the follicle grows. At the point of maturity the follicle bursts open, releasing the ovum for possible later fertilization.

After the ovum is released, the ruptured follicle develops into the *corpus luteum,* under the stimulation of LH (luteinizing hormone) from the anterior pituitary. The corpus luteum (corpus = body, *luteum* = yellowish or golden), or "golden body," is a secretory body that produces and secretes estrogens and progesterone during the second half of the menstrual cycle, and during pregnancy if fertilization occurs.

The two major functions of the ovaries, then, are development and expulsion of the female ova, and production and secretion of female sex hormones, the estrogens and progesterone.

What Are the Uterine Tubes?

The *uterine,* or *fallopian* (after the Italian anatomist Fallopius), *tubes* serve as the ducts to carry the ova from the ovaries to the uterus. The uterine tubes differ from the corresponding ducts for the male testes in that the uterine tubes are *not actually connected* to the ovaries. Rather, when an ovum is expelled from the ovary the finger-like projections from the ends of the

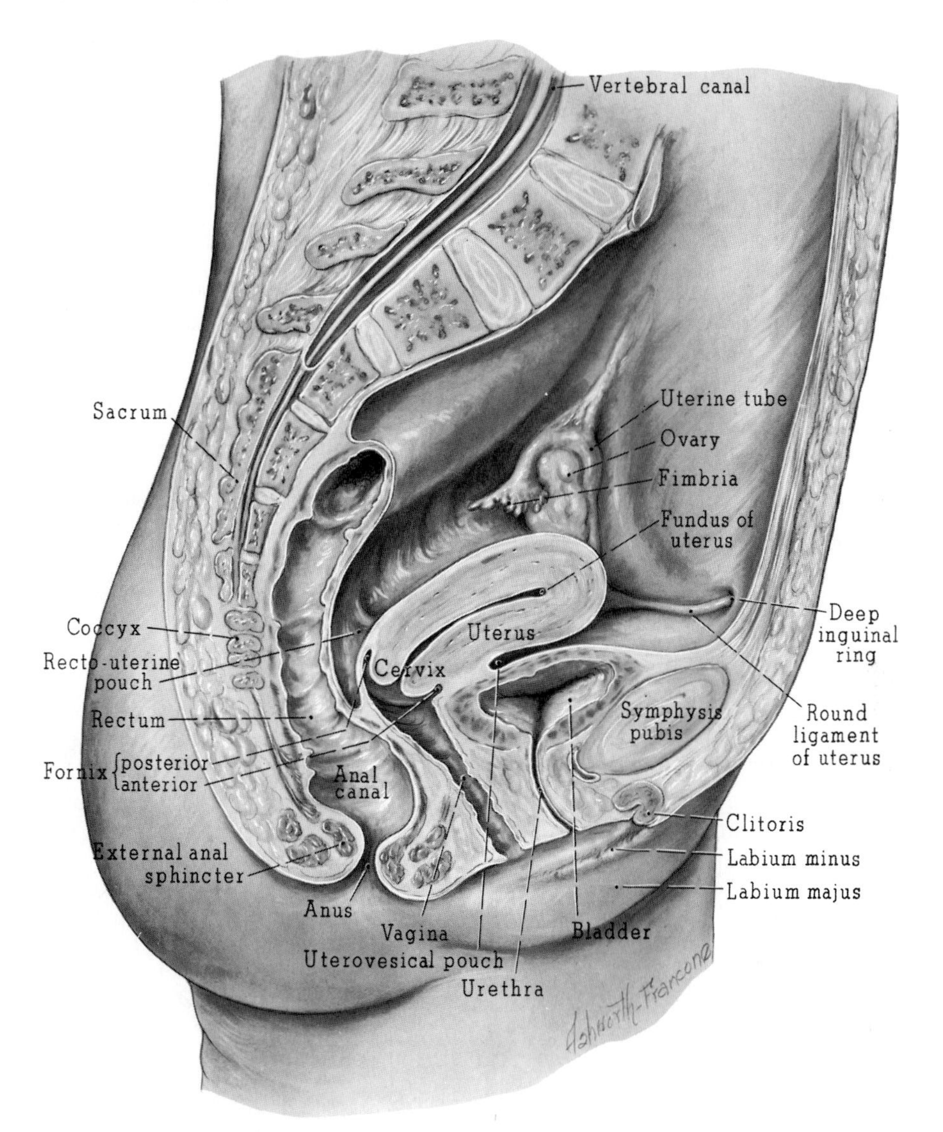

Figure 221. Mid-sagittal section of the female pelvis.

uterine tubes draw the ovum into the tube, and then transport it to the uterus.

Occasionally, an ovum becomes fertilized without entering the uterine tube. The term *ectopic pregnancy* means a pregnancy that develops outside of its proper place in the cavity of the uterus. The term "ectopic" is from the Greek, *ek,* meaning out of, and *topos,* meaning place; together they mean, literally, out of place.

What is the Uterus?

The *uterus* is a pear-shaped, thick-walled, muscular organ suspended in the pelvic cavity above the bladder and in front of the rectum. In its normal state it measures about 3 inches in length and 2 inches in width. The uterine tubes enter into its upper end, one into each side, and the lower end projects into the vagina. The lower portion, called the *cervix* (Latin for neck) corresponds to the neck of an inverted pear. The upper main portion of the uterus is known as the *fundus* (Latin for bottom, or portion farthest from the opening).

The wall of the uterus consists of three layers. The outer layer, called the peritoneal layer, is continuous with the broad liga-

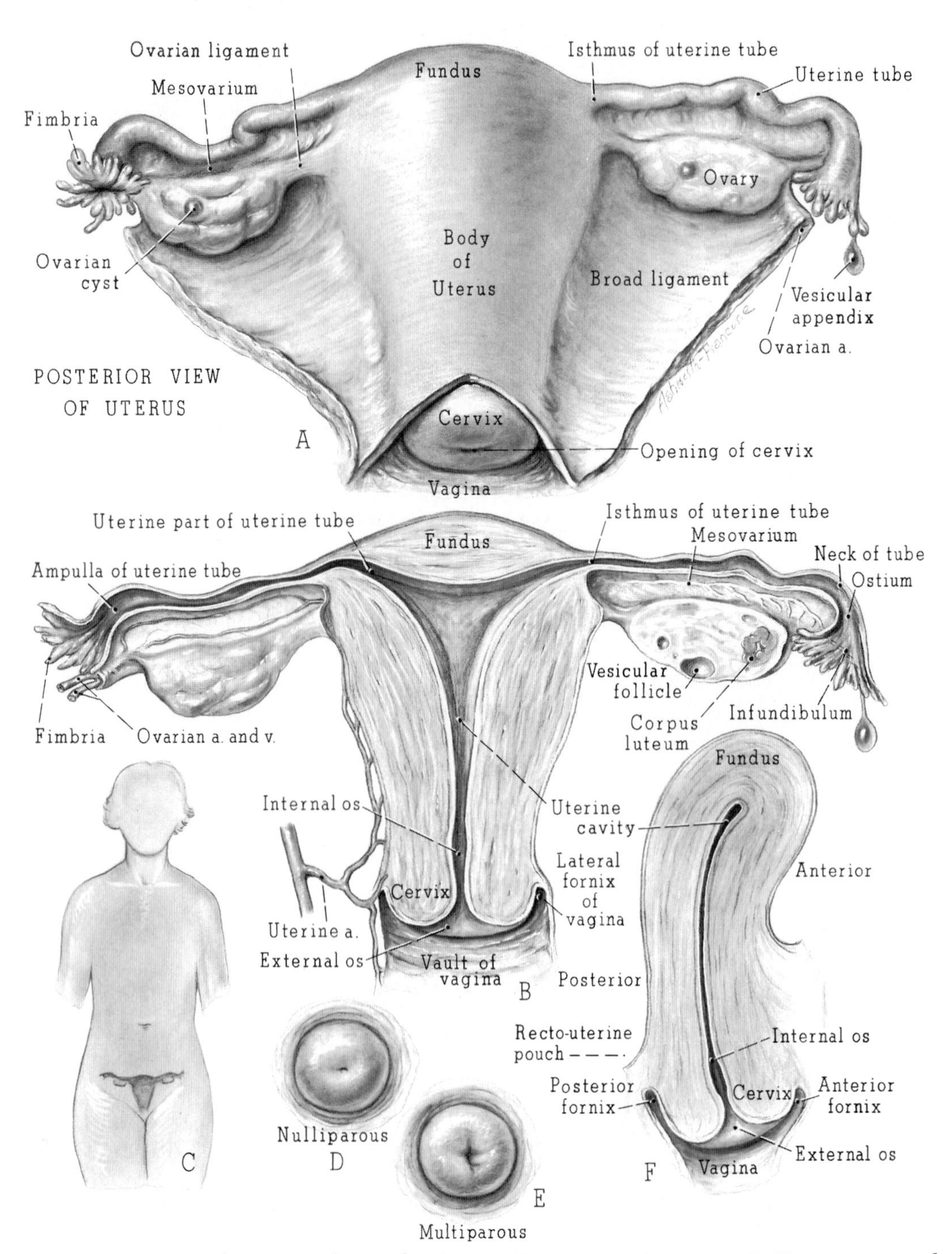

Figure 222. Female organs of reproduction. *A*, Uterus, posterior view. *B*, Uterus sectioned to show internal structure. *C*, Position in body. *D* and *E*, Shape of cervix before and after childbirth. *F*, Right lateral sagittal view.

ments that suspend the uterus. The middle layer, the *myometrium* (Greek *myo* = muscle, and *metra* = uterus), is a thick muscular layer that greatly increases in thickness during pregnancy.

The inner coat of the uterine wall is the mucous membrane, or *endometrium* (Greek *endo* = inside + *metra* = uterus). It consists of an epithelial lining and connective tissue. In *menstruation* (from the Latin word for monthly), the superficial portion of the endometrium pulls loose, leaving torn blood vessels underneath. Blood and bits of endometrium trickle out of the uterus into the vagina and out of the body. Regeneration of a new endometrial lining begins immediately after menstruation.

What Is the Vagina?

The *vagina* is a tubular canal 4 to 6 inches in length, directed upward and backward and extending from the external vestibule to the uterus. It is situated between the bladder and the rectum. The vaginal wall consists of an internal membranous lining and a muscular layer capable of constriction and enormous dilation (enlargement), separated by a layer of erectile tissue. The mucous membrane forms thick transverse folds and is kept moist by cervical secretions (the cervix is the lower part of the uterus).

The vagina serves as part of the birth canal and represents the female organ of copulation.

THE MAMMARY GLANDS

What Are the Mammary Glands?

The two *mammary* (from the Latin word for breast) *glands*, or *breasts*, are accessory reproductive organs. The breasts of pregnant women secrete milk for nourishment of the newborn. The *nipples*, containing the openings of the milk ducts, are located near the center of the breasts. A

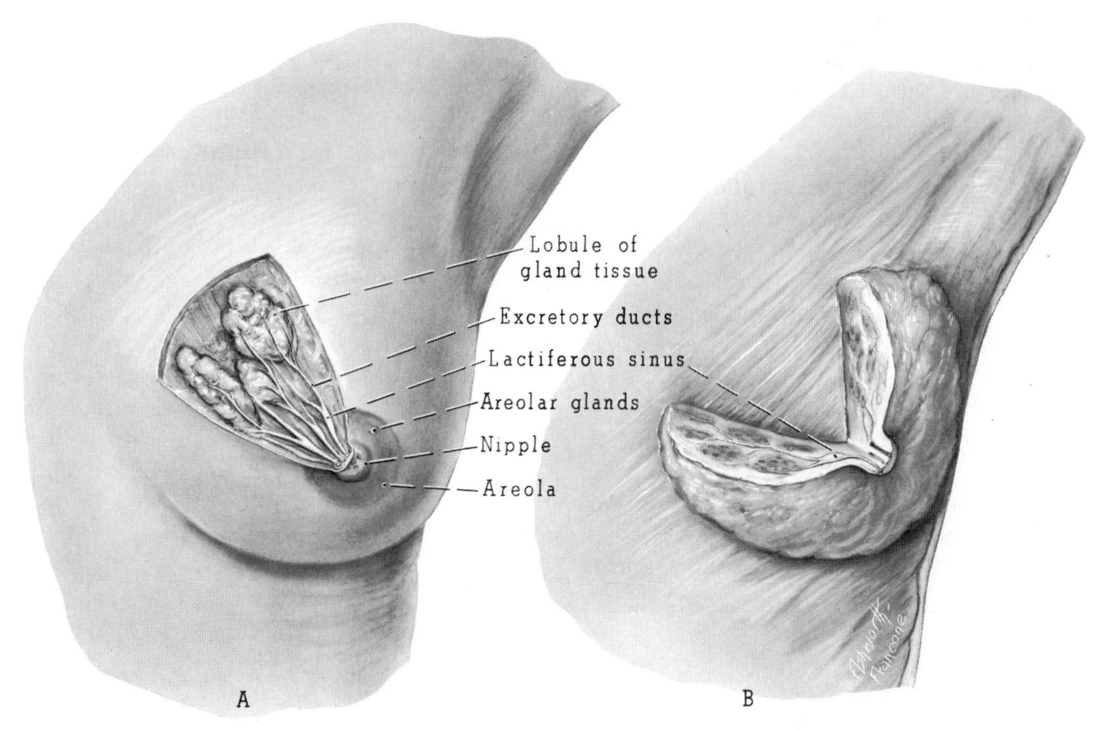

Figure 223. The female breast. *A*, The skin has been partly removed to show the underlying structures. *B*, A section has been removed to show the internal structures in relation to the muscles.

wider, circular area of pigmented skin, known as the *areola* (ah-re′o-lah), surrounds each nipple. There are from 15 to 20 lobes of glandular tissue arranged radially within the breast (Fig. 223).

What Are Some Factors that Affect Milk Production?

Ovarian hormones exert specific control over growth and development of the breast. Estrogen stimulates the development of the ducts; progesterone influences the growth of the milk-producing, glandular tissue.

Lactation, the production of milk, is a complex process requiring the interplay of various hormonal and nervous factors. *Prolactin,* which is secreted by the anterior pituitary, appears to be the prime factor. Suckling of the newborn stimulates release of *oxytocin* from the posterior pituitary, which stimulates release of milk from the glandular cells into the ducts, making it available to the infant.

THE OVARIAN HORMONES

What Are Estrogens? What Is Progesterone?

Estrogens and progesterone are essential ovarian hormones. Estrogens (Greek, *oistros* = mad desire + *gen* = to beget) are secreted by the maturing graafian follicles, the sacs in the ovary that contain the ova. Progesterone is secreted by the *corpus luteum,* the "golden body" that forms each month from the follicle that has matured and discharged its ovum.

The secretion of the estrogens and progesterone occurs in response to two specific hormones produced by the anterior pituitary gland—the follicle-stimulating hormone (FSH), and the luteinizing hormone (LH).

Estrogens are responsible for the development of female characteristics at puberty. They stimulate the growth of the uterus and the vagina; they also assist in the development of secondary sex characteristics, such as the figure, including the formation of ducts in the mammary glands.

During the menstrual cycle, estrogens cause thickening of the endometrium of the uterine wall in preparation for implantation of the fertilized egg. They also stimulate repair of the endometrium following menstruation.

Progesterone is secreted by the corpus luteum, and also by the *placenta* during pregnancy. It helps to prepare the uterine wall for implantation of the fertilized ovum, and is necessary for the process of implantation.

Progesterone maintains the development of the placenta and prevents the ovary from producing ova during pregnancy. It is responsible for enlargement of the breasts during pregnancy and for development of the milk-secreting cells of the mammary gland.

Diminished secretion of progesterone leads to menstrual irregularities in nonpregnant women and spontaneous abortion in pregnant women.

THE MENSTRUAL CYCLE

What Is Menarche? What Is Menopause?

The first menstrual cycle or *menses* (from the Latin word meaning monthly) occurs around 13 years of age, the age of puberty. This beginning is known as *menarche* (me-nar′-ke) (from the Greek *men-*, meaning monthly, and *arche,* meaning the beginning). From this beginning the menstrual cycle continues on a fairly regular basis, every 28 days, for about 35 years until *menopause* (Greek *meno* = monthly + *pausis* = cessation).

The day of onset of the menstrual flow is considered the *first day* of the cycle. The cycle ends on the last day prior to the next menstrual flow. Typically, the cycle is 28 days in duration, but it can vary from 22 to 35 days.

What Are the Three Phases of the Menstrual Cycle?

Three phases of the menstrual cycle are distinguished—the menstrual phase, the proliferative phase, and the secretory phase (Figs. 224 and 225).

Menstrual Phase. The menstrual phase lasts from the first to the fourth day of the cycle. When the ovum is not fertilized, the corpus luteum regresses. This leads to a decrease in the ovarian (corpus luteum portion) secretion of both estrogen and progesterone, which leads to the disintegration of the endometrial lining of the uterus. With some resultant bleeding, the endometrium drains out of the uterus, through the vagina, and out of the body.

Proliferative Phase. This phase, characterized by estrogen stimulation, begins about the fifth day of the cycle and extends through the ovulation, which occurs near the midpoint of the cycle—around the fourteenth day. This stimulation of estrogen secretion by the ovaries is begun by the secretion of FSH from the anterior pituitary. If synthetic estrogen is introduced at the beginning of this phase, FSH secretion is inhibited. As a result, the ovaries are not stimulated to grow and no ovum will be

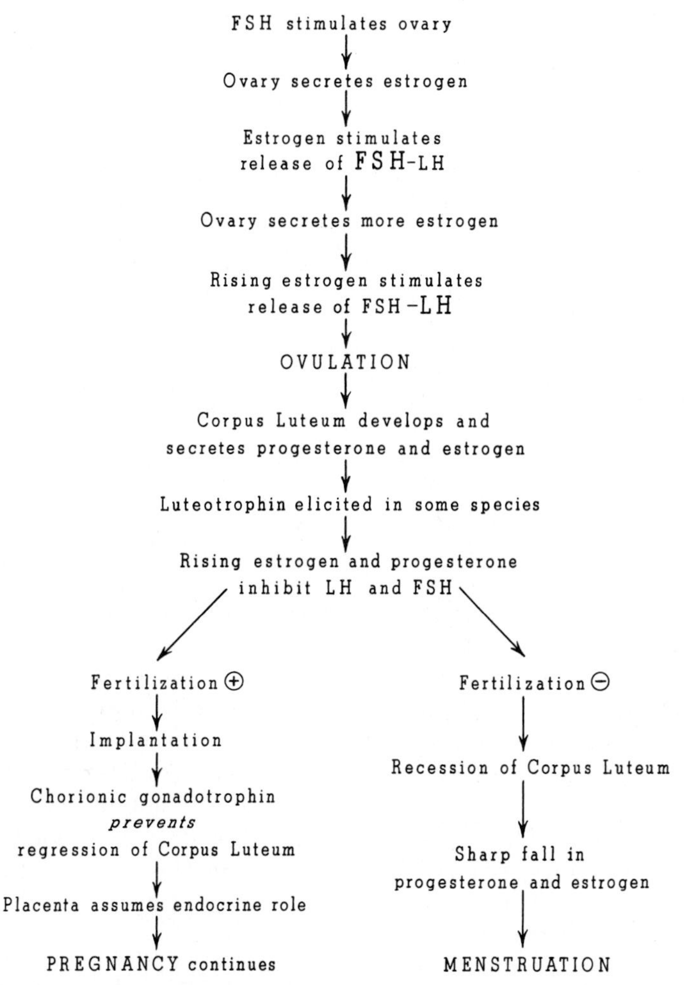

Figure 224. Diagrammatic relationships between hypophyseal and ovarian hormones. (From Tepperman: *Metabolic and Endocrine Physiology.* Chicago, Year Book Medical Publishers, 1962.)

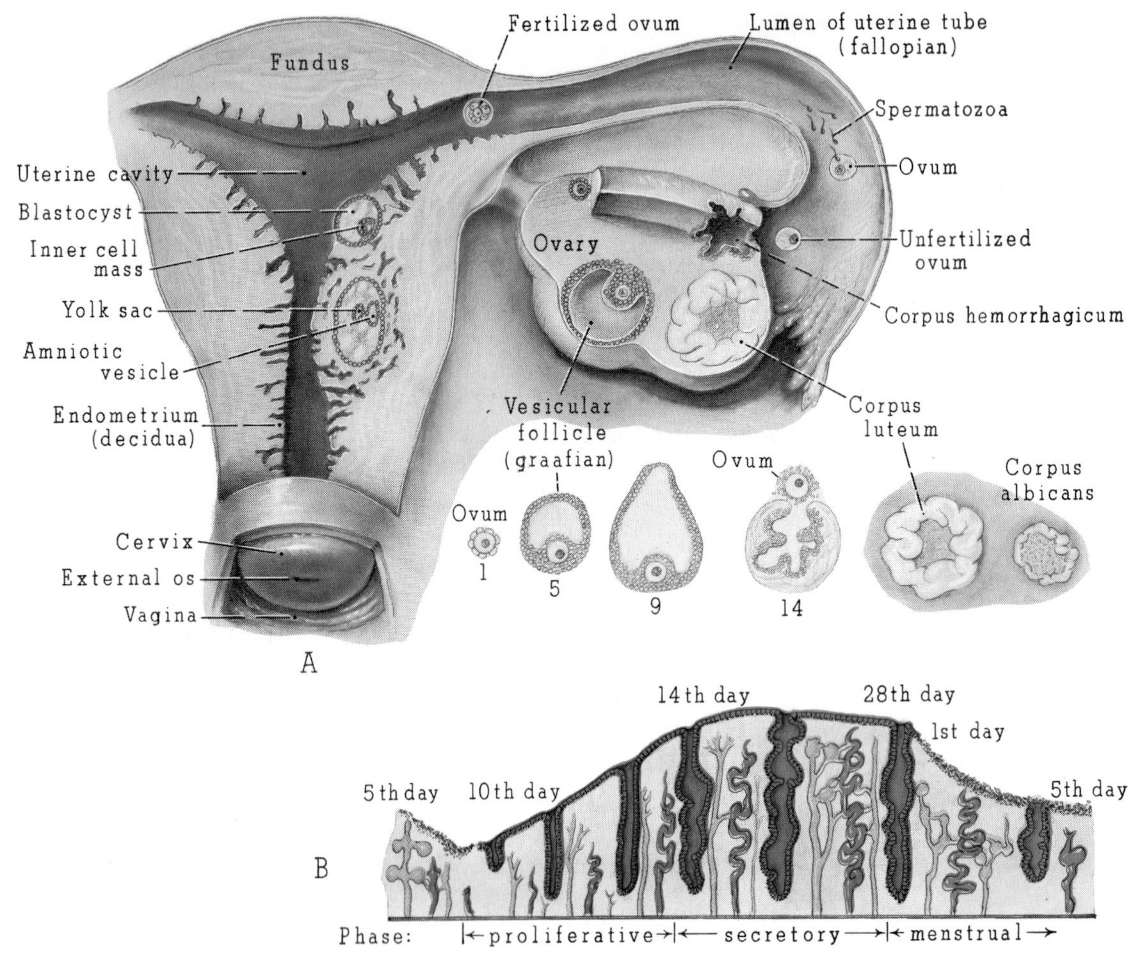

Figure 225. *A*, Physiologic processes of the ovary and uterus, showing ovulation, transportation of the ovum, and implantation. *B*, Cyclic menstrual changes in the uterine endometrium.

produced. This is a simplified version of how oral contraceptives work.

Secretory Phase. The normal *secretory phase* begins at ovulation. The rupture of the mature follicle and release of the ovum are stimulated by a rapid increase in the secretion of luteinizing hormone (LH)—sometimes called the "ovulation hormone." LH continues to act on the ruptured follicle, transforming (luteinizing) it into the secreting body, the *corpus luteum*. The corpus luteum secretes estrogens and, for the first time during the menstrual cycle, begins a significant secretion of progesterone. The secretions of the corpus luteum also further prepare the uterine wall for implantation, should the ovum become fertilized.

If implantation does not occur, the corpus luteum decreases its activity, which leads to disintegration of the endometrium and menstruation—the beginning of the menstrual cycle again.

At this point, you should reread the description of the menstrual phase.

FSH and LH

For summary purposes, one can consider the function of FSH to be stimulation of growth of the graafian follicles until one follicle matures and releases its ovum, on about the fourteenth day of the menstrual cycle.

The actual ovulation (release of the ovum from the mature follicle) is stimulated by a large secretion of LH. Subsequent LH secretions develop and maintain the corpus luteum.

CLINICAL CONSIDERATIONS

Pregnancy Testing

In the early diagnosis of pregnancy attention has recently been concentrated on tests designed to detect the presence of a substance peculiar to the pregnant state, human chorionic gonadotrophic hormone, HCG. This hormone is secreted by the developing placenta in large amounts beginning on about the twenty-sixth day of pregnancy.

Previously, the most widely used tests for the presence of HCG by examination of the patient's urine involved injecting the urine into mice or other experimental animals. After five days the animal was examined for stimulation of its ovaries, which HCG affects. If the animal's ovaries had been stimulated, this indicated that HCG was present in the patient's urine, and that the patient was in fact pregnant. The accuracy of this test has been relatively good, although it is time-consuming and expensive.

More recent procedures for detection of HCG involve blood reaction tests and various procedures to detect it chemically in the urine. These procedures are fast (about two hours), inexpensive, and over 99 per cent accurate when performed by trained laboratory personnel. The test is most accurate during the second and third months of pregnancy.

Symptoms and signs indicating that a pregnancy test should be given are largely subjective, and are appreciated by the patient herself. The signs include cessation of menses, changes in the breasts, alteration of the color of the mucous membranes, and increased pigmentation. The early symptoms include nausea, with or without vomiting, and frequency in urination.

Contraception and Abortion

There are many and varied methods utilized in the prevention of pregnancy. The mechanical means by which contraception is achieved with varying degrees of success include the use of the condom, the diaphragm, and, more recently, the intrauterine device (IUD). This latter device is simply a coil, or loop, placed within the uterus, the function of which now appears to be to interfere with sperm migration or the migration of the fertilized egg through the tube.

Physiologic or chemical means of contraception include the rhythm method, douching, suppositories, foams, and, more recently the contraceptive pill.

Oral contraceptives rely upon the hormone balance of the menstrual cycle for their effectiveness. Most birth control pills are taken 20 or 21 consecutive days of each cycle. Beginning on the fifth day after the initiation of the menstrual flow, a pill containing both estrogen and progesterone is taken. (Some recent contraceptives vary this protocol slightly.) As long as these two hormones are taken orally, secretion of FSH does not begin and the follicles of the ovaries are not stimulated to grow. Otherwise, the normal menstrual cycle proceeds and menstrual flow will begin a few days after the pill is discontinued. The pill must be taken for each cycle, but its efficacy as a contraceptive is insured even on the days that dosage is omitted and the menstrual flow permitted.

An *abortion* is any interruption of pregnancy prior to the period when the fetus is viable. (When infection occurs, the process is known as a septic abortion.) The fetus is considered viable when it weights 500 grams (just over one pound) or more and the pregnancy is over 20 weeks in duration. The term miscarriage is used when spontaneous loss occurs after 20 weeks.

Tumors of the Breast

Fibrocystic breast disease is characterized by fibrous growth in the breast tissue. It is quite common but does not occur before adolescence, and rarely, if ever, develops after menopause. A breast biopsy (removal and examination of tissue for diagnosis) is needed to distinguish between benign fibrous tumors and cancerous tumors.

When a malignant tumor is found in the breast, one of three surgical procedures is usually employed: a *lumpectomy* (-ectomy means removal), removing the tumor but leaving the breast intact; a *simple mastectomy,* or removal of the whole breast; or a *radical mastectomy,* removal of the breast, the associated lymph node chains, and any potentially involved muscles.

Venereal Diseases

Gonorrhea (gon″o-re′ah). This is a contagious inflammation of the genital mucous membrane transmitted chiefly by sexual intercourse, and caused by the bacteria *Neisseria gonorrhoeae.* The disease is marked by pain, difficulty in urinating, and a discharge of mucus and pus. Complications can occur, such as infection of other tissues and organs in the genital area. It may also produce arthritis and endocarditis (inflammation of the endothelial lining of the heart). This disease has reached nearly epidemic proportions in many areas of the U.S., principally in persons between 14 and 20 years of age.

Syphilis. A venereal disease that is almost invariably transmitted by sexual contact, syphilis is of particular clinical importance because the initial lesion and later widespread invasion are often not accompanied by disturbing signs or symptoms. The initial, *primary,* stage consists of the development of a local chancre (sore) in the male—frequently not detected in the female—between the end of the first week and the subsequent three months following contact. In from two to twelve weeks, the secondary stage becomes manifest in the form of a generalized skin rash, sometimes accompanied by involvement of the mucous membranes. An asymptomatic period lasting up to several decades follows the secondary stage. A final tertiary stage may develop in later life involving primarily the tissues of the brain and heart. The development of the disease process beyond the secondary stage is quite unpredictable; medical treatment is indicated in all cases.

LEARNING EXERCISES

1. Construct and label a diagram showing both the internal and external male reproductive organs and/or their parts.
2. Construct and label a diagram of the external female genitalia.
3. Construct and label a sagittal section diagram of the internal female reproductive organs.

INDEX

Note: In this index, page numbers in *italic* type refer to illustrations; page numbers followed by the letter t refer to tables.

Abdominal cavity, *4*, 5
  deep muscles of, *79*
Abductor muscles, 63
Abortion, 239
Absorption, 158
  of nutrients, abnormal, 175
Acidosis, definition of, 142
Acinar tissue, of pancreas, 165
Acne, 35
Acoustic meatus, 205, *205*
Acromegaly, 222
  with gigantism, *215*
ACTH. See *Adrenocorticotrophic hormone.*
Addison's disease, 224
Adductor muscles, 63
Adenine, in DNA, 23
Adenohypophysis, disorders of, 222
  function of, 212–213
Adenosine triphosphate, 20
  formation of, *21*
ADH. See *Antidiuretic hormone.*
Adipose tissue, functions of, 30
Adrenal cortex, disorders of, 222
  function of, 218–219
Adrenal glands, structure and function of, 218
  views of, *219*
Adrenal medulla, disorders of, 224
  function of, 219–220
Adrenocorticotrophic hormone, action of, 213
Age, in control of respiration, 143
Albumin, plasma, function of, 91
Aldosterone, in urine production, 153
  secretion of, 218
Aldosteronism, 222
Alkalosis, definition of, 142
Alpha cells, 220
Alveoli, description and function of, 136, *137*
Amino acids, code words for, 23–24, 24t
  in protein digestion, 168
  in protein manufacture, 23
Amniotic sac, fetus in, *226*
Anal canal, 164, *164*, *165*
Anaphase, in mitosis, 27, *27*
Anatomy, definition of, 1
  gross, 1
  microscopic, 1
  relation of physiology to, 2–4
  structural units in, 5–6
Androgens, actions of, 218
Anemia, blood count in, 97
  types of, 97
Angina pectoris, 124
Antibodies, in blood transfusion, 94–96
Antidiuretic hormone, 153
  action of, 214
  function of, 212
Anuria, 155
Aorta, *101*, 102
  and major branches, *111*
  coarctation of, 121
Appendicitis, 164
Appendicular skeleton, 41, *42*, *43*, 44t–45t
  parts of, 48–51
Appendix, 164, *164*
Arachnoid membrane, 183, *183*
Areola of breast, *235*, 236
Areolar tissue, functions of, 30
Arm, bones of, 48
  lower, arteries of, *116*
    muscles of, *73–75*
    veins of, *117*
  upper, arteries of, *114*
    muscles of, *70*, *71*
    veins of, *115*
Arrector pili muscles, 34, *34*
Arterioles, 106
Arteriosclerosis, definition of, 123
Artery(ies). See also specific arteries; e.g., *Pulmonary artery.*
  component parts of, 106
  description and function of, 106, *106*
  of lower arm and hand, *116*
  of pelvis and leg, *118*
  of shoulder, *114*

Artery(ies) (*Continued*)
  of upper arm, *114*
  to head and neck, *108*
  to internal organs, *113*
  to thyroid gland, *110*, 216
Arthritis, 55
Articular cartilage, 52
Articulations. See *Joints*.
Asthma, definition of, 143
Astigmatism, correction for, *204*
ATP. See *Adenosine triphosphate*.
Atrioventricular node, 103–104, *104*
Auditory tube, *205*, 206
Auricle of external ear, 205, *205*
Autonomic nervous system, diagram of, *200*
  structure and function of, 194–195, 201t
  subdivisions of, 195
Axial skeleton, 41, *42*, *43*, 44t–45t
  parts of, 46–48
Axilla, muscles in region of, *68*
Axons, 177

Back, muscles of, *69*
Baroreceptors, 181
Basal metabolic rate, definition of, 172
Basophils, *90*, 91
Bed sores, 35
Beta cells, 220
Bile, secretion of, by liver, 165
Bile ducts, 166, *166*
Birth control, methods of, 239
Blood, as connective tissue, 31
  composition of, *89*
  formed elements of, functions of, 90–91
  function of, 88
  major constituents of, 88–90
  stroke volume of, 105
  supply of to long bone, *40*
Blood cells, formation of, 92
  types of, *90*
Blood count, definition of, 96
  in anemias, 97
  in infectious mononucleosis, 97
  in leukemia, 96
Blood loss, mechanisms in stopping, 92–93
Blood pressure, definition of, 120
  diastolic, 121
  factors influencing, 121–123
  in various areas of circulatory system, *122*
  normal, 121
  regulation of, baroreceptors in, 122
    chemoreceptors in, 122–123
  systolic, 120
  vasomotor center of brain in, 121–122
Blood serum, 91
Blood transfusion, blood types in, 93–96, 94t
  immune response in, 94
  Rh factor in, 96
Blood types, 93–96, 94t
  cross matching of, *95*
  relationships between, 94–96
Blood vessels. See also *Arteries*, *Veins*, and specific arteries and veins; e.g., *Pulmonary artery* and *Pulmonary vein*.
  types of, 106–107
Body, anatomic positions of, *3*, 4
  coordination of, nervous and endocrine systems in, 16
Body (*Continued*)
  main cavities of, 4, *4*, 5
  water distribution in, 28–29
Body fluids, calcium concentration in, control of, 40–41
  chemical composition of, 28t
  extracellular, 28t, 29
  intracellular, 28, 28t
Boils, 35
Bone(s), as connective tissue, 31, *32*
  calcium in, 39–40
  cancellous, 38
  carpal, 48
  diaphysis of, 38
  epiphysis of, 38
  growth of, 37, *38*
  Haversian canals of, *39*
  long, anatomy of, *39*
    blood supply to, *40*
  medullary canal of, 38
  metacarpal, 48
  metatarsal, 51, *54*
  nature of, 38–39
  of arm, 48
    anterior view of, *51*
  of cranium, 46, *46*
  of face, 46, *46*
  of fingers, 48–50
  of hand, anterior view of, *51*
  of leg and foot, anterior and posterior views of, *53*
  of lower extremities, 50
  of skeleton, 41, *42*, *43*, 44t–45t
  of torso, 46
  of upper extremities, 48
  of wrist, 48
  ossification of, 37
  tarsal, 51, *54*
  typical, appearance of, 38
Bone marrow, function of, 38–39
  red, 38
  yellow, 39
Bone tissue, common disorders of, 52
Bowman's capsule, 148
Brachial artery, blood pressure in, 120
Brachial plexus, skeletal relations of, *197*
Bradycardia, 104
Brain, circulation of cerebrospinal fluid in, *186*
  description of, 185–186
  left half, sagittal view, *187*
  ventricles of, 186, *186*
Brainstem, 188
Breast(s). See also *Mammary glands*.
  areolae of, 236
  nipples of, 235
  structure and function of, 235–236, *235*
  tumors of, 240
Breathing, mechanics of, 136–137
  process of, 136
Bronchi, description and function of, 134–136, *137*
Bulbourethral glands, functions of, 230
Burns, degrees of, 36, *36*
  determining extent of, 35, *36*
Bursae, function of, 52, *57*
Bursitis, 58

Calcitonin, action of, 216–217
Calcium, deposition of, 40, *41*
  function of in body, 39–40, *41*
  reabsorption of, 40, *41*

Calcium phosphate, in bone, 31
Capillaries, description and function of, 106–107, *108*
  lymphatic, 124
Capillary bed, *107*
Carbaminohemoglobin, in gas exchange mechanism, 141
Carbohydrates, digestion of, 170, *171*
  metabolism of, 174
Cardiac arteries, 99
Cardiac cycle, sequence of events in, atrioventricular node in, 103–104, *104*
    sino-atrial node in, 103, *104*
Cardiac muscle, structure and function of, 61
Cardiac output, factors determining, 104–106
  force of contraction in, 106
  heart rate in, 105
  stroke volume of blood in, 105
  venous return in, 105
Cardiac veins, 99
Carpal bones, 48
Cartilage, articular, 52
  function of, 31
  thyroid, 133, *135*
  types of, *32*
Cavity(ies), abdominal, *4*, 5
  cranial, *4*, 5
  dorsal, *4*, 5
  pelvic, *4*, 5
  spinal, *4*, 5
  thoracic, *4*, 5
  ventral, 4, *4*
Cecum, 164, *164*
Cell(s), components of, relative sizes of, 26
  composition of, 5
  daughter, 26
  definition of, 18
  formation of ATP in, *21*
  in nervous system, 177
  in pancreatic islets, 220
  internal environment of, 16
  interstitial, 221
  liver, vitamin storage in, 168
  olfactory, 132
  reproduction of, 26
  three main divisions of, 18–19
Cell membrane, 5
  description of, *18*, 19, *19*
  function of, 20
  size of, 26
Cell nucleus, description of, 18, 19, *19*
  function of, 21–23
  in protein manufacture, 21
  size of, 26
Cellular physiology, 2
Centrioles, in mitosis, 27, *27*
Cerebellum, description and function of, *187*, 188
Cerebral aqueduct, 186
Cerebral cortex, lobes of, 189, *189*
Cerebral medulla, 189
Cerebrospinal fluid, 183
  circulation of in brain and spinal cord, *186*
Cerebrum, structure and function of, *186, 187*, 189, *189*
Cervical vertebrae, 48
  relationship of to face, *46*
Cervix of uterus, 233, *233*, *234*
Chemoreceptors, in control of respiration, 143
Childbirth, oxytocin in, 214
Cholecystitis, 175–176
Choroid of eye, 202, *203*
Chromosomes, formation of, 27, *27*
Chyme, 159
Cilia, nasal, 132
Circulatory system, blood pressure, velocity and cross section of vascular tree in, *122*
  clinical conditions in, 123–124
  function of, 98
  in homeostasis, 16
  portal flow in, 110
  pulmonary circulation in, 110
  relationship of lymphatic system to, *125*
  structures in, 98–99
  systemic and portal, relationships between, *121*
  systemic and pulmonary, relationship between, *120*
  systemic flow in, 110
Circumcision, 227
Cirrhosis, 175
Clavicle, 48
Clitoris, 231, *231*
Coagulation, mechanism of, abnormalities in, 93
  phases of, *94*
  process of, 93
Coccygeal vertebrae, 48, *49*
Coccyx, 48, *49*
Cochlea, *205*, 206
Collar bone, 48
Colliculi, of midbrain, 188
Colon, arteries to, *113*
  description of, *164*, 165
Compound fracture, 56
Concussion, definition of, 207
Conjunctiva of eye, 202, *203*
Connective tissues, 5
  dense, 30, *32*
  function of, 30
  loose, 30, *32*
Contraception, methods of, 239
Contracture, of muscle, 63
Cornea of eye, 202, *203*
Coronary sinus, 100, *102*
Corpus callosum, 189
Corpus luteum, 232, *234*
  estrogen secretion in, 238
  in progesterone secretion, 236
Cortex, adrenal, 218–219
  of kidney, 147–148
Cortisol, 218
Costal breathing, 136
Cowper's glands, functions of, 230
Cranial cavity, *4*, 5
Cranial nerves, 185
  description of, 191, *191*
  distribution of, *190*
  table of, 192t
Cranium, bones of, 46, *46*
Cushing's disease, 222, *223*
Cyanosis, 143
Cystic duct, 166, *166*
Cystitis, 156
Cytoplasm, 5
  description of, *18*, 19
  function of, 20
Cytoplasmic organelles, definition of, 20
Cytosine, in DNA, 23

D agglutinogen, 96
Daughter cells, 26

Dendrites, 177
Dental caries, 172
Dentin, *32*
Deoxyribonucleic acid, 21
  definition of, 22
  language of, 23–24
Deoxyribonucleic acid helix, 23, *23*
Deoxyribonucleic acid molecule, chemical structure of, 22
  size of, 26
Deoxyribose, in DNA, 23
Dermis, 33, *33*
Detrusor muscles, of urinary bladder, 155
Diabetes, definition of, 214
Diabetes insipidus, 222
  diuresis in, 214
Diabetes mellitus, 224
  insulin secretion in, 220–221
Diaphragm, relationship of to rib cage, *99*
Diaphragmatic breathing, 137
Diaphysis, 38
Diencephalon, structure and function of, 188
Diffusion, passive, in absorption mechanism, 170
Digestion, 158
  of carbohydrates, 170, *171*
  of proteins, 168–170, *171*
  process of, stages of, 168
Digestive system, disorders of, 172–176
  functions of, 158
  in homeostasis, 16
  structures of, 158–159, *159*
Digestive tract, upper, functions of, 159
Disease, definition of, 27–28
  types of, 28
Diuresis, 153
  in diabetes insipidus, 214
Diuretics, in water retention, 154
DNA. See *Deoxyribonucleic acid.*
Ductus deferens, functions of, 230
Duodenum, description of, 163
  relationship of pancreas to, *166*
  ulcer of, 173
Dura mater, 183, *183*
Dwarfism, 222

Ear, description and function of, 204–206
  external, parts of, 205, *205*
  in frontal section, *205*
  inner, parts of, *205*, 206
  middle, parts of, 205–206, *205*
  three divisions of, *205*
Eardrum, *205*, 206
Ectopic pregnancy, 233
Edema, diuretics for, 154
  in pneumonia, 143
  pulmonary, definition of, 146
Electrocardiogram, *105*
Electron microscope, 1, 2
Elimination, 158
Embolism, 93
Emphysema, definition of, 143
Endocardium, 100, *100*
Endocrine glands, 165
  function of, 210, *211*
Endocrine system, description and function of, 210
  in body coordination, 16
  structures in, 210–211
Endometrium, 235
  cyclic menstrual changes in, *238*
Endoplasmic reticulum, *19*, 21
Enzymes, 21
Eosinophils, *90*, 91
Epicardium, 100, *100*
Epidermis, 33, *33*
Epididymis, functions of, 230
Epiglottis, 134, *135*
Epilepsy, seizures in, 207
Epinephrine, production of, 219
Epiphysis of bone, 38
Epithelial tissues, 5
  function of, 30
  types of, *31*
Erythrocytes, *90*
  function of, 90
Esophagus, anatomic position of, *161*
  description and function of, 159
Estrogens, definition of, 236
  in milk production, 236
  production of, 221, 231
Ethmoid sinus, 133
Eustachian tube, 133, *205*, 206
Exercise, and muscle tone, 62
Exocrine glands, 165, 210
  types of, *31*
Exophthalmos, 222
  in hyperthyroidism, 216, *217*
Extensor muscles, 63
Eye, description and function of, 202–204
  external appearance of, *202*

Face, arteries to, *108*
  bones of, 46, *46*
  relationship of skull and cervical vertebrae to, *46*
  superficial muscles of, *65*
  venous drainage of, *109*
Facial nerve (VII), branches of, *193*
Fallopian tubes, structure and function of, 232–233, *234*
False ribs, 48
Farsightedness, correction of, *204*
Fats, digestion of, 170
  neutral, digestion of, *172*
Femur, 50, *52*
Fetus, in amniotic sac, *226*
Fibrin clot, formation of, 92, *93*
Fibrinogen, 93
  function of, 92
Fibrosis of lung, in tuberculosis, 146
Fibula, 51, *54*
Fingernails, 34, *34*
Fingers, bones of, 48–50
Flexor muscles, 63
Floating ribs, 48
Fluid, cerebrospinal, 183, *186*
Follicle stimulating hormone, 236
  action of, 213
Foot, arteries to, *118*
  bones of, anterior and posterior views of, *53*
  nerves of, *199*
  superficial muscles of, *84*, *85*
Forearm, arteries of, *116*
  muscles of, posterior view, *75*
  nerves of, *198*
  venous drainage of, *117*

Forebrain, structure and function of, 189–190
Fractures, types of, 56
Froehlich's syndrome, 222
Frontal lobe of cerebral cortex, 189, *189*
Frontal sinus, 133
FSH. See *Follicle stimulating hormone.*
Fundus, of stomach, 159
  of uterus, 233, *233*, *234*

Gallbladder, digestive functions of, 165–166
  disorders of, 175–176
  external view of, *168*
Gamma globulin, function of, 92
Ganglia, of autonomic nervous system, 194
Gas exchange, in lungs, 138–142, *141*
    simple diffusion in, 141
  in tissues, 142
Genetic code, 23
Gland(s). See also specific glands; e.g., *Adrenal glands.*
  description and function of, 30, *31*
Genital system, relationship of urinary system to, *148*
Genitalia, male, external, midsagittal section of, *227*
GH. See *Growth hormone.*
Gigantism, 222
  with acromegaly, *215*
Globulin, function of, 91
Glomerular filtration, 149
  normal pressure in, *152*
Glomerulonephritis, 156
Glomerulus, 148
Glucagon, effect of on blood glucose concentration, 221
  production of, 220
  secretion of, by pancreas, 165
Glucocorticoids, action of, 218
Gluconeogenesis, definition of, 168
  process of, 219
Glucose, in digestion of carbohydrates, 170
  metabolism of, liver in, 166–168
Glycogen, 166
  breakdown of, 221
Glycogenosis, 168
Glycogenolysis, 168
Goiter, 216
Gonads. See also *Testes.*
  function of, 221, 227, 228
Gonorrhea, 240
Gout, 55
Graafian follicles, 232, *234*
Grand mal seizure, 207
Gray matter, of cerebellum, 188
Growth hormone, action of, 213
  overproduction of, and acromegaly, 213, *215*
  underproduction of, and Simmond's disease, 213, *214*
Guanine, in DNA, 23

Hair, parts of, 34, *34*
Hand, arteries of, *116*
  muscles of, posterior view, *75*
  nerves of, *198*
  veins of, *117*
Haversian canals, 39
HCG. See *Human chorionic gonadotrophic hormone.*
Head, arteries to, *108*
  venous drainage of, *109*
Heart, atria of, 100, *100*, 102
  chambers of, 100–102
  conducting system of, *104*
  normal rate of, 104
  relationship of lungs to, *139*
  relationship of to rib cage, *99*
  septum of, 100
  Starling's Law of, 106
  stroke volume of, 105
  structures of, 99–100, *100*
  valves of, function of, 102
    mechanism of operation, 103
    types of, 101, *101*, 102, *102*, *103*
  ventricles of, 100, *100*, 102
Heart sounds, causes of, 103
Hematocrit, 88, *90*
Hematopoiesis, 39
Hematopoietic tissue, *32*
Hemoglobin, 90
  in gas exchange mechanism, 141
Hemophilia, 93
Hepatic artery, 110
Hepatic ducts, 166, *166*
Hepatic jaundice, 176
Hilus, of kidney, 147
  of lung, 136
Hindbrain, structures and functions of, 186–188
Hip joint, with attachments, *56*
Hives, 35
Hemodialysis, definition of, 156–157
Homeostasis, 16
Hormone(s), 16, 210. See also specific hormones; e.g., *Adrenocorticotrophic hormone, Luteinizing hormone.*
  definition of, 211
  function of, 211–212
  in urine production, 153
  ovarian, and hypophysial, relationship between, *237*
Human chorionic gonadotrophic hormone, tests for, 239
Human physiology, 2
Humerus, 48
Hunchback, 48
Hunger, sensation of, sensory receptors for, 202
Hydrocephalus, 186
  definition of, 206
  head enlargement in, *207*
  operative procedure for, *208*
Hydrocortisone, 218
Hymen, 231, *231*
Hyperopia, correction of, *204*
Hyperpnea, 143
Hyperthyroidism, 222
  exophthalmos in, 216, *217*
Hyperventilation, alkalosis in, 142
Hypophysis, description of, 212
Hypothalamus, structure and function of, 188
Hypothyroidism, 222
  myxedema in, 216, *217*
Hypoxia, 143

ICSH. See *Interstitial cell stimulating hormone.*
Ileocecal valve, 164, *164*
Ilium, 50
Incus, *205*, 206

Infectious mononucleosis, blood count in, 97
Injuries, classification of, 28
Insulin, lack of, progressive symptoms in, *220*
  production of, 220
  secretion of by pancreas, 165
Interbrain, structure and function of, 188
Internal organs, arterial supply and venous drainage of, *113*
Interneurons, function of, 179
Interphase, in mitosis, 27, *27*
Interstitial cells, 221
Interstitial cell stimulating hormone, 229
Intestine, large. See *Large intestine* and *Colon*.
  small. See *Small intestine*.
Intramuscular injections, locations for, 72, *86*, *87*
Intrauterine device for contraception, 239
Iodine, in production of thyroid hormone, 216
Iris of eye, *202*, 203
Iron deficiency anemia, blood count in, 97
Ischium, 50
Islets of Langerhans. See *Pancreatic islets*.

Jaundice, 176
Joint(s), description and function of, 51–52, *54*
  disorders of, 55, 58
  fixed, 51, *54*
  freely movable, movements permitted by, *55*
  hip, with attachments, *56*
  knee, in frontal section, *56*
  movable, 51–52, *54*
  structure of, 52, *56*
  types of, 51
Joint capsule, 52

Keratin, 34
Kidney(s), arteries to, *113*
  artificial, diagram of, *156*
  common disorders of, 155–156
  description of, 147
  function of, 147
    control of, 152–154
  gross structure of, *150*
  internal structure of, 147–149
  transplant of, *156*, 157
Knee joint, frontal section, *56*
  lateral views, *57*
Kneecap, 51, *56*, *57*
Kyphosis, 48

Labia majora, 230, *231*
Labia minora, 231, *231*
Lactation, 236
Lactogenic hormone, action of, 213, 214
Large intestine, function of, 163–165
  structure of, 163–165, *164*
Larynx, description and function of, 133–134, *135*
Leg, arteries of, *118*
  bones of, anterior and posterior views of, *53*
  lower, superficial muscles of, *84*, *85*
  lumbar and sacral nerve plexuses of, *199*
  upper, superficial muscles of, *81*, *82*, *83*
  veins of, 119
Lens of eye, 203, *203*
Leukemia, blood count in, 96
Leukocytes, *90*
  functions of, 90–91
LH. See *Luteinizing hormone*.
Ligaments, 52
  ovarian, 231, *234*
Lingual tonsil, *128*
Lipids, metabolism of, *175*
Liver, arteries to, 110, *113*
  digestive functions of, 165–166
  disorders of, 175–176
  normal relationships of, *167*
  role of in glucose metabolism, 166–168
Lordosis, 48
Lumbar plexus, of leg and foot, *199*
Lumbar vertebrae, 48
Lumpectomy, 240
Lung(s), arteries to, *113*
  description and function of, 136, *137*
  functional unit of, microscopic, *140*
  gas exchange in, 138–142, *141*
    simple diffusion in, 141
  medial aspect of, *139*
  relationship of to heart and pulmonary vessels, *139*
  respiratory capacities and volumes of, *140*
  tubercle of, 145
  venous drainage of, *113*
Lunula of nail, 35
Luteinizing hormones, 236
  action of, 213
Lymph, definition of, 124
Lymph nodes, description of, 124, *126*
  location of, 126
Lymphatic capillaries, 124
Lymphatic ducts, 124, *125*
Lymphatic system, and drainage, *127*
  functions of, 126
  relationship of to circulatory system, *125*
  structure of, 124
Lymphocytes, *90*, 91
Lymphoid tissue, *32*
Lysosomes, *19*, 20
  size of, 26

Malleus, *205*, 206
Mammary glands. See also *Breast(s)*.
  structure and function of, 235–236
Mastectomy, 240
Maxillary sinus, 133
Mediastinum, 132
Medulla, cerebral, 189
    description of, 186
    in control of respiration, 142
  of kidney, 147
Medullary canal, of bone, 38
Melatonin, secretion of, 221
Menarche, definition of, 236
Meninges, 183, *183*
Meningitis, definition of, 207
Menopause, definition of, 236
Menstrual cycle, 236–239
  phases of, menstrual, 237
    proliferative, 237
    secretory, 238
Menstruation, 236
  endometrium in, 235
Messenger ribonucleic acid, 21
  description of, 24, *25*

Metabolism, basal, 172
of carbohydrate, *174*
of lipids, *175*
of protein, *174*
Metacarpal bones, 48
Metaphase, in mitosis, 27, *27*
Metatarsal bones, 51, 54
Microscope, electron, 1, *2*
light, limits of, 1, *2*
Midbrain, structure and function of, 188
Mineralocorticoids, action of, 218
Mitochondria, *19*, 20
size of, 26
Mitosis, detail of division in, *27*
main stages of, 27, *27*
process of, 26
Monocytes, *90*, 91
Motor neurons, function of, 179
Mouth, 159
roof of, *161*
Mucous membrane, function of, 30
Multiple sclerosis, definition of, 206
Muscle(s). See also specific muscles; e.g., *Adductor muscles* and *Extensor muscles*, etc.
atrophy of, 72
contracture of, 63
disorders of, symptoms of, 63, 72
female, first layer of, *7*
hypertrophy of, 72
in region of axilla, *68*
insertion of, 63, 66t–67t
male, first layer of, *6*
second and third layers of, *8*
of abdominal, thoracic, and pelvic cavities, deep, *79*
of back, *69*
of face, superficial layer, *65*
of hand and forearm, palmar aspect, *73*, *74*
posterior view, *75*
of lower leg and foot, superficial, *84*, *85*
of neck, superficial, *65*
of shoulder and upper arm, *70*, *71*
of upper leg, superficial, *81*, *82*, *83*
origin of, 63, 66t–67t
skeletal, location and action of, 63, *64–65*, 66t–67t
movement in, 63
smooth, 60–61
striated, 59–60
types of, differences in, 59–61, *60*
Muscle contraction, isometric, 61, *62*
isotonic, 61, *61*, *62*
mechanics of, 60
Muscle fibers, 59
Muscle tone, definition of, 62
exercise and, 63
Muscular dystrophy, 72
Muscular system, function of, 59
Muscular tissues, 5, 32–33, *32*
Myasthenia gravis, 72
Myelin, in insulation of nerve fibers, 178
Myocardial infarction, 124
Myocardium, 100, *100*
Myometrium, 235
Myopia, correction of, *203*
Myxedema, 222

Nails, 34, *34*
Nasal cavity, 46, *46*
description of, 132
Nasal cilia, 132
Nasal septum, 132
components of, *133*
Nasopharyngeal tonsils, *128*
Nearsightedness, correction of, *203*
Neck, arteries to, *108*
superficial muscles of, *65*
venous drainage of, *109*
Neisseria gonorrhoeae, 240
Nephritis, 155
Nephrons, 148, 150
detailed diagram of, *151*
function of, 149
functions of parts of, 153t
Nephrosis, 156
Nerve fibers, insulation of, myelin in, 178
Nerve plexuses, 194
brachial, skeletal relations of, *197*
lumbar, of leg and foot, *199*
sacral, of leg and foot, *199*
Nerves. See also *Neurons.*
of brachial plexus, *197*
of forearm and hand, *198*
of spinal cord, *196*
sympathetic, pathways for distribution of, *197*
Nervous system. See also specific system; e.g., *Autonomic nervous system*, *Central nervous system*, etc.
cell types in, 177
central, 182–190
functions of, 177
in body coordination, 16
two main divisions of, 182, *182*
Nervous tissues, 5
function of, 33
Neural synapse, *179*, 181
Neuroglia, 177
Neurohypophysis, 212
disorders of, 222
function of, 214
Neurons, *32*, 33. See also *Nerves.*
all-or-none rule in, 182
classes of, 178–179
description and function of, 177–178, *178*
impulse transmission in, 180–181, *180*
of autonomic nervous system, 194–195
of reticular activating system, 188
stimulus sensitivity of, 181–182
synapse in, *179*
Neutrophils, *90*, 91
Nipples of breasts, 235, *235*
Norepinephrine, production of, 219
Nose, description of, 132, *133*
Nucleus of cell. See *Cell nucleus.*
Nutrients, absorption of, 170–172, *173*
abnormal, 175
specific, utilization of, 172, *174*, *175*

Occipital lobe of cerebral cortex, *189*, 190
Olfactory cells, 132
Oliguria, 155
Oral contraceptives, 239
Organ(s), composition of, 5–6
of Corti, 206
Organ systems, composition of, 6
Organelles, cytoplasmic, 20
Os coxae, 50
Ossification, process of, 37

Osteoarthritis, 55
Osteoporosis, 52
Ovarian ligament, 231, *234*
Ovaries, description and function of, 231–232, *234*
  endocrine function of, 221
Ovulation, process of, *238*
Ovum(a), 221, 232
  and sperm, size relationships, *226*
  transportation and implantation of, *238*
Oxytocin, action of, 214
  in milk production, 236

Pain, sensation of, sensory receptors for, 202
Palatine tonsil, *128*
Pancreas, acinar tissue of, 165
  digestive function of, 165
  disorders of, 224
  function of, 220–221
  relationship of to duodenum, *166*
  section of, *220*
Pancreatic islets, description of, 220, *220*
  function of, 220–221
Paralysis, 72
  in spinal cord injury, 206
  in stroke, 208
Paranasal sinuses, description of, 133, *134*
Parasympathetic nervous system, diagram of, *200*
  structure and function of, 195–202
Parathormone, action of, 217
Parathyroid glands, description and function of, 217
  disorders of, 217–218, 222
  in control of calcium concentration, 41
  location of, *218*
Parathyroid hormone, 217
Parietal lobe of cerebral cortex, 189, *189*
Parkinson's disease, definition of, 208
Patella, 51, *56*, *57*
Pectoral girdle, bones of, 48
Pelvic cavity, *4*, 5
  deep muscles of, 79
Pelvic girdle, bones of, 50
Pelvis, arteries of, *118*
  renal, 147
  veins of, *119*
Penis, description and function of, 225, *227*, *228*, *229*
  prepuce of, 227
Peptic ulcer, 173
Pericardium, 99, 100, *100*
Perineum, female, skin and superficial fascia removed, *232*
Peripheral nervous system, main subdivisions of, 191
Peristalsis, definition of, 163
  in ureters, 154
Peritoneum, description and function of, 158
Pernicious anemia, blood count in, 97
Petit mal seizure, 207
Phagocytosis, 20, *20*
Pharynx, 159
  description of, 133
  in digestion, 159
Phosphoric acid, in DNA, 23
Physiology, definition of, 2
  relation of anatomy to, 2–4
Pia mater, 183, *183*
Pineal gland, function of, 221
Pinna of external ear, 205, *205*
Pinocytosis, 20, *20*
Pituitary gland, description of, 212
Plasma, 88
  composition of, 91
Plasma proteins, functions of, 91–92
Platelets, 91
Pleural cavities, 132
Pleurisy, 132
Pneumonia, definition of, 143–145
Polyuria, 155
Pons, description of, 188
Portal vein, 110
Pregnancy, ectopic 233
  position of fetus in, *226*
  prevention of, 239
  tests for, 239
Prepuce of penis, 227
Progesterone, production of, 221, 231
  secretion of, 236
Prolactin, action of, 213–214
  in milk production, 236
Prophase, in mitosis, 27, *27*
Prostate, function of, 230
Protection, skin function in, 35
Protein, digestion of, *171*
  manufacture of, cell nucleus in, 21–23
  metabolism of, 174
  synthesis of, 25–26
Prothrombin, function of, 92
Psoriasis, 35
Pubis, 50
Pulmonary artery, 102, 110
  angiograph of, *138*
Pulmonary edema, definition of, 146
Pulmonary veins, 110
  angiograph of, *138*
Pulse, taking of, pressure points for, *125*
    reasons for, 123
Pupil of eye, *202*, 203, *203*
Pyelonephritis, 156
Pyloric sphincter, 161
Pylorus of stomach, 159
Pyorrhea, 172

Radius, 48
Rectum, arterial supply, 113
  description of, 165, *165*
  function of, 164
Reflex arc, 179, *179*, *180*
Renal pelvis, 147
Renal tubule, 148
Reproductive organs, male, external, 225–227
    internal, 227–230
Reproductive system, female, disorders of, 239–240
    external structures of, 230–231
    internal structures of, 231–235, *234*
  function of, 225
  male, structure and function of, 225–230
Respiration, control of, age in, 143
    chemoreceptors in, 143
    medulla in, 142
Respirator, artificial, *145*
Respiratory system, disorders of, 143
  functions of, 131
  in homeostasis, 16
  structures of, 131–143
Resuscitation, mouth-to-mouth, *144*

Reticular activating system, 188
Retina of eye, 202, *203*, 204
  rods and cones of, 204
Rheumatoid arthritis, 55
Rh factor, in blood transfusion, 96
Rib cage, anterior view of, *47*
  posterior view of, *47*
  relationship of heart and diaphragm to, *99*
  relation of humerus, scapula, and clavicle to, *50*
  relationship of spleen to, *130*
Ribonucleic acid, 21
  definition of, 24
  messenger, 21
  ribose in, 24
  ribosomal, 22, 25
Ribosomes, *19*, 21
  size of, 26
Ribs, false, 48
  floating, 48
  true, 48
Rickets, 52
RNA. See *Ribonucleic acid.*

Sacral plexus, *199*
Sacral vertebrae, 48
Sacrum, 48
Scapula, *47*, 48
Sclera of eye, 202, *202*, *203*
Scoliosis, 48
Scrotum, 221
  description and function of, 225, *227*, *228*
Sebaceous gland, 34, *34*
Sebum, 34
Secretion, skin function in, 35
Seizures, convulsive, 207
Semen, 225, 228
Semicircular canal of inner ear, *205*, 206
Semicircular duct of inner ear, *205*, 206
Seminal vesicles, functions of, 230
Sensation, skin function in, 35
Sensory neurons, function of, 179
  special receptors of, 202
Sex glands. See *Gonads, Testes,* and *Ovaries.*
Shaking palsy, 208
Shock, definition of, 123
Shoulder, arteries of, *114*
  muscles of, *70*, *71*
  veins of, *115*
Shoulder blade, 48
Shoulder girdle, 48
Sickle cell anemia, blood count in, 97
Simmond's disease, 222
Simple fracture, 56
Sino-atrial node, 103, *104*
Sinus(es), coronary, 100, *102*
  ethmoid, 133
  sphenoid, 133
Skeletal muscles, 60
Skeletal system, articulations of, 51–52, *54*
  function of, 37
Skeleton, anterior view of, *42*
  appendicular, 41, *42*, *43*, 44t–45t
  axial, 41, *42*, *43*, 44t–45t
  development of, 37
  lateral and posterior views of, *43*
Skin, appearance of, importance of, 35
  appendages of, 34–35
  clinical aspects of, 35–36
  definition of, 33
  functions of, 35
  three dimensional view of, *33*
  tissues of, 33–34
Skin disorders, common, 35
Skull, description of, 46
  relationship of to face, *46*
Small intestine, description and function of, 163
  diagram of, *162*
  in absorption of nutrients, 171
Smooth muscle, structure and function of, 60–61
Somatotrophin, action of, 213
Sound, conduction of, 206
Special senses, receptors of, 202
Sperm, and ovum, size relationships, *226*
  formation of, 221
  production of, 228
Sphincter, pyloric, 161
Spinal cavity, *4*, 5
Spinal cord, brain communication function of, 184–185, *185*
  circulation of cerebrospinal fluid in, *186*
  description of, 182–184
  formation of spinal nerve in, 183
  injury to, results of, 206
  major anatomical divisions of, *182*
  major ascending and descending tracts of, *184*
  nerves from, *196*
  reflex functions of, *180*, 184–185
  relation of to vertebra, *183*
Spinal nerves, 185
  description of, 191–194
  relation of to vertebrae, *183*
Spirometer, in measuring lung capacity, 138
Spleen, arteries to, *113*
  description and functions of, 129–130
  relationship of to stomach and rib cage, *130*
Stapes, *205*, 206
Sternum, 48
STH. See *Somatotrophin*, 213
Stomach, anatomical position of, *161*
  description and function of, 159–163
  relationship of spleen to, *130*
Stomach ulcer, 173
Striated muscle, description and function of, 59–60
Stroke, definition of, 208
Sunburn, 35
Suprarenal glands, structure and function of, 218
Sweat glands, 34, *34*
Sympathetic nervous system, diagram of, *200*
  structure and function of, 195–202
Syphilis, 240
Systole, ventricular, 103

Tachycardia, 104
Tachypnea, 143
Tarsal bones, 51, 54
Teeth, in digestion, 159, *160*
Telophase, in mitosis, 27, *27*
Temperature regulation, skin function in, 35
Temporal lobe of cerebral cortex, 189, *189*
Testes, description of, 228
  functions of, 221, 227, 228
  interstitial cells of, 229

Testes *(Continued)*
section of, *229*
seminiferous tubules of, 228, *229*
Testosterone, 218
production and secretion of, 221, 227, 229
Tetany, 222
TH. See *Thyroid hormone* and *Thyroxin.*
Thalamus, 188
Thirst, sensation of, sensory receptors for, 202
Thoracic cavity, *4*, 5
deep muscles of, *79*
description of, 132
Thoracic vertebrae, 48
Thorax, 46–48, *47*
Thrombosis, 93
Thymine, in DNA, 23
Thymus, description and function of, 129
location of, and relationship to lungs, *129*
Thyroid cartilage, 133, *135*
Thyroid gland, arteries to, *110*, 216
description and function of, 216–217
disorders of, 222
venous drainage of, 110, *216*
Thyroid hormone, action of, 216
Thyroid stimulating hormone, action of, 213
Thyroxin, action of, 216
Tibia, 51, *53*
Tidal volume, definition of, 137
Tissue(s), acinar, of pancreas, 165
adipose, functions of, 30
bone, common disorders of, 52
composition of, 5
in percentages, 28t
connective, 5, 30, *32*
definition of, 30
epithelial, *31*
gas exchange in, mechanism of, 142
hematopoietic, *32*
lymphoid, 32
muscular, 5, 32–33, *32*
nervous, 33
skin, 33–34
types of, 5, 30–33
Tongue, in digestion, 159
views of, *160*
Tonsillitis, chronic, 128
Tonsils, function of, 128
Torso, bones of, 46
female, all viscera removed, *13*
first muscle layer exposed, *7*
visceral relations exposed, *11*
male, anterior, chest and abdominal muscles removed, *9*
first muscle layer exposed, *6*
second and third layers of abdominal muscles exposed, *8*
muscles of, anterior superficial, 78
anterior surface of, *77*
posterior, deep muscle layers exposed, *15*
first muscle layer exposed, *14*
rib cage and omentum removed, *10*
stomach, small bowel, most of colon, and anterior part of lungs removed, *12*
Trachea, description and function of, 134, *136*
Tracheotomy, 134, *136*
Transfer ribonucleic acid, 22
description of, 24
Transport, active, in absorption of nutrients, 170
Trigeminal nerve (V), branches of, *193*
True ribs, 48
TSH. See *Thyroid stimulating hormone.*
Tubercle of lung, 145
Tuberculosis, definition of, 145
Tumors of breast, 240
Tympanic cavity, parts of, 205–206, *205*
Tympanic membrane, *205*, 206

Ulcer, 173
duodenal, *161*
Ulna, 48
Uracil, in RNA, 24
Uremia, 155
Ureters, 147
description and function of, 154
Urethra, description and function of, 155, *155*
male, 225
function of, 230
Urethral orifice, 231
Urinary bladder, description and function of, 147, 154–155, *155*
Urinary system, function of, 147
in homeostasis, 16
relationships to genital system and great vessels, *148*
Urine, composition of, 153t
formation of, glomerular filtration in, 149
tubular reabsorption in, 149–150
tubular secretion in, 150–152
Uterine tubes, structure and function of, 232–233, *234*
Uterus, structure and function of, 233–235, *234*

Vagina, structure and function of, 235
vestibular glands of, 231
vestibule of, 231
Vaginal orifice, 231, *231*
Vagus nerve (X), branches of, *195*
Vas deferens, function of, 230
Vasectomy, 230
Veins, bicuspid valves in, 107, *107*
component parts of, *106*
description and function of, 107
of head and neck, *109*
of lower arm and hand, *117*
of pelvis and leg, *119*
of shoulder and upper arm, *115*
of thyroid gland, *110*, 216
varicose, 107, *107*
Vena cava, and tributaries, *112*
inferior, 100, *101*
superior, 100, *101*
Venereal disease, types of, 240
Ventricles, of brain, 186, *186*
of heart, 100, *100*, 102
Ventricular systole, 103
Venules, 1–7
Vermis, of cerebellum, 188
Vertebrae, cervical, *46*, 48
coccygeal, 48
function of, 48
lumbar, 48
relation of spinal cord and nerves to, *183*
sacral, 48
thoracic, 48

Vertebral column, curvature of, 48
  function of, 48
  in relation to body outline, 49
Vestibule of inner ear, *205*, 206
Vestibulocochlear nerve (VIII), branches of, *194*
Visceral pericardium, 100, *100*
Vision, abnormalities of, corrections for, *204*, *205*
Vital capacity, definition of, 138
Vitamin D, in control of calcium absorption, 41, *41*
Vitamins, table of, 169t
Vocal cords, 134, *135*
Vulva, 230, *231*, *232*

Wrist, bones of, 48